Water Science and Technology Library

Volume 84

The aim of the Water Science and Technology Library is to provide a forum for dissemination of the state-of-the-art of topics of current interest in the area of water science and technology. This is accomplished through publication of reference books and monographs, authored or edited. Occasionally also proceedings volumes are accepted for publication in the series.

Water Science and Technology Library encompasses a wide range of topics dealing with science as well as socio-economic aspects of water, environment, and ecology. Both the water quantity and quality issues are relevant and are embraced by Water Science and Technology Library. The emphasis may be on either the scientific content, or techniques of solution, or both. There is increasing emphasis these days on processes and Water Science and Technology Library is committed to promoting this emphasis by publishing books emphasizing scientific discussions of physical, chemical, and/or biological aspects of water resources. Likewise, current or emerging solution techniques receive high priority. Interdisciplinary coverage is encouraged. Case studies contributing to our knowledge of water science and technology are also embraced by the series. Innovative ideas and novel techniques are of particular interest.

Comments or suggestions for future volumes are welcomed.

Vijay P. Singh, Department of Biological and Agricultural Engineering & Zachry Department of Civil Engineering, Texas A&M University, USA
Email: vsingh@tamu.edu

More information about this series at http://www.springer.com/series/6689

Arup K. Sarma · Vijay P. Singh
Rajib K. Bhattacharjya · Suresh A. Kartha
Editors

Urban Ecology, Water Quality and Climate Change

Springer

Editors
Arup K. Sarma
Department of Civil Engineering
Indian Institute of Technology Guwahati
Guwahati, Assam
India

Rajib K. Bhattacharjya
Department of Civil Engineering
Indian Institute of Technology Guwahati
Guwahati, Assam
India

Vijay P. Singh
Department of Biological and Agricultural
  Engineering, Zachry Department
  of Civil Engineering
Texas A&M University
College Station, TX
USA

Suresh A. Kartha
Department of Civil Engineering
Indian Institute of Technology Guwahati
Guwahati, Assam
India

ISSN 0921-092X   ISSN 1872-4663   (electronic)
Water Science and Technology Library
ISBN 978-3-030-09006-7   ISBN 978-3-319-74494-0   (eBook)
https://doi.org/10.1007/978-3-319-74494-0

Printed on acid-free paper

This Springer imprint is published by the registered company Springer International Publishing AG part of Springer Nature
The registered company address is: Gewerbestrasse 11, 6330 Cham, Switzerland

Preface

The twentieth century witnessed massive urbanization at a rapid pace in developing countries, and this pace is continuing in the twenty-first century. Currently, nearly half of the world's population lives in urban areas. The rapid urbanization is a direct result of economic growth and people's urge toward higher comfort and access to education, health, and travel facilities. Urbanization, more often, causes strain on natural as well as man-made resources that are meant for the well-being of people. One direct and most perceptible consequence of urbanization is the change in land use and land cover which, in turn, impact the ecological balance, including the hydrological cycle. This impact is witnessed through changes in surface and groundwater levels, surface runoff patterns, diurnal variations in temperature, humidity, cloud cover, smog, pollution, radiation, and wind. Therefore, scientific studies on urban hydrology are vital for appropriate design of urban landscapes and civil infrastructure works.

A center of excellence for "Integrated Landuse Planning and Water Resources Management" sponsored by Ministry of Urban Development, Government of India, was established at Indian Institute of Technology Guwahati in the year 2010 to undertake studies on urban hydrology of the northeast region of India. As part of the mandate, the center conducted an international conference ENSURE and invited authors to submit articles related to urban hydrology. A number of authors presented highly thoughtful research papers on the topic and seeing their value, it was decided to publish these conference articles in book form. Consequently, we revisited the conference themes and articles and selected some of them for inclusion in this volume.

Different issues of sustainable urban development are being addressed in this book, and the subject matter of the book is divided into five parts. Rapid and unplanned urbanization to meet the short-term need and without considering the long-term sustainability of the ecosystem, particularly in the eco-sensitive areas like hills and wetlands, is now becoming a major concern. Such development is leading to multiple hazards, like flood, landslides, soil erosion, and water quality degradation in almost all urban areas of the developing world.

Part I, comprising eight chapters, deals with the urbanization and ecological sustainability. This part aims at giving a broad idea of what is meant by a sustainable urban ecosystem. The need of having an environment-friendly construction, including the material used, is discussed in one of the chapters. The emission in various forms caused by construction activities and also to meet transportation needs of an urban area needs special attention. Several in-depth studies on these issues are discussed with some case studies in this part. The scope of adopting a green road approach in planning and design of highways to have better urban environment is also presented in the last chapter of this part.

Climate change issues in the urban sector are elaborately discussed with several case studies across six different chapters in Part II. As renovating a fully developed unplanned urban area is extremely difficult, the emphasis is also given in considering climate change issues in planning and development of semi-urban areas to have a better future, and the first chapter of this part is dedicated to this aspect. Facts of having increasing extreme temperatures and the effect of heat island in urban setup are discussed with a case study on Guwahati City of Northeast India. The trend in sunshine duration in the humid climate of Northeast India is also presented in one of the chapters. The scope of projecting important hydrological parameters, like precipitation and temperature, in climate change scenario is discussed with a case study of Kashmir Valley of India.

Part III, comprising five chapters, deals water quality concerns in the urban environment. While rainwater harvesting is advocated strongly worldwide, that equal importance should be given to the quality of rooftop harvested water is discussed in the first chapter of this part. The effect of effluent of a paper mill in the receiving water body is discussed in one of the chapters. The scope of bioremediation of heavy metal pollution is also discussed in this part. An additional advantage derived in assessing river water quality by multivariate analysis is presented with a case study of Iran. Construction of an embankment to protect urban areas is a common practice in India. As the failure of embankment may have several concerns about water quality, a chapter of stability analysis of earthen embankment is also included in this part.

Developing countries, because of their economic constraints, generally cannot afford costly measures to prevent industrial pollution. Therefore, the developing world suffers from such pollution, and research need in this direction is enormous. Studies related to the effect and remediation of industrial pollution, including heavy metals, are discussed in eight different chapters of Part IV.

Part V comprises five chapters, which deal with groundwater contamination and its management. Seawater intrusion is one of the major challenges faced by the coastal cities and is discussed in the first chapter of this part. Drinking water need in most of the cities is met from groundwater storage, which is depleting day by day due to unplanned exploitation. Reduced groundwater recharge resulting from urbanization is farther aggravating the problem. All these issues are addressed in this concluding part.

Water is one of the essential elements of the ecosystem, disturbance of which, both in terms of quality and quantity, can lead to an irreversible ecological disturbance. Raising awareness on the importance of a sustainable ecosystem, particularly in an urban environment, is highlighted through theoretical analyses and different case studies. This book is therefore expected to provide not only a theoretical basis for sustainable development but will also provide examples of practical cases highlighted through various case studies from the developing world.

Guwahati, India	Arup K. Sarma
College Station, USA	Vijay P. Singh
Guwahati, India	Rajib K. Bhattacharjya
Guwahati, India	Suresh A. Kartha

Contents

About the Editors

Prof. Arup K. Sarma Professor and former Head of Civil Engineering Department, Indian Institute of Technology Guwahati, was honored with the prestigious *B. P.Chaliha Chair Professor* position by Ministry of Water Resources, Government of India in 2009. Professor Sarma also served as Visiting Professor in the Asian Institute of Technology, Bangkok, Thailand during 2015. The NPTEL video course on hydraulic engineering developed by Prof. Sarma is receiving wide appreciation from different parts of the globe and entered into the top 5 most visited course. Till date, he has completed 24 sponsored research project and 52 consultancy projects from India and abroad. Working with 22 Ph.D. research scholars and 56 Master's students, he has published more than 100 technical papers in reputed national and international journals, books, and in conference proceedings. Reviewer of several reputed international journals, Prof. Sarma has also served as member and adviser of various national and international scientific committees. He has also devoted himself to the promotion of scientific temperament in the society through music and drama. As an approved lyricist of All India Radio, Prof. Sarma has composed several songs, drama, and musical features for All India Radio and Television to bring scientific awareness among masses.

Prof. Vijay P. Singh is a University Distinguished Professor, a Regents Professor, and Caroline and William N. Lehrer Distinguished Chair in Water Engineering at Texas A&M University. He received his B.S., M.S., Ph.D., and D.Sc. degrees in engineering. He is a registered professional engineer, a registered professional hydrologist, and an honorary diplomate of ASCE-AAWRE. He has extensively published in the areas of hydrology, hydraulics, groundwater, irrigation, and water resources. He has received three Honorary Doctorates from University of Basilicata, University of Waterloo, and University of Guelph; and has received 87 national and international awards, including James M. Todd Technological Achievement Medal; Arid Lands Hydraulic Engineering Award, Ven Te Chow Award, Richard R. Torrens Award, Lifetime Achievement Award, and Norman Medal, from ASCEs; Ray K. Linsley Award and Founder's Award, from AIH; Gold Medal, from KSCE; Crystal Drop Award and Ven Te Chow Award from

IWRS; Sigma Xi Outstanding Distinguished Scientist Award; and Jiangsu Provincial Friendship Award, and Distinguished Award from China. He is a Distinguished Member of ASCE, and a fellow EWRI, AWRA, IWRS, ISAE, IASWC, and IE. He is member of 10 international engineering academies. He has also served as President of the American Institute of Hydrology (AIH). He is editor-in-chief of three journals and two book series and serves on editorial boards of more than 20 journals and three book series.

Prof. (Dr.) Rajib K. Bhattacharjya is a professor in the Department of Civil Engineering, Indian Institute of Technology Guwahati, India. He received his Bachelor's and Master's degree in Civil Engineering from Gauhati University, India in the year 1993 and 1995, respectively, and Ph.D. in Civil Engineering from Indian Institute of Technology Kanpur, India in the year 2004. His current research interests include coupled simulation–optimization modeling for groundwater management, identification and management of virus sources in groundwater aquifers, management of saltwater intrusion in coastal aquifers, optimization methods, and artificial neural networks. He has more than 15 years of teaching and research experiences and has authored more than 70 peer-reviewed scientific publications in various reputed international journals and conference proceedings. He has also been a visiting professor at other institutes, including the Dalhousie University, Halifax, Canada and École Centrale de Nantes, France. More about his research and academic activities can be found at http://www.iitg.ac.in/rkbc.

Dr. Suresh A. Kartha is presently an Associate Professor in the Department of Civil Engineering, IIT Guwahati. He joined IIT Guwahati in the year 2007 as a Senior Lecturer in the specialization Water Resources Engineering and Management in the Department of Civil Engineering. As a faculty, Dr. Kartha has taught courses on numerical methods, hydrology, subsurface hydrology, fluid mechanics, etc., and his research focus is in the area of flow and transport processes in the porous media. His internet web profile can be accessed through the link http://www.iitg.ac.in/kartha/homepage/index.html.

Abbreviations

AAS	Atomic absorbance spectrophotometer
ACs	Air conditioner
AFEX	Ammonia fiber expansion
AFOLU	Agriculture, forestry and other land use
AM	Arbuscular mycorrhizal
AMI	Advanced meter infrastructure
AMR	Automated meter reading
AMTRON	Assam Electronics Development Corporation Ltd.
ANN	Artificial neural network
APHA	American Public Health Association
ARSAC	Assam Remote Sensing Application Centre
ASTEC	Assam Science Technology and Environment Council
ATP	Adenosine triphosphate
BET	Brunauer–Emmett–Teller
BJH	Barrett–Joyner–Halenda
BNR	Biological nitrogen removal
BOD	Biological oxygen demand
BRTS	Bus rapid transit system
BSA	Bovine serum albumin
BT	Brightness temperature
CA	Cluster analysis
CAAA	Clean Air Act Amendments
CGCM3	Canadian third generation climate model
CGI	Corrugated galvanized iron
CGWB	Central Groundwater Board
CL	Methane emission from cropland
CNS	Central nervous system
COD	Chemical oxygen demand
CPM	Cachar Paper Mill
CRS	Coordinate reference system

CTMABr	Cetyl trimethyl ammonium bromide
CWW	Coke oven wastewater
DGM	Directorate of Geology & Mining
DNPAOs	Denitrifying phosphate accumulating organisms
DO	Dissolved oxygen
EC	Electrical conductivity
EDX	Energy dispersive X-ray
EF	Emission factor
EPA	Environmental Protection Agency
EPS	Extracellular polysaccharides
ETCCDMI	Expert Team on Climate Change Detection, Monitoring and Indices
FA	Factor analysis
FCC	False color composite
FE-SEM	Field emission scanning electron microscopy
FOLU	Forestry and other land use
FS	Deforestation and conversion of forest to settlement
FSI	Forest Survey of India
FTIR	Fourier transform infrared
FW	Burning of biomass
GCM	Global climate model
GF-AAS	Graphite furnace atomic absorption spectrometry
GH	Glycoside hydrolases
GHG	Greenhouse gas
GI	Galvanized iron
GIT	Gastrointestinal tract
GMA	Guwahati city metropolitan area
GMC	Guwahati Municipal Corporation
GMDA	Guwahati Metropolitan Development Authority
GP	Gaon panchayats
GPR	Ground penetrating radar
GRIHA	Green Rating for Integrated Habitat Assessment
GWP	Global warming potential
HCN	Hydrogen cyanide
HPLC	High performance liquid chromatography
HVAC	Heating, ventilation and air conditioning
ICP-OES	Inductively coupled plasma optical emission spectrometry
ICT	Information and communication technology
IGBC	Indian Green Building Council
IMD	India Meteorological Department
IonE	Institute on the Environment
IPCC AR4	Fourth Assessment Report of the Intergovernmental Panel on Climate Change
IPCC	Intergovernmental Panel on Climate Change
IPTG	Isopropyl-β-D-thiogalactopyranoside
IQ	Intelligence quotient

KMO	Kaiser–Meyer–Olkin
LARS-WG	Long Ashton Research Station Weather Generator
LB	Luria broth
LRT	Light rail transit
MLR	Multiple linear regression
MoRTH	Ministry of Road Transport and Highways
MSL	Mean sea level
MSM	Minimal salt medium
MSM	Mineral salt medium
MVA	Multivariate statistical analyses
NCAR	National Center for Atmospheric Research
NCEP	National Center for Environmental Prediction
NDVI	Normalized difference vegetation index
NGOs	Nongovernment organizations
NIH	National Institute of Hydrology
NIL	NIL
NIUA	National Institute of Urban Affairs
NOx	Nitrogen oxides
NUTP	National Urban Transport Policy
nZVI	Nano zero valent iron
O&M	Operation and maintenance
OD	Optical density
OH$^-$	Hydroxyl ions
PAOs	Phosphorous accumulating organisms
PCA	Principal component analysis
PHA	Poly hydroxyalkanoate
PNS	Peripheral nervous system
PRBs	Permeable reactive barriers
PRBW	Permeable reactive barrier wall
PSD	Phosphate treated sawdust
PTFE	Polytetrafluoroethylene
PYE	Peptone yeast extract
RBF	River bed filtration
RCM	Regional climate model
RF	Reserved Forests
RFFP	Rainfed floodplain
ROS	Reactive oxygen species
RUSLE	Revised Soil Loss Equation
RWH	Rainwater harvesting
RWH	Rooftop rainwater harvesting
RWS&S	Rural Water Supply and Sanitation Department
SAR	Sodium adsorption ratio
SAT	Soil aquifer treatment
SBR	Sequencing batch reactor
SBS	Sick building syndrome

SDSM	Statistical downscaling model
SDSS	Spatial decision support system
SEM	Scanning electron microscopic
SIF	Seasonally integrated flux
SMB	Stabilized mud blocks
SOI	Survey of India
SSF	Simultaneous saccharification and fermentation
SUHI	Surface urban heat island
TCLP	Toxicity characteristic leaching procedure
TDS	Total dissolved solids
TEOS	Tetraethyl orthosilicate
TGA	Thermogravimetric analyzer
TM	Thematic mapper
TNPCB	Tamil Nadu Pollution Control Board
UAE	United Arab Emirates
UBL	Urban boundary layer
UCL	Urban canopy layer
UHII	Urban heat island intensity
UHIs	Urban heat islands
UV	Visible detector
VFA	Volatile fatty acids
VOCs	Volatile organic compounds
VWSC's	Village Water and Sanitation Committees
WHO	World Health Organization
WL	Net CO_2 emission from wetlands
ZVI	Zero valent iron

Part I
Urbanization and Ecological Sustainability

Sustainable Urban Ecosystems: Problems and Perspectives

Vijay P. Singh

Abstract In urban ecosystems, the main problems are drinking water supply, energy supply, drainage, waste disposal, land use change, pollution, mitigation of natural disasters, and protecting the integrity of ecosystems. These problems are being compounded by population explosion including migration, sociocultural upheaval, and climate change. In order to discuss these problems from a hydrologic perspective, this paper first revisits ecology, ecosystems, ecosphere, and biosphere. It then looks at ecosystems in the hierarchy of biological organization. This then leads to stating the sciences that are allied with ecology and ecosystems. The next question that needs to be addressed is the one of sustainability. Defining sustainability is, some aspects of sustainable development and sustainability imperatives are enumerated. With this background in hand, urban ecosystems and their components are formulated. This leads to stating the main urban hydrology problems and challenges. These problems and challenges are being complicated by growing global population, particularly urban population and especially in India and its cities. Emphasizing water as the source of life, food, fiber, and energy, the question of meeting the challenges of water security, energy security, and food supply arises. To that end, water use, including individual, virtual water use, water withdrawals, and water use in different sectors, is discussed. The use of water in energy production is then highlighted. There is a dualism of energy use in India, and this gives rise to India's energy challenge. Without sustainable energy supply in the long term, development will be impeded, for this is one of the sustainability imperatives. The problem of drinking water which is still plaguing cities in India is then examined. Like energy, it is related to human health and development. These challenges must be met with an integrated approach to water management under the specter of looming climate change. The discussion is concluded with a personal perspective on water, food, and energy security and on development itself.

V. P. Singh (✉)
Department of Biological & Agricultural Engineering, Zachry Department of Civil
Engineering, Texas A&M University, College Station, Texas 77843-2117, USA
e-mail: vsingh@tamu.edu

© Springer International Publishing AG 2018
A. K. Sarma et al. (eds.), *Urban Ecology, Water Quality and Climate Change*,
Water Science and Technology Library 84,
https://doi.org/10.1007/978-3-319-74494-0_1

3

Keywords Ecology · Ecosystem · Ecosphere · Biosphere · Urban hydrology
Water–energy–food nexus · Development · Sustainability

1 Introduction

Urbanization is occurring at a rapid pace all over the world, particularly in developing countries, such as India. For example, in 1970, about 23% of the population of India lived in cities and this percentage grew to 33% in 2000. Given India's population, this is a massive growth, largely fueled by migration from rural areas. Associated with urbanization come problems such as drinking water supply, energy supply, drainage, waste disposal, land use change, pollution, and mitigation of natural disasters. These problems are being compounded by population explosion and migration, sociocultural upheaval, and climate change and they threaten the integrity of ecosystems. Little is often done to address or plan beforehand and the government administration has to grapple with these problems after people have already migrated. In a democratic society where short-term political considerations often take on primacy over long-term societal good, it is almost impossible to plan for such migration.

  With growing urbanization, an important question that arises is if urban areas that are becoming so humongous are sustainable and what should be done to make them sustainable. There are basic needs that must be met on a sustainable basis, that is, there have to be water security, food security, energy security, air quality, healthcare facilities, schools, transportation facilities, housing, waste disposal, recreation, and environmental security. These are essentially sustainability imperatives and without ensuring these resources to the common man, sustainability and sustainable development will not be assured. These are huge challenges but are fundamental to the maintenance of ecosystems, ecosphere, and biosphere. To meet these challenges, there has to be an integrated approach, especially under the specter of looming climate change. This paper first revisits ecology, ecosystems, ecosphere, and biosphere, then visits sustainability and sustainable development. The discussion is concluded with a personal perspective on water, food, and energy security and on the development itself.

2 Ecology and Ecosystems

Environment means one's house and consists of two components: biotic and abiotic. Biotic means living that includes all organisms, and abiotic means nonliving or physical, such as temperature, sunlight, and precipitation. Ecology is comprised of two words: eco which means house and logy which in Greek means study. Thus, ecology means the study of one's house. Then, ecology can be defined as the study of interactions among organisms and between organisms and their abiotic environment. There is a biological organization or hierarchy that organisms follow. From the smallest scale to the largest scale, the hierarchy in a sequential order comprises atoms, molecules, cells, tissues, organs, body systems, individual multicellular organisms, population, and communities. Populations of different species interact to form a community. Thus, organisms occur in populations and populations are organized into communities.

A community together with its physical environment defines an ecosystem. It encompasses not only all the biotic interactions of a community but also the interactions between organisms and their abiotic environment. Thus, an ecosystem is studied at the community level and can be inferred as a community and its abiotic environment.

All communities of organisms on the Earth comprise the biosphere, that is, Earth's communities are organized into biosphere. Organisms of the biosphere depend on one another and on the Earth's physical environment that comprises atmosphere (the gaseous envelope surrounding the Earth), hydrosphere (Earth's supply of water—liquid and frozen, fresh and salty.), and lithosphere (soils and rocks of Earth's crest). Biosphere extends to ecosphere which encompasses the biosphere and its interactions with the atmosphere, hydrosphere, and lithosphere. Thus, a study of biosphere or ecosphere becomes essential for the development of any paradigm for sustainability, because it examines the complex interactions among the Earth's atmosphere, land, water, and organisms.

Now ecology can be defined as a study at the highest levels of biological organisms comprising population, communities, ecosystems and the biosphere, and ecosphere. The study of ecology comprises a number of other sciences—biological, physical and chemical as well as socioeconomic. Examples of allied sciences include biology, chemistry, physics, hydrology, meteorology, geology and Earth sciences, economics, politics, and population dynamics.

3   Urban Ecosystems and Their Problems

Some of the problems plaguing urban ecosystems in developing countries even today are drinking water supply, energy supply, flooding, drainage systems, sanitation, waste disposal, transportation, land use change, population explosion, air pollution, natural disasters, health care, schools, housing, food security, social and cultural conflicts, and environmental security. Compounding these problems are global warming and climate change and are threatening the integrity of ecosystems. These problems are not going to go away and looking at the growing world population, trends of urbanization, and sociocultural upheaval, they are going to get more complicated in the decades ahead and their solution will become much more daunting.

Fundamental to the sustainability of urban ecosystems are water security, energy security, and food security (Singh 2017). Water security is built on (1) demand for and use of quality water, (2) access to quality water, (3) supply of quality water, and

(4) availability of water. In a given country, certain parts are water secure but other parts may be water scarce. Likewise, a region can be water secure in one part of the year but not in other parts.

Water is used for a variety of uses, such as for agriculture, horticulture, livestock, fisheries, domestic consumption, industry, energy generation, waste disposal, and recreation (Singh 2008). Water consumption significantly changes from country to country and from city to city. An individual needs 15–20 L of water per day to survive. Besides drinking, water is needed for cooking, washing, bathing, sanitation, cooling, and heating.

Agriculture is the largest consumer of freshwater, however, in urban areas agriculture is not directly the largest water user, but it is the industry that uses the most water. Five liters are needed to produce 1 L of bottled water, 960 L for 1 loaf of bread, 50 L for 1 whole orange, 170 L for 1 glass of orange juice, 70 L for 1 whole apple, 190 L for 1 glass of apple juice, 2400 L for 1 dozen eggs, 3900 L for 1 kg of chicken meat, and 4800 L for 1 kg of pork. It is often not appreciated that to produce our daily diet takes a lot of water.

Electricity production is one of the largest users of water. On average about 25 gallons of water are used to produce 1 kilo Watt hour (kWh). For a 60-Watt incandescent light bulb burning for 12 h a day for a year in 111 million houses, a power plant would consume about 655 billion gallons (2.5 km^3) of water (Singh et al. 2014).

In many countries, such as India, there is dualism of energy use. The consumption basket of the rich consists of goods and services which have greater intensive use of fossil fuel, minerals, chemicals, etc., than that of the poor. A vast majority of the rural and urban poor have to depend on unclean, unconverted, and highly inefficient biomass fuel for cooking. There is still significant lack of connectivity with electricity and/or its reliable supply for, particularly, the rural households. Energy sustainability entails removal of current energy dualism, modern energy for all, reducing cost of connectivity to modern energy for the poor, reliable supply of electric power in rural areas, reducing the cost of renewable, e.g., solar, aggressive research and development investment, technology collaboration with developed nations, and coal gasification and liquefaction.

Water and human health are intimately connected. About 80% of diseases in developing countries are caused by the lack of access to clean potable water. Pathogens transmitted through water kill 25 million people every year by amoeba linked diarrhea, cholera, and typhoid, and about 3,900 children die every day (World Health Organization 2004). The joint report of WHO/UNICEF (2012) provides some interesting and hopeful but also stark facts. For example, in 2010, 89% of the world's population, or 6.1 billion people, used improved drinking water sources, exceeding the MDG target (88%); 92% are expected to have access in 2015. Between 1990 and 2010, two billion people gained access to improved drinking water sources. However, 11% of the global population, or 783 million people, are still without access. In 2015, the WHO/UNICEF JMP projects that 605 million will still not have access.

Sub-Saharan Africa accounts for more than 40% of the global population without access to improved drinking water. Sub-Saharan Africa is not on track for meeting the drinking water target, but some countries have already met the target: Malawi, Burkina Faso, Ghana, Namibia, and Gambia. Liberia is on track to meet the target. 593 million in China and 251 million in India gained access to improved sanitation since 1990. China and India account for just under half the global progress on sanitation.

India still has 626 million people who practice open defecation, which is more than twice the number of the next 18 countries combined. It accounts for 90% of the 692 million people in South Asia who practice open defecation; accounts for 59% of the 1.1 billion people in the world who practice open defecation; and has 97 million people without access to improved sources of drinking water, second only to China.

The per capita availability of water depends on population, rainfall, and surface and groundwater resources. On a per capita per year, the countries with most freshwater resources are Suriname with 479,000 m^3 and Iceland with 605,000 m^3, and Egypt has the lowest with 25 m^3, followed by United Arab Emirates (UAE) with 61 m^3. The rainwater has a high degree of variability from one region to another. Likewise, runoff also varies from place to place and hence the same applies to flow variability. The land can be dry in one year but completely deluged in another year. The high rainfall variability poses a challenge for water resources management and to ensure water security.

4 Sustainability and Sustainable Development

The U.S. Environment Protection Agency provided a consensus "straw man" definition of sustainability as "Sustainability occurs when we maintain or improve the material and social conditions for human health and the environment over time without exceeding the ecological capabilities that support them (National Research Council 2017)." Earl Beaver from Institute for Sustainability defined sustainability as "Sustainability is a path of continuous improvement, wherein the products and services required by society are delivered with progressively less negative impact upon the Earth." Along similar lines, Brundtland Commission (1987) defined sustainable development as "Sustainable development is development that satisfies the needs of the present without compromising the needs of the future." Thus, sustainability can be viewed as the goal (or the state) and sustainable development as the process by which we achieve that goal (or reach that state) (Clift 2000).

Proceeding further, environmental sustainability can be defined by the ability of the environment to function indefinitely without degradation due to stresses imposed by human beings on natural systems, such as soil, water, air, and biological diversity. This suggests that sustainability can be visualized at the community, regional, national, and global levels. In the long term, sustainability should be viewed as global and the sustainability of any country should not be viewed in isolation. There can be no global sustainability without sustainability of all countries. In reality, sustainability is not being realized at all levels. In some countries, sustainability is far from achieved at any level. The reasons that some countries or regions that are not operating sustainably are that they are using non-recoverable resources as if they are infinite, using renewable resources at a rate faster than the rate of their replenishment, polluting the environment with toxins as if the environment has a limitless capacity to absorb them, or/and growth in population despite the area's limited capacity to support them. Sustainability is the integral product of environmental, social, and economic factors, and interactions of these factors are visible, bearable, equitable, and sustainable.

5 Engineering Sustainability

Fundamental to engineering sustainability is an integrated approach that must encompass integrated development, integrated demand and use management, integrated supply management, and integrated overall management. The integrated management must be multi-criteria, multi-objective, and multi-constraint; and must be the basis for decision making. Social, cultural, political, legal, environmental, and economic considerations will constitute constraints within which the integrated management must be practiced. It is vital that people, processes, and policies are brought together in a seamless manner.

Integrated management applies to water, energy, food, land use, industrial development, as well as all other sustainability factors. That should be done with utmost care and with full support of the stakeholders. Put succinctly, there must be a well-formulated management policy or law that those who manage and those for whom management is done understand the policy unambiguously. Integrated management should be practiced at the basin scale, regional scale, or national scale.

Integrated management requires that management at the watershed scale is integrated with the entire ecosystem and watersheds are managed in an integrated manner. The health of a watershed is the responsibility of the management and must be maintained so that it is sustained for generations to come. Integrated supply and demand management must consider demand management, requirement management, use management, and supply management. The integrated management must also consider the requirements and demands of different stakeholders, order of priorities, valuation or pricing of resources, and social and economic considerations. More diverse portfolios for supply must be employed.

Since resources are limited, they should not be wasted. Integrated management will have to simultaneously consider reduction in consumption; efficient use; conservation; and recycling, reuse, and treatment of resources, wherever possible. Appropriate incentives must be provided to do all these. Development of additional resources may be necessary where justified.

Planning for new projects must include climate change impacts and changing priorities. Existing infrastructure facilities were designed that did not reflect the effect of climate change. Therefore, a systematic revision of existing infrastructure facilities is needed. For example, larger floods overwhelm existing control structures. Reservoirs do not get enough water to store for people and plants during droughts. There are more swings between floods and droughts. In many areas, people are now willing to support construction of dams. Food waste is waste of water. Likewise, sanitation and electricity grids were not planned for the rapid increase in consumers.

In order to develop a water, food and energy security system, partnership among academia, government sector, nongovernment organizations (NGOs), and private sector (farmers) must be developed. The educational system that trains the trainers and teaches the teachers must be emphasized.

6 Key Challenges

To ensure and sustain urban ecosystems is one of the grand challenges of the twenty-first century. The key to ensure sustainability is to ensure energy security, food security, and ecosystem security. Sustainable development is not possible without the water–energy–food security. Fundamental to this security is proper management which involves both technical as well as nontechnical aspects. Technical issues are relatively easy to handle, because there is enough technical knowhow to deal with most technical issues. It is the nontechnical issues that are difficult to handle, especially in a democratic society. There are too many conflicting interests and political considerations so unique solutions acceptable to all parties are hard to come by.

Compounding the security challenge is climate change. Although it is widely accepted that climate change is real and its causes are primarily human-induced, not enough is being done to arrest the change. This is because of a number of reasons. First, climate has a global scale and it is not easy to have an international consensus on concrete measures and their implementation in a time-bound manner. Second, economic development and measures for controlling climate change are perceived to be not in alignment, especially in the near term. Third, democratic systems of government are not always conducive to making tough decisions and are more motivated by election cycles. Fourth, there are legal and environmental issues that have to be grappled with.

Social conflicts and wars are becoming a problem to development and to ensuring security. These days, conflicts are becoming only too frequent and we are losing a sense of who we are and who we want to be. Imposition of social, political, and religious ideologies and beliefs is the root cause of many of the problems we are facing today.

Abatement of pollution of water bodies and of air in urban areas developing countries is a huge challenge. Having laws and policies on books is as good as not having them, if they cannot be implemented. This is one of the major problems in developing countries. To enforce the laws is not easy. Waste management is another serious issue. In developing countries, littering is commonplace and solid waste management can no longer be neglected if water security is to be achieved (Ahmad 2017).

It is well known that management for security is as much nontechnical as technical. Therefore, integration of engineering and technology with socioeconomic and political sciences is urgently needed, but it is a challenge that we do not yet know how to cope with or we have not yet been able to successfully deal with.

Another major challenge is how to live in peace. We spend far too much money, man power, and resources on fighting each other, dominating each other, resolving conflicts and social ills, and creating problems in the first place than developing the environment in which we all can live in peace and share resources so we can enjoy the quality of life, help each other develop together, and truly grow as one world family.

7 Reflection

The question we must ask is: Where are we headed as a society? Our approach from development to management calls for a paradigm shift. We have achieved tremendous scientific and technological progress. However, it is not certain if the same can be said about social progress. Our values seem to be changing rapidly not necessarily for the better. We are witnessing enormous changes in global demographic landscape. We are having too many conflicts and wars. We are constantly in the race of competition and consumption.

We are often forgetting that fertile lands and resources are finite, and our standard of living does not have to be unsustainable. After all, the objective of life is far more than consumption, competition, and comfort. Natural resources are God's Gift. We must use them wisely. We owe it to our future generations that we leave them in a better shape than the shape we got them in.

8 Conclusions

From this discussion, it is concluded that we have to strive hard for sustainability and sustainable development, even harder under climate change and social conflicts. We need integrated approaches to sustainable development which are as much nontechnical as technical. No sustainable development is possible without ensuring water security, energy security, and environmental security. Management of these resources requires management of supply, management of demand and requirement, and management of use as well as efficiency, conservation and elimination of waste. Environmental pollution is threatening ecosystems and must be arrested without delay. There must be well-tested laws and policies that can bring people, policies, and processes together.

References

Ahmad R (2017) US water regulations and India's water challenges. J AWWA 109(3):65–67

Brundtland Commission (1987) Report of the World Commission on Environment and Development: Our Common Future, by G.H. Brundtland, Oslo, Norway

Clift R (2000) Forum on sustainability. Clean Prod Process 2(1):67 (Verlag, Berlin)

National Research Council (NRC) (2017) Sustainability and the U.S. EPA, 21 June 2017

Singh VP (2008) Environment, engineering, religion, and society. J Hydrol Eng 13(3):118–123

Singh VP (2017) Challenges in meeting water security and resilience. Water Int 42(4):349–359. https://doi.org/10.1080/02508060.2017.1327234

Singh VP, Khedun CP, Mishra AK (2014) Water, environment, energy, and population growth: implications for water sustainability under climate change. J Hydrol Eng 19:667–673

World Health Organization (2004) Water, sanitation and hygiene links to health: facts and figures. (www.who.int/water_sanitation_health/publications/factsfigures04/en)

WHO (UNICEF) (2012) Joint statement-integrated community management (iCCM): an equity focused strategy to improve access to essential treatment services for children. Geneva, Switzerland

Construction in Nature Versus Nature of Construction

Aabshar U. K. Imam and M. S. Indu

Abstract Unplanned urbanization often leads to environmental degradation which is also related to economic development. The present study highlights the contemporary issues arising out of urban environmental degradation such as urban heat islands, air pollution, and sick building syndrome. Solutions are sought through natural habitats found in nature and attempts at their adaptation and implementation are highlighted. Green building concepts are proposed as a viable alternative.

Keywords Energy efficiency · Green buildings · Natural habitats
Sick building syndrome · Urban heat islands

1 Introduction

'*The control man has secured over nature has far outrun his control over himself,*' as stated by Ernest Jones befits the urban scenario of the present times. The rapid urbanization of cities has resulted in over-densification and haphazard development of built structures. The impetus has shifted to accommodation of the burgeoning population instead of comfortable housing for the residents.

Over-densification, urbanization, use of 'warm' construction materials and decreasing quantum of green areas disturb the balance in natural environment. Increase in the expanse of cities necessitates the use of automobiles. Gaseous

A. U. K. Imam (✉)
Department of Architecture and Regional Planning, Indian Institute of Technology
Kharagpur, Kharagpur, India
e-mail: ar.aabsharimam@gmail.com

M. S. Indu
Department of Civil Engineering, Indian Institute of Technology Kharagpur,
Kharagpur, India
e-mail: indumsiitkgp@gmail.com

© Springer International Publishing AG 2018
A. K. Sarma et al. (eds.), *Urban Ecology, Water Quality and Climate Change*,
Water Science and Technology Library 84,
https://doi.org/10.1007/978-3-319-74494-0_2

13

emissions from industries and automobiles result in continuous worsening of the gaseous balance in the atmosphere. Increasing percentage of the greenhouse gases like carbon dioxide, nitrous oxide and methane exacerbates global warming. The cumulative effect of these factors has endangered the ecology of cities and urban centres. One of the most predominant and widespread consequence of the disturbed urban natural environment is the generation of urban heat islands.

This study attempts to highlight the misgivings of unplanned urbanization and proposes biomimicry and the use of green buildings as probable solutions. It takes a brief discourse through the consequences of unchecked urbanization which are discussed under the heads of urban heat islands, health-related issues, air pollution, energy and cost expenditure, use of building materials and sick building syndrome. A solution is sought through the design and construction of sustainable wildlife habitats found in nature and the modern attempts at biomimicry. Green buildings are proposed as a viable alternative.

2 Urban Heat Islands (UHIs)

Built structures and the construction materials used in the urban fabric of the city determine its thermal behaviour at the micro- and macro-scale (Santamouris et al. 2011; Akbari et al. 2001). High-rise dense neighbourhoods built with low albedo construction materials like concrete and asphalt tend to trap heat. The diminished sky view factor and multiple reflections between the built surfaces prevent the escape of heat into the atmosphere. This results in the formation of heat islands where the temperature remains higher than those of the adjacent rural surroundings. It has been shown that the extent of an area affected by urban heat island is directly proportional to urban sprawl and grows accordingly (Yang et al. 2010). The UHIs have been shown to result in temperature differences of up to 7 °C as in the case of large conurbations and the surrounding rural areas in UK (Smith and Levermore 2008).

Summertime heat islands give rise to a number of problems such as, increased residential cooling loads, increased electricity consumption, exacerbated thermal discomfort, and detrimental effects on human health (Yang et al. 2010).

2.1 Health-Related Issues

The thermal conditions in heat islands get further worsened during conditions of heat waves. For instance, the heat waves of 2003 resulted in about 35,000 European deaths in a span of 2 weeks and more than 1,900 deaths in India. Besides increased mortality, heat stroke, heat exhaustion, infectious diseases, cardiovascular and respiratory problems also increase during the summer seasons (Harlan et al. 2006).

The heat-induced stress appears to fall disproportionately upon the marginalized groups, i.e. the poor, the physically weak and the elderly (Harlan et al. 2006), thus highlighting the increased need for environmentally sustainable cities.

2.2 Air Pollution

The UHIs increase air pollution due to increase in air temperatures. As per studies conducted in California, at maximum daily temperatures below 22 °C, ozone concentration of 90 ppb is acceptable; however, at temperatures above 35 °C, almost all days are smoggy (Akbari et al. 2001). The level of air pollution and overall environmental degradation varies with the nature of urbanization and depends on the level of economic development reached. It is depicted by the Kuznets curve as shown in Fig. 1.

2.3 Energy and Cost Expenditure

Dense neighbourhoods and improperly designed buildings in hot climates may result in indoor thermal discomfort, compelling the user to resort to the use of air conditioners (ACs). While the ACs help to maintain the thermal comfort indoors, they release waste gases such as carbon dioxide to the external environment which is a major factor responsible for urban warming. Anthropogenic heat emissions and the resulting urban warming aggravate the issues arising out of global warming.

As per Hulme et al. (2002) and IPCC (2007), temperature increase of 0.1–0.5 °C per decade is predicted in Europe during the twenty-first century (Smith and Levermore 2008). Typically, the electricity demands in cities increase by 2–4% for each 1 °C increase in temperature (Akbari et al. 2001) thus aggravating the economic burden.

Fig. 1 The environmental Kuznets curve: a development–environment relationship. *Source* http://staging.unece.org/fileadmin/DAM/ead/pub/032/032_c2.pdf

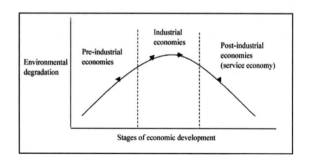

2.4 Building Materials

'The use of materials defines the global albedo of cities'. The urban areas in general have much lower albedo than the surrounding rural areas. The typical albedo for European/American cities of 0.15–0.30 stands in stark contrast to that of North African cities, i.e. 0.45–0.60 (Doulos et al. 2004).

The urban environment is often characterized with the use of materials like concrete, asphalt, pebbles and stone, which are poor reflectors and absorb much of the incoming radiation and lead to heating up of the neighbourhood. Experiments have shown the thermal performance of these materials to be poor and not satisfactory. Instead, use of 'cooler' materials like marble and mosaic which absorb minimal insolation and minimize heat gains has been advocated (Doulos et al. 2004). They simultaneously reduce energy consumption for heating, ventilation and air conditioning (HVAC) requirements (Akbari and Taha 1992). However, the cost of materials like marble and mosaic acts as a limiting factor to their usage. Alternatively, a change in the colour of urban surface from darker to lighter shade will help to reduce heat gains (Akbari and Taha 1992).

3 Sick Building Syndrome (SBS)

Unhygienic indoor environmental conditions have a detrimental effect on human health which is commonly referred to as 'building-related illness'. As per the Environmental Protection Agency USA, a building-related illness is one in which the occupant shows symptoms of diagnosable illness that can be attributed directly to airborne building contaminants. A very common building-related illness is the sick building syndrome where the occupants experience acute health and comfort effects that are linked to the time spent in the building and no other specific illness or cause is identifiable (Sick buildings). A sick building forms an environment that does not answer to the human needs for (Sick or Health?):

- Absence of pathogens, biotic, chemical or physical pollutants;
- Thermal, lighting and acoustic comfort;
- Sufficient space, privacy, peace, quiet, control and
- Contact with outdoors through a sense of vision, smell, sound and air quality.

Given the numerous environmental and health-related problems being faced due to the erroneous designs of buildings and cities, it is but logical to turn to nature for ideas aimed at sustaining the micro- and macro-climates in urban areas.

4 Built-in Nature

4.1 Habitat Design

4.1.1 Magnetic Termite Nests

Nature provides us with numerous instances of sustainable and energy efficient habitats acclimatized to the microclimatic conditions. One such example is the '*magnetic termite nests*' by the *Amitermes meridionalis* as witnessed in Cape York peninsula of Queensland, Australia (Hansell 2007) (Refer Fig. 2). The longer axes of the mounds are oriented in the north–south direction which explains their name of '*compass termites*' (Termite). The flat face of the mound faces east–west direction which enables the mound to quickly reach temperatures of 33–34 °C, which is maintained throughout the day with minimal variation, thus exhibiting *thermoregulation* (Hansell 2007). The top face of the wedge-shaped mound is a sharp edge which minimizes intake of heat during the mid-day.

The mounds of some termites such as, *Macrotermes subhyalinus* exhibit an advanced system of ventilation. The mound has numerous pores—the hot air escapes through the central pore having an opening at the top while cooler air enters through the pores at the sides as shown in Fig. 3.

The Eastgate Centre at Harare in Zimbabwe uses this ventilation system as an inspiration emphasizing Albert Einstein's words '*we shall require a substantially new manner of thinking if mankind is to survive*'. The structure employs passive cooling using a continuous row of chimneys placed on top of the building and uses only 10% of the energy that would be used by a similar conventionally cooled building (Eastgate Centre, Harare) (Refer Fig. 4).

Fig. 2 High-rise profiles of magnetic nests in Australia. *Source* Hansell (2007)

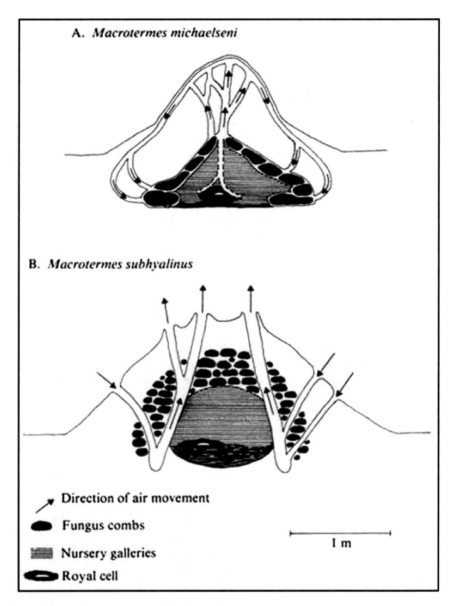

A. *Macrotermes michaelseni*

B. *Macrotermes subhyalinus*

➤ Direction of air movement

● Fungus combs

▨ Nursery galleries

⬭ Royal cell

1 m

Fig. 3 Two contrasting designs of termite mounds showing varying nature of air movement. *Source* http://inhabitat.com/building-modelled-on-termites-eastgate-centre-in-zimbabwe/wp-content/blogs.dir/1/files/termitemound_cross.jpg

Fig. 4 Eastgate Centre,
Harare, Zimbabwe. *Source*
http://en.wikipedia.org/wiki/
Eastgate_Centre,_Harare

4.1.2 Leafcutter Ants Nest

The South American leafcutter ants build massive nests, which extend up to 6 m underground and house around 8 million adults and 2–3 million larvae and eggs. The colony is thus comparable to the largest of human cities in terms of population.

The nest is a buildup of a network of chambers, spaces separated for domestic and horticultural needs and is supported by a complex ventilation system required for the sustenance of the massive structure as shown in Fig. 5. At the surface of the mound, there are various turrets providing openings for air currents, which stabilize the partial vacuum created due to wind movement on top, enabling underground ventilation (Hansell 2007).

4.2 Construction Techniques

4.2.1 Trigonopsis Mud Dauber Wasp

It is not only the design but the technology used for construction of the natural habitats which is innovative. Some clays exhibit *thixotropic* properties and their behaviour is of great importance to the strength of the structure, especially in earthquake-prone zones [Composites Basics: Materials (Part 4)]. Thixotropicity is well tackled by the solitary mud dauber wasp *Trigonopsis* during the construction of its dwelling. It places the newly brought pellet to the cell and simultaneously ejects water from its crop, emitting a soft buzzing sound. The vibrations of the sound liquefy the mud, spreading it in order to build the cell. The junctions between the newly introduced pellet and the nest vibrate together due to the shared water resulting in a common consistency, and thereby producing better structural quality (Hansell 2007).

Fig. 5 Leafcutter ants nest in Argentina showing the chambers and highways after the nest had been flooded with a mix of cement and water. *Source* Built by animals—the natural history of animal architecture, By M. Hansell

4.2.2 Nest of the Oriental Hornet

The thermoregulation observed in the nest of oriental hornet is due to the material properties. As per Ishay and Barenholz-Paniry (1995) and Ishay et al. (2002) in Hansell (2005), the silk caps covering the pupal cells of the hornet *Vespa orientalis* act as insulator and also a thermostatic regulator, maintaining its temperature between 28 and 31 °C. Some of the heat input during the day is stored as electrostatic charge in the nest, while the excess is removed through evaporation. During the cold nights, water is absorbed and its conveyance helps to release the electric charge as current, generating heat (Hansell 2005).

5 Modern Green Building Concepts

Although nature has bestowed us with a lot of examples on 'green architecture', the quest for modernization and urbanization has resulted in unplanned and haphazard development which has led to the aforesaid environmental and health issues.

To overcome these issues, energy-efficient building concept was introduced in 1970s in the US which was later refined in 1990s by including indoor environmental quality criteria, leading to green building concepts (Atlee 2011). The main objectives of green building technologies are the following (Cheng-Li 2003; Paul and Taylor 2008; Chau et al. 2010):

- Efficient use of resources like water and energy;
- Improving occupant comfort and productivity by improved indoor environmental conditions and
- Reducing wastage and reuse of resources wherever possible.

These goals could be achieved by utilizing better architectural concepts, technology and building materials. The role and the effect of these three factors are explained in the following sections.

5.1 Green Architecture

A major part of the energy requirements of buildings comes from heating and cooling of the indoor spaces, lighting and heating of water (energy efficiency in buildings, sustainable energy regulation and policy making for Africa, http://www.unido.org/fileadmin/import/83276_Module19.pdf). The architecture of green building allows natural lighting for a better quality of illumination, operable windows and fans that enable personal control over ambient conditions (Paul and Taylor 2008). In green architecture, placement of windows, walls and landscaping is done in such a way as to shade the windows during summer and to expose them to direct insolation in winter.

To reduce artificial heating and cooling requirement of a building, extra heat input to buildings during summer and heat loss during winter is limited, using green building concepts. To reduce heat exchange between the indoor and outdoor environment, layered construction like cavity walls consisting of cladding, cavity, insulation, vapour barrier, masonry structure and finish is used in green architecture (Wanga et al. 2005). Selective glazing and shading of the walls help in reducing the artificial heating and cooling requirements. This along with use of smooth and light coloured finishing materials in the building interiors yields better reflection of light, thus reducing the lighting requirements.

5.2 Green Technologies

It is reported that buildings account for 40% of the natural resources extracted in the industrialized countries, the consumption of 70% of the electricity and 12% of potable water, and the production of 45–65% of the waste disposed to landfills (Chau et al. 2010; Franzoni 2011).

A method to reduce heat exchange, apart from the architectural features discussed before, is the use of green roofs. Green roofs protect buildings from solar radiation through the combined effects of *shielding, insulation and evapotranspiration-induced cooling* (He and Jim 2010). The essential architecture of green roofs includes an upper layer of vegetation followed by a drainage layer, root barrier and waterproof membrane (Sailor 2008). Such roofs minimize heat fluxes through roof slab by reducing the temperature fluctuation and increasing the thermal capacity at the outer surface of the roof (He and Jim 2010). The other environmental benefits of green roof include reduced flood risk, improved rainwater runoff quality, UHI mitigation, energy saving, and providing urban habitats for birds (Sheng et al. 2011). A reduction in temperature by 2–7 °C (Jaffal et al. 2012) is reported by the use of green roofs, which reduces the load on the cooling requirements of the building, thereby enhancing energy savings. Another prospect for green roof is its integration with rainwater harvesting system. This system has the added advantage of natural cleaning of the runoff before it gets collected in the rainwater harvesting system and its utilization for irrigating green roofs during summer (Sheng et al. 2011).

Another feature in green technology is the use of readily available renewable energy sources like solar power and wind power. Zahi et al. (2007, 2008) reported the use of solar-powered integrated energy system used for heating, air conditioning and hot water supply. This technology was applied with success in green buildings of Shanghai Research Institute of Building Science. This helps in improving the comfort level inside the building without increasing stress on the non-renewable energy sources (Zahi et al. 2007, 2008). Generation of electricity using photovoltaic cell is also an established technology used for trapping solar energy. The percentage roof area to be occupied by solar panels and green roof could be decided based on the urban heat island effect and renewable energy potential of the area (Sheng et al. 2011).

Water conservation in green buildings is achieved by reuse of treated grey water and use of harvested rainwater after primary treatment. Also, use of water-saving devices like new-style water taps, water-saving toilets, two-sectioned water closets and water-saving shower nozzles are used in green buildings to reduce water loss (Cheng-Li 2003).

About 4% of energy use in domestic buildings and 30% of energy use in commercial buildings is for lighting. The use of technologies like intelligent lighting system, where the light is dimmed and brightened based on occupant presence and absence in a region, can reduce wastage of energy for lighting (Zhang and Cookie 2010).

5.3 Green Materials

As mentioned before, the indoor quality of energy efficient buildings was modified to overcome the 'sick building syndrome'. The main objective in improved indoor environment is the reduction in Volatile Organic Compounds (VOC) and microbial contamination. This could be achieved by the use of chemically stable materials with lesser VOC emission. Methods like red listing of chemicals, green screen and green spec project guide are used for identifying greener building materials (Atlee 2011).

The selection of green building materials is based on three main criteria (Hoang et al. 2009):

- Materials certified by a third party or listed on a well-established directory of green materials;
- Materials prevalent in residential buildings and schools, and commonly used for ceiling, flooring, wall coverings or cabinetry and
- Materials available as both unfinished and finished products (tinted or UV-coated).

The use of these materials along with natural ventilation and heat control enhances the indoor environmental quality of green buildings.

Another goal of green buildings is to reduce the wastage of resources and to promote reuse as far as possible. To achieve this goal, locally available recycled materials and construction materials from demolition sites are used in green buildings.

6 Green Buildings in India

In India, the urge to convert to green is increasing with newer environmental regulations. US Green Building Council estimates a 20% growth in the green building sector in India by 2018 (in comparison to 2015) (http://economictimes. indiatimes.com/wealth/real-estate/green-building-to-grow-by-20-in-india-by-2018-usgbc/articleshow/51045482.cms). GRIHA (Green Rating for Integrated Habitat Assessment) and IGBC (The Indian Green Building Council) act as apex bodies for certification of green buildings in India. Both the agencies certify buildings based on a fixed set of criteria including site selection and planning, sustainable architecture and design, water conservation and energy savings, construction management, building materials, indoor environmental quality, performance of the building and so on. (http://www.grihaindia.org/files/GRIHA_V2015_May2016. pdf, https://igbc.in/igbc/redirectHtml.htm?redVal=showGreenNewBuildingsno sign#CertificationLevels).

India being a tropical country has the advantage of abundant solar radiation and hence the design of green buildings is carried out considering this factor. GRIHA recommends a window to wall ratio of 60% to maximize the use of daylight, at the same time an orientation with longer facades facing north and south and well-insulated walls and windows are suggested for improving the thermal comfort. The use of construction materials with less embodied energy is also encouraged for better thermal comfort. Use of recycled materials is promoted, however the material selection is to be carried out carefully to avoid adverse impact on health of the inhabitants, projecting the importance of indoor environmental quality. Water conservation is promoted in green buildings with emphasis on reuse of treated wastewater for irrigation. Another major aspect highlighted is the use of renewable energy system and minimization of wastes (http://www.firstpost.com/business/challenging-environment-how-green-buildings-in-india-will-enhance-your-quality-of-life-2739872.html).

India's first platinum rated green building is the CII-Sohrabji Godrej Green Business Centre in Hyderabad, Telangana (Refer Fig. 6), which was certified in 2004. This building incorporates many green features like green roof, solar panels, wind tower and zero water discharge. The green roof covers 55% of the total roof area. With an extensive courtyard, 90% of this building is lighted up with sunlight. An extensive array of photovoltaic cells generates 16% of the building's energy demand. The wastewater from the building is subjected to root zone treatment and is utilized for irrigation and low flush toilets, achieving zero water discharge along with 35% reduction in the use of municipally supplied potable water (Thüring 2012).

Another platinum-rated green building in India is Suzlon 'One Earth' in Pune, Maharashtra covering an area of 41,000 m^2 with a capacity to house 2,300 people and is one of the largest green buildings in India. The major features of this building include use of low energy construction materials (70% of materials used in interiors); renewable energy based LED street lighting (25% reduction in total lighting load); daylight utilization and occupancy sensors controlled task lighting

Fig. 6 CII-Sohrabji Godrej Green Business Centre. *Source* http://www. greenroofs.com/content/ guest_features005.htm

(20% reduction in energy cost); efficient ventilation system and stormwater–rainwater management system (Suzlon global headquarters 'Óne Earth' receives 'LEED Platinum' certification).

7 Conclusion

Social and economic development post-industrialization has led to haphazard development of towns and cities. As a result of this unplanned development, many environmental and health issues like urban heat islands, air pollution and sick building syndrome have been observed. Probable solutions can be sought in the habitat design of the wildlife. Simultaneously, innovative architectural and environmental concepts can be explored. Green buildings provide an established pathway to sustainable development. They encourage the efficient use of material resources for creation of indoor comfort while minimising the wastage of resources. Therefore, there is an acute need to harmonise man made and natural environment which entails a detailed understanding of the *nature of construction while constructing in nature.*

Bibliography

Akbari H, Taha H (1992) The impact of trees and white surfaces on residential heating and cooling energy use in four Canadian cities. Energy 17(2):141–149

Akbari H, Pomerantz M, Taha H (2001) Cool surfaces and shade trees to reduce energy use and improve air quality in urban areas. Sol Energy 70(3):295–310

Atlee J (2011) Selecting safer building products in practice. J Cleaner Prod 19:459–463

Chau CK, Tse MS, Chung KY (2010) A choice experiment to estimate the effect of green building experience on preferences and willingness to pay for green building attributes. Build Environ 45(11):2553–2561

Cheng-Li C (2003) Evaluating water conservation measures for green building in Taiwan. Build Environ 38:369–379

Composites basics: materials (Part 4). Accessed 20 Jan 2012. Available from World Wide Web. http://www.mdacomposites.org/mda/psgbridge_cb_materials4_other_constituents.html

Doulos L, Santamouris M, Livada I (2004) Passive cooling of outdoor spaces. The role of materials. Sol Energy 77:231–249

Eastgate Centre, Harare. Accessed 19 Jan 2012. Available from World Wide Web. http://en.wikipedia.org/wiki/Eastgate_Centre,_Harare

Energy efficiency in buildings, Sustainable energy regulation and policy making for Africa. Accessed 6 Jan 2017. http://www.unido.org/fileadmin/import/83276_Module19.pdf

Franzoni E (2011) Materials selection for green building: which tools for engineers and architects? Procedia Eng 21:883–890

Hansell M (2005) Animal architecture. Oxford University Press, Oxford

Hansell M (2007) Built by animals: the natural history of animal architecture. Oxford University Press, Oxford

Harlan SL, Brazel AJ, Prashad L, Stefanov WL, Larsen L (2006) Neighbourhood microclimates and vulnerability to heat stress. Soc Sci Med 63:2847–2863

He H, Jim CY (2010) Simulation of thermodynamic transmission in green roof ecosystem. Ecol Model 221:2949–2958

Hoang CP, Kinney KA, Corsi RL (2009) Ozone removal from green building materials. Build Environ 44:1627–1633

Hulme M, Jenkins GJ, Lu X, Turnpenny JR, Mitchell TD, Jones RG, Lowe J, Murphy JM, Hassell D, Boorman P, McDonald R, Hill S (2002) Climate change scenarios for the United Kingdom: the UKCIP02 scientific report. Tyndall Centre for Climate Change Research, School of Environmental Sciences, University of East Anglia, Norwich

http://economictimes.indiatimes.com/wealth/real-estate/green-building-to-grow-by-20-in-india-by-2018-usgbc/articleshow/51045482.cms. Accessed 6 Jan 2017

http://www.grihaindia.org/files/GRIHA_V2015_May2016.pdf. Accessed 6 Jan 2017

https://igbc.in/igbc/redirectHtml.htm?redVal=showGreenNewBuildingsnosign#CertificationLevels. Accessed 6 Jan 2017

http://www.firstpostcom/business/challenging-environment-how-green-buildings-in-india-will-enhance-your-quality-of-life-2739872.html. Accessed 6 Jan 2017

IPCC (2007) Climate change 2007: impacts, adaptation and vulnerability. In: Parry ML, Canziani OF, Palutikof JP, van der Linden PJ, Hanson CE (eds) Contribution of Working Group II to the Fourth Assessment Report of the Intergovernmental Panel on Climate Change. Cambridge University Press, Cambridge, pp 541–580

Ishay JS, Barenholz-Paniry V (1995) Thermoelectric effect in hornet (Vespa orientalis) silk and thermoregulation in a hornet's nest. J Insect Physiol 41(9):753–759

Ishay JS, Litinetsky L, Linsky D, Lusternik V, Voronel A, Pertsis V (2002) Hornet silk: thermophysical properties. J Thermal Biol 27(1):7–15

Jaffal I, Ouldboukhitina SE, Belarbi R (2012) A comprehensive study of the impact of green roofs on building energy performance. Renew Energy 1–8

Paul WL, Taylor PA (2008) A comparison of occupant comfort and satisfaction between a green building and a conventional building. Build Environ 43:1858–1870

Sheng LX, Mari TS, Ariffin ARM, Hussein H (2011) Integrated sustainable roof design. Procedia Eng 21:846–852

Sailor DJ (2008) A green roof model for building energy simulation programs. Energy Build 40:1466–1478

Santamouris M, Synnefa A, Karlessi T (2011) Using advanced cool materials in the urban built environment to mitigate heat islands and improve thermal comfort conditions. Sol Energy 85:3085–3102

Sick Buildings. Accessed 24 Jan 2012. Available from World Wide Web. http://www.phe.bwk.tue.nl/Master_Courses/Lectures/7y900-HealthComfort/7y790-Sick%

Sick or Health? Accessed 24 Jan 2012. Available from World Wide Web. http://www.phe.bwk.tue.nl/WetPub/Documenten/Onderwijs_diversen/SickorHealth.ppt

Smith C, Levermore G (2008) Designing urban spaces and buildings to improve urban sustainability and quality of life in a warmer world. Energy Policy 36:4558–4562

Suzlon global headquarters 'One Earth' receives 'LEED Platinum' certification. Accessed 24 Jan 2012. Available from World Wide Web. http://www.suzlon.com/images/Media_Center_News/166_LEED%20Press%20Release%20Draft_042910_120pmCST_Final.pdf

Termite. Accessed 19 Jan 2012. Available from World Wide Web. http://en.wikipedia.org/wiki/Termite

Thüring C (2012) Green buildings. Accessed 24 Jan 2012. Available from World Wide Web. http://www.greenroofs.com/content/guest_features005.htm

Wanga W, Zmeureanu R, Rivard H (2005) Applying multi-objective genetic algorithms in green building design optimisation. Build Environ 40:1512–1525

Yang F, Lau Stephen SY, Qian F (2010) Summertime heat island intensities in three high-rise housing quarters in inner-city Shanghai China: building layout, density and greenery. Build Environ 45:115–134

Zahi XQ, Wang RZ, Dai YJ et al (2007) Solar integrated energy system for a green building. Energy Build 39:985–993

Zahi XQ, Wang RZ, Dai YJ et al (2008) Experience on integration of solar thermal technologies with green buildings. Renew Energy 33:1904–1910

Zhang F, Cookie P (2010) Green buildings and energy efficiency. Accessed 24 Jan 2012. Available from World Wide Web. http://www.dime-eu.org/files/active/0/Cooke-2010-Fang-Green-building-review.pdf

The Use of Vegetation as Expression of a Happy Memory in Jean Nouvel's Recent Projects

Smaranda Maria Todoran

Abstract The abundance of studies on the memory of architecture in recent years as well as a growing preoccupation for ecological approaches represent the premises of the present study. In particular, French architect Jean Nouvel's work of the last two decades, the two issues intermingle in an apparent paradox. On one hand, his preoccupation for context as a major condition for the creative process breaks from the classical tradition of visual and historical references, and in general, from any form of art history and aesthetics. The use of vegetation appears as a natural solution. On the other hand, the temporal dimension, intrinsic to a memory value attributed to the idea of monument, operates profoundly in the possibility of this superposition of context and vegetation. The paper tries to establish a common ground of temporality between the two which may become relevant for a new contemporary understanding of the memory of architecture and to determine its relation with the past. This has been done following his theoretical discourse as well as his architectural projects, also comparing the grounds of his thesis with reflections on the concept of context expressed by Frampton, Rossi, Eisenman, and Sola Morales. Paul Ricouer's work on memory fundaments the idea of a happy memory, unchallenged by the burden of the past, perceived as an anteriority that *has been*, rather than one that *no longer is*. The study wishes to reveal a form of understanding memory, specific to weak architecture and Deconstructivism, based on the idea of becoming and integrating forgetting as an acceptance of the fragile condition of city, architecture, and vegetation alike.

Keywords Memory · French contemporary architecture · Ephemeral Vegetation · Context · Weak architecture

S. M. Todoran (✉)
Department of Architecture, Faculty of Architecture and Urbanism,
Technical University of Cluj-Napoca, Cluj-Napoca, Romania
e-mail: arh.smaranda.todoran@gmail.com

© Springer International Publishing AG 2018
A. K. Sarma et al. (eds.), *Urban Ecology, Water Quality and Climate Change*,
Water Science and Technology Library 84,
https://doi.org/10.1007/978-3-319-74494-0_3

1 Introduction

This research aims to study, inside the extended field of the memory of architecture, one of its contemporary forms, related to the innovative use of vegetation in architectural works. We shall refer to the work of Jean Nouvel, whose extensive critical discourse questions the problematic nature of contemporary architecture—the very possibility of its existence, inside or out of what he sees as a too fulfilled tradition. In this perspective, we propose the understanding of memory as formulated by Paul Ricoeur, in his vast work, *La memoire, l'histoire, l'oubli*, according to whom, memory is "about the past".

This assumption creates an apparent difficulty. Contemporary philosophical discourse questions the issue of memory (the abundance of writings on memory since the 90s stands as proof), but it has not entirely untied memory from architecture as one of its basic functions. Thus, although a philosopher like Jacques Derrida reevaluates and deconstructs the fundamentals of thought, he states however that architecture's role as guardian of the memory of the city remains one of its four defining constants (Derrida 1987). A tight relationship, an invariant, architecture seems to remain memory and memory, architecture: memory is a form of architecture (Bourgeois 1999). Without conveying this second premise with an axiomatic value, we shall note that it provides an identification of the aporia itself: if memory refers to the past, then the difficulty seems to be situated, in the contemporary context, in the understanding of the past as value. Because it is within the past that the doubts appear: the crisis of modernity emerges here, whether we understand it as modern movement or as an entire western tradition. The present paper discusses two fundamental theses which propose solutions for this crisis (Aldo Rossi's *The architecture of the City* and Kenneth Frampton's *Critical Regionalism*), but also, Peter Eisenman's more radical approach, presented in his essay *The End of The Classical: The End of the Beginning, the End of the End*. The latter exposes the failure of the classical system, with its temporal values, built upon the idea of a linear time, which presupposes the idea of a beginning as origin and of an end as purpose. The recent eras of great confidence in the future and respectively great respect for the past, before and after the 60s, prove both to be, for Jean Nouvel, disappointing. The general motivation of the paper presents itself at this point as a questioning on how we can address memory as a fundamental issue in architecture, inside the new paradigm, which integrates the recognition of the crisis.

The use of vegetation in Jean Nouvel's architecture constitutes the premise for a more particular motivation of the study for two reasons. On one hand, its recurrence during the last decade and a half in the French architect's work is linked to the idea of context (expressions such as "hyper-contextuality" and "hyper-specificity" are granted a particular place in his discourse). On the other hand, vegetation introduces a temporal dimension to architecture, stressing an idea of a past which no longer has to do with the linearity of a classical architecture of memory, but rather with the cyclical rhythms of nature. This first and brief observation on the different temporalities comprised by nature and architecture remains to be nuanced throughout the paper, but it also reformulates the motivation of the study as a question on how an architecture which integrates vegetation at such extent remains a memory-fundamented architecture, and, if memory refers to the past, as stated by Ricoeur, what kind of memory is at stake here?

The paper first asserts that there is a common ground where the interest for both context and vegetation meet in recent Jean Nouvel projects. We try to determine the characteristics of this commonality, showing what it is not: it is not a common ground of materiality, of imagery, of form, of general aesthetics, of art history. Further on, we advance the hypothesis of a common ground of temporality. The recent works of the architect thus seem to be situated in a particular paradigm involving an idea of the past. Could they then reveal a conception of architecture as memory? This is the question we try to answer to as a conclusion of the paper.

2   Use of Vegetation and Context: Common Ground

The recurrence of vegetation in recent Jean Nouvel projects stretches from such varied situations as the Opera House in Seoul or Dubai, The Guggenheim Museums in Guadalajara, Rio de Janeiro or Tokyo, the Icelandic National Concert and Conference Hall in Reykjavik or the Parisian projects (of which The Cartier Foundation and the Quai Branly Museum are probably the best known). Each time, the use of vegetation is linked to the importance of the local context—both of interest to the architect in a specific way.

An interview with the architect published in number 112/113 of *El Croquis* magazine, dedicated to Nouvel's work between 1994 and 2002, stresses the importance of the context in his projects and also launches a debate on the use of vegetation as architectural material. Throughout the creative process, the analysis of the "specific" context is of major importance, whether it represents a given situation of the site or the very finality of the project (as is the case of the Danish Radio Concert Hall, where the preexisting data is absent, thus forcing the architect to design a project that can itself generate context.) In both situations, the idea of context, or hyper-context, as Nouvel prefers, signifies a force that must be understood in its specificity, beyond generalities and above all, in a nonvisual code of references. New architecture adopts its pre-existing context, marking it, making it reverberate, without, however, continuing a historical and visual tradition. Matter is not by itself—as it is visually structured in the historical city—a model, just as the buildings that are meant to create context do so by generating symbolical power rather than by exercising a visual presence.

The interview further develops the recent interest of the architect's work in vegetation as architectural material. The shift from artificial to living matter is not— in the same way as it is not for the context—a shift in terms of visual references. On one hand, it represents a rebounding with the a priori urban context which is always, after all, also a geography, a territory, the true founding element (Nouvel 2010). Towns have drifted away from nature, but nature is not an invented element;

it reappears from the desire to tighten the relationship between town and nature without however considering modernity and technology as enemies (Nouvel 2010). Vegetation itself is in the architect's understanding, following the tradition of French landscaping, a structured, organized material. We may observe at this point that vegetation is used above all as something *from* the context—something vital, at the same time leaving aside the architectural context, which is historically exhausted. The interest for context and nature are distinctive from an interest in visual effects, in scale, proportions, composition; the aim is rather, as we have seen, to acquire symbolical force, or to reveal "the essence" of the building, as Nouvel asserts, with confessed ambiguity (Baudrillard and Nouvel 2000).

Eliminating the interest for the visual, a common ground for the use of vegetation and the attention to context in Nouvel's work may be the temporal dimension. Before fully embracing this hypothesis, let us first evaluate the specificity of the temporal dimension introduced by vegetation, and further on, by Nouvel's understanding of context.

2.1 Vegetation and Temporality: Disappearance and Becoming

Jean Nouvel stresses the importance of the ephemeral dimension that vegetation expresses. The fugacious becomes relevant as opposed to the perennial—temporal value which produces repetitive architectures, clones, models. While these are contrary to poetical emotions and fail to be authentic (Baudrillard and Nouvel 2000), the ephemeral values the very moments that permanence misses. Emotion resides in the passing of things, in the glimpse of light through the leaves, in the short breeze blowing in the forest (Nouvel 2001). The passing of things does not refer to the visual perception, it does not intervene with scenic effects, following the model of Baroque compositions, but it is rather the composition itself that integrates the ephemeral dimension. As suggested by Anne Cauquelin, we have to do with a renunciation with the laws of perspective that govern the tradition of western landscape painting, and thus, with a rupture from the classical relation established between art and nature (Cauquelin 2000). The disruption marks a diminished interest for "what is seen", for the materiality and weight of things, toward the very disappearance of form and space, which leaves behind a kind of essence. This essence is a complement of matter, seized by an uncertain eye (Nouvel 2001), supported by architectural and poetical elements that Nouvel calls "nothings", "mouvences", or vibrations (Nouvel 2001). The economy of visual effects can be traced back to classical Greek culture, following Aristotle, for whom the visible— while it is capable of seducing the public—is foreign to art and poetics. It is possible to draw a parallel with theater and cinematography, who draw poetic effect from the ephemeral, the sequential and passing image, which reverberates even after it has disappeared. The association between movement and image related to cinema are to be found in Gilles Deleuze's writings who develops a semiology of force. Ephemeral images may produce the event and the event reveals meaning—an ambiguous, ambivalent one. There is no true meaning—sense and nonsense communicate and change places continuously (Deleuze 1985). In this light, architecture is pure event (Baudrillard and Nouvel 2000) and the use of vegetation offers architecture the occasion for the revelation of meaning. The ephemeral of the architectural effect relies on a special relation with disappearance—a natural and familiar relation which is also a condition of appearance and reappearance of things, ultimately, of becoming. This apparently stands against the contemporary pro-preservation discourse who associates disappearance to an irremediable lose, and demands a sustained effort to avoid its occurrence. What is at stake in the fugacity of vegetation is, unlike with other materials, the very positivity of accepting disappearance as a precondition of becoming.

2.2 Context: A Quest for the Origins and Values of Architecture

In the description of his work, on at least two occasions, Jean Nouvel chooses vegetation as a solution to exist in a too fulfilled context: the project for the Belfiore hotel in Florence and the Guggenheim Temporary Collection in Tokyo. Because in certain cities, the idea of architecture is so perfect that the very challenge of confronting it is discouraging, artifice and nature are chosen as part of an indispensable strategy (Nouvel 2001).

The too fulfilled context has reached the limits of form. Thus, analyzing the existing parameters of the site as part of the creative process, toward the creation of a specific architecture, does not aim at the analysis of form as historical tradition. The specific situation is found within those elements that can retrace a primitive space, where nature is the true inspiration. This is the case of the forest on the Branly quay, which, although the architect intends as a counterpoint to the close-by Champs de Mars, is, first of all, a sacred wood, proper for the collection hosted by the museum, but outside a presumptuous western conception of monument and monumentality.

However, the question remains whether the preference for a natural context engages a specific temporality or temporal preference. In order to establish this aspect, we propose a comparison between the idea of context as defined by Nouvel and the view of two major authors, Kenneth Frampton and Aldo Rossi.

Frampton's interest for the tectonic, the topographic, the climatic is manifest as the fifth point in the program of critical regionalism. Observing the late modernists' failure to seize the complexity of the site, which results in a sense of placelessness, the topography of the region is embraced, following Heidegger's thesis, as the ultimate resource for specificity. This regionalism functions by exclusion of everything that does not belong, i.e., technology and the megalopolis.

A well-defined boundary of the region is the very premises of an architecture of resistance (Frampton 1983). However, this confinement proves contradictory when it comes to buildings like Utzon's Bagsvaerd church, which is praised not only for its local characteristics but also for its universality (Paterson 1995). For Jean Nouvel, there is no such conflictuality—we have previously stressed a sense of harmony he aims at in combining vegetation and technology. His interest is territorial rather than regional and his architecture is not one of resistance but rather, one of existence in a context oversaturated with cultural references. Frampton's proposed rupture is only with recent development of modernism, which has been led astray; his tectonics aims at rebounding with an earlier modern tradition which goes back to the very purity of the origins of architecture. Tectonics, following Karl Bötticher, is for Frampton nothing else but form adequate to function, as expressed by Greek columns (Frampton 1983). Turning to tectonics is a return to the origins. But for Jean Nouvel, it is not an origin which is at stake. The exclusion does not concern a recent past but rather the entire cultural tradition. The idea of context does not, in his case, imply a value system that would place nature at the origins of architecture—which would imply the same classical evolutionary process over again.

Another idea of the context is developed by Aldo Rossi. Unlike Frampton, he gives up the quest for the origin and purity of modernism, taking a step away from the Italian Tendenza. The author of *Architecture of the city* is a critic of the concept of context, observing the apparently unsolved conflict between place and design, between the specific nature of place and the rational character of design (Rossi 1984). Context finds an opposite in the idea of monument—which has a reality that can be analyzed, and that becomes, by extension, the town itself. The architectural form of the city is constituted by the summing up of singular, well-identified monuments. They are just like dates in the calendar—a representation of the passing of time, their importance resides in the very way they reflect the moment of their creation. But Rossi also asserts that this reflection of time manifests itself with a certain constancy, in a permanent way—thus monuments or rather cities, as sum of monuments, developing into types (Rossi 1984). Typology is possible when history ends and memory begins (Eisenman 1984), transforming human artefacts into products of collective memory. Introducing the idea of memory, the object becomes the holder of itself and of a previous self, thus preserving a certain authenticity and sense of origin.

The town appears as a large house of memory, a house of illusion and death where history and function are over where the memories of individual childhood homes are gone; collective memory is what remains (Eisenman 1984). A crisis of history signals a contradiction inside the concept of context—founded on the appreciation of the monument as unique and singular. Aldo Rossi's thesis, even if it stays within a humanist–modernistic paradigm, admits the end of eternal rebounding with the origins of architecture. However, his discourse is still run by a belief in permanence, in a persistence of things—the buildings that, even drained of function, remain in the city being intrinsic to its very existence (Rossi 1984).

In this way, an idea of context is possible, as expression of the persistence which resides within the idea of a collective memory.

Both for Frampton and Rossi, there is the evidence of a crisis within the architectural tradition, manifest as crisis of the modernity. But the solution to the crisis is always found inside this tradition, whether it is in its origin or in its permanence. Origin and permanence as valid architectural values maintain a place of relevance for the past. But they are not so in the case of Nouvel's architecture. The interest for vegetation does not reveal a quest for a pure origin nor does permanence represent a value—we have seen that it is rather its counterpart, fugacity, which is at stake. At this point, we shall try to clarify what type of relation Nouvel's architecture establishes with the past and implicitly, if there is a sense of persistence—or rather, duration intrinsic to his work.

3   Memory and Sense of the Past in a Vegetation Using Architecture

Because neither origin nor permanence represent a major interest in Jean Nouvel's work, we have further asked on what terms there may be a discussion in terms of "value" on the past. Following the thesis of Eisenman (1996), we are now at the end of the classical and thus, outside the paradigm of permanence. The architect speaks of an end of the beginning (as there is no intrinsic value, there is no need for an origin to found it); he also speaks of an end of the end (architecture is free of the idea of a purpose, and thus of the myth of the end following progress and the linearity of history) (Eisenman 1996). Modern utopia was, from the point of view of finality, a failure to keep alive the expectations regarding the future, a future which is no longer threatened by its value load. Like Nouvel, Eisenman defines architecture after the classic tradition as text, no longer preoccupied with scale, image, and proportions. Time is then freed of its burden: a timelessness space in the present unpressured by an idealized past or an ideal future (Eisenman 1996).

The fact that architecture no longer wishes to retain the past (at least not a historical, classical past as represented by art history) and avoids the vocation of permanence, belonging first of all to a passing, changing present, always *becoming*, can be translated by a renunciation to the value paradigm. The very idea of monument functions however, within this paradigm, followed by a set of institutional means in the service of the concept. Preservation is what is destined for the classical monument: an ambition strange to the nonclassical model to which Nouvel adheres.

Yet, a type of ambition still resides within the architect's approach. In Florence, "a town so perfect", he adopts a courageous position: it is in such a town that the architect must prove at least some ambition (Nouvel 2001). More than *some* ambition is required in the case of the Phare Tower of Paris which is to match the monuments of the past, more particularly, Madame Eiffel. There is a manifest desire to equal the past, to exist alongside its monuments. Unlike Anthony Vidler's thesis on the crisis of confidence in monumentality expressed by the glass architecture of recent Parisian presidential projects (Vidler 1998), the ambition proclaimed by Nouvel seems compatible with a certain idea of monument.

The ephemeral is linked to emotion, familiarity, poetical effects. A sensible stake is sought for, not as a substitute for intelligence, as suggested by Lyotard, but by integrating it outside a classical tradition. Achieving poetical emotion is an intellectual job (Baudrillard and Nouvel 2000) as far as it breaks away from the classical way of thought with its fictions of representation, simulating meaning, of reason, simulating truth and historicism, simulating permanence (Eisenman 1996). The preservation of monuments of the past and the future preservation of the *singular object* is based not on these values, or at least not exclusively, but rather on the emotion they may still produce (Baudrillard and Nouvel 2000). Small gestures, drops of rain and rays of light—these are at the origin of what makes emotion last and offer a support to recollection.

Ignasi de Sola Morales finds that it is within these small gestures resides the very force of what he names, following Giani Vatimo, "weak architecture". For him, the idea of monument relevant for the architectural object and for its memory is an idea related to an openness toward a more intense reality; it reverberates like the sound of bells which continue to echo long after they have ceased to toll. This echo is a form of residue, a form of remembering. Sola Morales further distinguishes between the meaning of the monument as defined by Aldo Rossi—which implied a sense of permanence, intrinsic to his static understanding of the city—and a dimension of the monument which relies on its resonance, like poetry once it has been heard, like architecture, once it has been seen. The power of weak architecture is thus within this mild, tangential position, unaggressive, and un-dominant (de Sola Morales 1998).

Recollection is for the author the only thing that the classic and the contemporary monument have in common. Architecture's ambition to exist can thus be explained today as an ambition not to last, to reveal meaning, or to say the truth but rather the ambition of this familiarity which is in fact another form of lasting, through the power of recollection. Since we have identified recollection (following Eisenman and Sola Morales) as the unifying element on the terrain of both the classical and the contemporary, this kind of memory also implies a new relation to the past.

Ricoeur offers a possible answer. Following Bergson, he centers on the ideas of recognition and survival as central to Bergon's *Matiere et memoire,* and to the phenomenon of remembering. The inscription of a memory represents the very survival of the contemporary image of the original experience which can some time be recognized again. Survival itself is but a form of forgetting, a fundamental form, which designates the very persistence of a memory outside the limits of consciousness: forgetting thus preserves. This result clarifies a distinction between a type of anteriority that *was* and one that *no longer is* or between a type of forgetting as resource for memory and one as inexorable destruction (Ricoeur 2000).

Recollection in the proximity of such a constructive form of forgetting becomes relevant for Nouvel's work with vegetation. The memory implied by his architecture is on the terrain of this fragility of remembrance where the existence of disappearance as forgetting is essential to a reappearance as recognition. Because the idea of survival works inside this mechanism allowing the possibility of becoming, what really lasts in the process is the idea of this becoming, something that does not produce clones and still invites to recollection. Such memory implies a relation to a past that is no longer "absence to recover" but an event of "happy recognition" assuming the idea of forgetting as a fragility of the town and city.

4 Conclusion

We have seen that Jean Nouvel's recent work testifies for a recurrent interest regarding two major issues, context and the use of vegetation. The context resists through its force, through its symbolic power—away from visual or historical references. The use of vegetation promises to achieve this distance, toward a curing of a disappointed contemporary gaze.

From this point of view, the study was situated around the temporal dimension, which both context and vegetation exploit. Vegetal time is the ephemeral, the passing by, opposed to the permanence presumed by classical architecture. Disappearance is intrinsic to this temporal dimension, a disappearance as premises for becoming, following Deleuze and Baudrillard. The problem of a too fulfilled context—the problem of architecture's very existence—is solved in several recent Nouvelian projects by the use of vegetation. Unlike Frampton or Rossi who try to retrieve in the origin or permanence of architecture its salvation, with Nouvel, there is no such immovability at stake.

At this point, the question on the memory of contemporary architecture using vegetation as material becomes the question on its capacity to last. We have seen that there is an ambition to the monument status within Nouvel's approach. But we have proposed a type of memory and duration, following Ricoeur, in the proximity of forgetting, on a fragile ground. Its force comes from its acceptance of the disappearance of things as premises for their becoming. The use of vegetation as expression of a happy memory of architecture may open a discussion on a new approach to an era that reflects on the status of its inheritance, between respect and burden, duty and fear of danger. Vegetation is in this context a manifesto for living—for a vital memory.

References

Baudrillard J, Nouvel J (2000) Singular objects. University of Minesota Press, Mineapolis
Bourgeois L (1999) Interview with Josef Helfenstein, New York, 31 Mar 1999. Louise Bourgeois: memory and architecture. Museo Nacional Centro de Arte Reina Sofía. Aldeasa, Madridm, p 26

Casey E (1193) Getting back into place—toward a renewed understanding of the place-world. Indiana University Press, Bloomington

Cauquelin A (2000) L'invention du paysage. Quadrige, Puf

de Sola Morales I (1998) Weak architecture. In: Hays M (ed) Architecture theory since 1968. The MIT Press Cambridge, London

Deleuze G (1985) Cinema 2: l'image temps. Les Editions de Minuit, Paris

Deleuze G, Guattari F (1980) Capitalisme et Schizophrenie 2. Mille Plateaux. Editions de Minuit, Paris

Derrida J (1987) Maintenant l'architecture. In: Psyché. Inventions de l'autre. Galilée, Paris, pp 477–492

Eisenman P (1996) The end of the classical: the end of the beginning, the end of the end. In: Nesbitt K (ed) Theorizing a new agenda for architecture: an anthology of architectural theory 1965–1995. Princeton Architectural Press, New York

Frampton K (1983) Towards a critical regionalism: six points toward an architecture of resistance. In: Foster H (ed) The anti-aesthetic. Essays on postmodern culture. Bay Press, Port Townsend

Garcia GE, Moreno CD (2002) Interview [A conversation with Jean Nouvel]. El Croquis: Jean Nouvel 1994–2002, 112/113

Paterson S (1995) Critical analysis of "towards a critical regionalism" by Kenneth Frampton. http://home.earthlink.net/~aisgp/texts/regionalism/regionalism.html

Ricoeur P (2000) La mémoire, l'histoire, l'oubli. Seuil, Paris

Rossi A (1984) The architecture of the city. The MIT Press, Cambridge

Vidler A (1998) The architectural uncanny: essays in the modern unhomely. In: Hays M (ed) Architecture theory since 1968. The MIT Press, Cambridge

Soil Leaching Behaviour of Different Urban Landscapes

Banasri Sarma and C. Mahanta

Abstract Leaching behaviour of soil has an significant role in influencing composition of water and it is important to study the soil leaching behaviour to
understand their role in modifying water composition. Guwahati, the capital city of
Assam exhibits high soil erosion rate from its urbanized hilly catchments leading to
their subsequent role in leaching of ionic species to water. In this paper, soil and
sediment samples from 11 different sites representative of hilly and valley areas of
the Guwahati City were collected and their leaching behaviour was studied using
the standard Toxicity Characteristic Leaching Procedure (TCLP) developed by
US EPA (Townsend et al. 2003). The study evaluated the total leachable amounts
of some cations (Na, K and Ca) and trace metals of concern (Pb, Fe, Ni, Mn, Cr and
Cd) and their role in influencing the surface water chemistry. The average leaching
was observed in the following order: Ca > Fe > K > Na > Mn > Pb > Ni.
Leaching of Cr and Cd was below detectable limit. The study indicated that the
leaching of dissolved nutrients and trace metals varies with site condition, background condition and level of saturation. While there may not be a generalized
leaching behaviour for these soils, characterization at local level may be possible by
the current approach. Overall analysis of all the samples revealed that the soil
samples with high initial concentration of ions can leach out higher ionic species.
Since a more or less linear trend was observed between the ions present in soil and
their subsequent amount of leaching, it is likely that increased toxic contamination
of urban soils due to anthropogenic sources can lead to higher water contamination.

Keywords Soil leaching · TCLP · Water quality

B. Sarma (✉)
Water and Sanitation Support Organization, Public Health Engineering Department Assam,
Betkuchi Guwahati, Guwahati, India
e-mail: banasris@gmail.com

C. Mahanta
Department of Civil Engineering, Indian Institute of Technology Guwahati, Guwahati, India
e-mail: chandan@iitg.ernet.in

© Springer International Publishing AG 2018
A. K. Sarma et al. (eds.), *Urban Ecology, Water Quality and Climate Change*,
Water Science and Technology Library 84,
https://doi.org/10.1007/978-3-319-74494-0_4

1 Introduction

Soil erosion delivers significant amount of sediment to downstream water bodies. When soil erosion occurs in a watershed, sediment particles ultimately find way into the water bodies. Some parts of these sediments are deposited on the bed of the water bodies and some remain in suspension throughout the water column. These sediments release its exchangeable ionic species to the water through the process of leaching unless the saturation limit of the water is reached.

Soils and sediments significantly influence the composition of water coming in contact with them. Soil releases nutrients and heavy metals to surface water and groundwater through the process of leaching. This process occurs in three ways— (i) when the rainwater moves over soil surfaces as runoff (Schipper et al. 2008) (ii) when the rainwater/runoff percolates through the soil profile to groundwater (Hansen and Djurhuu 1997) and (iii) leaching from bed sediments of water bodies (Jain and Ram 1997; Jiang et al. 2008; Butler 2009).

Natural sediments are the source of dissolved minerals to the water bodies and govern the water quality dynamics in the water column. Sediments are also the site for growth and survival of aquatic flora and fauna, which in turn interferes with the water quality. However, when sediment gain in a water body becomes too high beyond the desirable limit, it releases dissolved nutrients and pollutants in excess amount. This over nourishment condition is often detrimental to the water body, because it may lead to conditions like eutrophication. Also contaminated sediments from anthropogenic sources often become undesirable for a water body due to their addition of toxic ionic species into the receiving water (Jain and Ram 1997; Butler 2009).

It is important to study the leaching behaviour of soil to understand their role in modifying water composition. In this paper, soil and sediment samples from different sites representative of hilly and valley areas of Guwahati City were collected and their leaching behaviour was studied using the standard Toxicity Characteristic Leaching Procedure (TCLP) developed by US EPA. We evaluated the total leachable amounts of cations and some trace metals of concern and their role in influencing surface water chemistry has been assessed.

2 Study Area

Soil and sediment samples from eleven different sites of Guwahati City, Assam, India were collected for studying the leaching behaviour of the soil to make the sampling representative of the entire city. Guwahati, the capital city of Assam exhibits high soil erosion rate from its urbanized hilly catchments leading to their subsequent deposition in the natural and manmade drainage systems. To understand the process of leaching, 32 soil samples from different (11) sites were collected and their leaching behaviour was studied. The collected samples also covered soils from commercial and industrial dumping sites (sample code from S1–S11). Samples from residential hilly areas (sample code from S12–S16) and totally natural unpolluted sites (sample code from S17-S28) were also collected. Four samples from river bed sediments (sample code from B1–B4) were collected too.

The locations of the sampling sites are shown in Fig. 1 and their key features are described in Table 1.

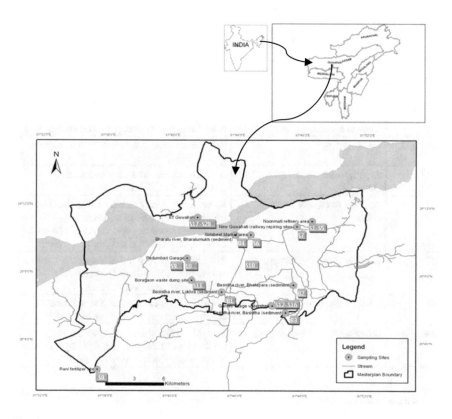

Fig. 1 Map of the study area with location of the sampling sites

Table 1 Description and code assigned for the soil sampling sites

Code	Place	Description
S1-S2	Noonmati refinery hospital road side and (near equalizer tank)	Outside the campus of an industrial area (refinery), low level of soil contamination is expected: surface soil
S3-S5	Refinery (near railway, tank area and new tank farm area)	Inside the campus of an industrial area (refinery), moderate level of soil pollution is expected: surface soil
S6	Solabeel market area	Fish market area, very high organic contamination is expected: surface soil
S7	New Guwahati Lokoshed area	An intense railway repairing site, soil contamination with oil and grease are expected: surface soil
S8	Rani fertilizer industry	Outside the campus of a fertilizer industry: surface soil
S9	Kiron automobile	Garage site, with various kind of vehicle repairing activities and standing site of automobiles: surface soil
S10	Podumbari garage	Small car repairing site: surface soil
S11	Boragaon waste dump	A municipal waste dumping site: surface soil
S12-S15	Games village	Hilly barren area due to human intervention: surface soil
S17-S27	Experimental watershed IIT Guwaahti	A vegetated area in a hilly undisturbed watershed: surface soil
B1	Basistha river, Kotabnari, Lokhra	River site with very turbid (almost dark) which receives a great deal of effluents from small-scale industries and domestic effluents, moderately polluted site: river bed sediment
B2	Basistha river, at Bhetapara	River site with very turbid water, the location receives domestic effluents, moderately polluted site: river bed sediment
B3	Basistha river, at Basistha	River site with clear water just at immediate downstream, unpolluted: river bed sediment
B4	Bharalu river, at Bharalumukh	A highly polluted river site just at the confluence of river Bharalu, dark-turbid water, highly waste loaded: river bed sediment

3 Methodology

3.1 Determination of Electrical Conductivity (EC) of Soil and Sediments

The standard BIS Code (IS 14767: 2000) was followed for determination of electrical conductivity of soil and sediment samples. The soil samples were air-dried and sieved through a 2 mm sieve. Then, 20 g of the sample was taken in a conical flask and 40 ml of water was added. The conical flasks were covered with cotton and placed in a horizontal shaking machine for 30 min. Then each sample was transferred into a beaker and the electrical conductivity of the samples was measured with a portable EC metre (Make: Wagtech). The reading of the blank solution (distilled water only) was subtracted from all the sample readings.

3.2 Soil Digestion Method

Extraction of trace metals from soil samples was carried out by following the standard ASTM methodology (D 3974-81, Reapproved 2003, Digestion Practice A) (ASTM-D 3974-81 2007). The soil samples were dried overnight at 105 °C until a constant weight was achieved and allowed to cool in a desiccator. The 4 g of the dried samples were taken into a beaker and 100 ml of distilled water was added. Then 1 ml of Nitric acid and 10 ml of Hydrochloric acid were added to the solution. The beaker was covered with a watch glass and heated on a hotplate at 95 °C. The beaker was removed from the hot plate when the solution remained within 10–15 ml. The solution was allowed to cool at the room temperature. The solution was transferred to a 50 ml volumetric flask and diluted to the volume of the flask. For determining concentration of elements for each dry sample, following formula was used:

$$C = \frac{(Q - S) \times V}{U} \tag{1}$$

where

Q Concentration of the elements in the digested soil sample, µg/ml
S Concentration of the element found in the reagent blank, µg/ml
V Volume of the extract = 50 ml
U Weight of the sample corrected to a dried sample at 105 °C = 4 g
C Trace element per g of the dry sample, in µg

For determination of the concentration of the metal in the wet sample, the following formula was used:

$$A = \frac{C \times B}{100} \tag{2}$$

$A = C * B/100$

where

A Metal per g of the wet sample, in µg
C Trace element per g of the dry sample, in µg
B Percent solid of the sample, which is determined as

$$B = \frac{M \times 100}{N} \tag{3}$$

where

M Dry weight of the sample obtained after drying overnight in oven at 105 °C
N Weight of the sample before drying

3.3   Soil Leaching Study

To study the leaching properties of soil, standard Toxicity Characteristic Leaching Procedure (TCLP) developed by US EPA (Townsend et al. 2003) was followed. The test involves extracting the elements from a 100 g size reduced sample (425 μ sieved) with distilled water. A specific L/S ratio (liquid to solid ratio); 20:1 was maintained and the mixture was rotated for 18 ± 2 h at 30 rpm. The sample was then filtered and the filtrate was used for the analysis of heavy metals and cations.

3.4   Determination of pH and Electrical Conductivity (EC) of Leachate Samples

The pH and EC of the leachate samples were tested immediately after filtration by the portable pH metre (Wagtech) and EC metre (Wagtech). These metres are pre-calibrated and the precession and accuracy of measurements were verified before analysis.

Analysis of Ionic Concentration of Leachate Samples

The digested soil samples and the leachate samples were analysed for Na, K, Ca and trace metals (Fe, Mn, Ni, Pb and Cd). The cations, sodium (Na), potassium (K) and calcium (Ca) were analysed in Flame Photometer (Make: Systronics) (APHA, WEF, AWWA 1998). Concentartions of Iron (Fe), Manganese (Mn), Lead (Pb) Cromium (Cr), Nickel (Ni) and Cadmium (Co) were determined in Atomic Absorption Spectrophotometer (AAS-Model: AA240—Varian Inc) (APHA, WEF, AWWA 1998).

3.5   Determination of Soil Loss from Watershed Area

Revised Soil Loss Equation (RUSLE) is applied to the Games Village watershed area to determine the annual soil loss from the watershed.

RUSLE states that the field soil loss in tonne per acre, A, is the product of six factors.

$$A = RKLSCP \qquad (4)$$

where

A Soil loss, tonnes/acre/year
R Rainfall and runoff erosivity index, in 100 ft * tonne/acre * in/h.
K Soil erodibility factor, tonne/acre per unit of R
LS Slope length and steepness factor, dimensionless
C Vegetative cover factor, dimensionless
P Erosion control practice factor, dimensionless

4 Results and Discussion

4.1 Leaching of Soil and Sediment

The mean, standard deviation and range of pH, EC along with the detectable ionic species as observed in the leachate samples are presented in Table 2. The average leaching was observed in the following order:

$$Ca > Fe > K > Na > Mn > Pb > Ni$$

As indicated by standard deviation values, the variability of the values is observed to be very high except for Pb and Ni. In this study, Cr and Cd was not detected in the leaching test.

The variation of pH in all the samples was almost consistent (Fig. 2), significant variation was observed for EC values (Fig. 3), indicating the variability in the ionic concentrations through all the samples.

In Fig. 4, the variation of Na, K and Ca in the leachate samples is shown. The leaching of Na, K and Ca is higher for the soil samples from the urban areas representing comparatively polluted sites (sample code S1-S11) and also for river bed sediments (sample code B1-B4) than the samples that are from the hilly watershed areas (sample code S12-S28), i.e. in Games Village watershed and experimental watershed. The leaching of Pb and Ni was more in the soil samples from the urban areas (sample code S12-S16) compared to the samples from the hilly watershed areas (Fig. 5), except for one sample from the experimental watershed of IIT Guwahati site. The leaching of Fe and Mn was much higher for the soil samples of experimental watershed of IIT Guwahati site (sample code S17-S28) than the

Table 2 Summary of the results of leaching study for the collected samples

	pH (L)	EC (L) (µS/cm)	Na (L) (µg/g)	K (L) (µg/g)	Ca (L) (µg/g)	Pb (L) (µg/g)	Ni (L) (µg/g)	Fe (L) (µg/g)	Mn (L) (µg/g)
Average	7.4	74.1	44.2	53.0	166.0	2.0	1.3	61.5	17.4
Standard deviation	0.60	88.89	99.05	88.26	200.80	1.02	0.74	74.11	23.50
Range	6.1–8.2	9.6–432	[a]BDL–571	1–394	1.2–687	0.2–3.4	0.08–2.36	BDL–188	0.06–64.5

[a]*BDL* Below detectable level

Fig. 2 pH values in the leachate samples

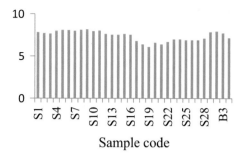

Fig. 3 EC values in the leachate samples

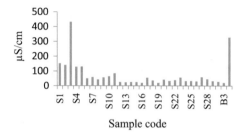

Fig. 4 Na, K and Ca concentrations in the leachate samples

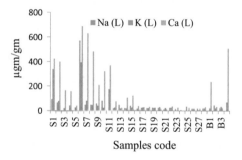

Fig. 5 Pb and Ni concentrations in the leachate samples

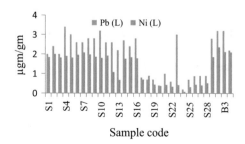

Fig. 6 Fe and Mn
concentrations in the leachate
samples

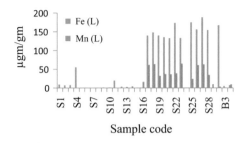

other sites (sample code S1-S16) (Fig. 6), except the Fe leachate for bed sediment
of Basistha river (sample code B2). Correlation coefficient of all soil and leachate
samples is presented in Table 3.

4.2 Leaching Behaviour of Soil of Experimental Watershed of IIT Guwahati

In Table 4, the leaching percentage and average leaching in experimental watershed
of IIT Guwahati are presented. For the samples of the experimental watershed of
IIT Guwahati, the percentage leaching is in the following order:

$$Na > Mn > Pb > Ni > Ca > K > Fe$$

The average leaching is observed in the order as follows:

$$Fe > Mn > Ca > Na > K > Pb > Ni$$

In Figs. 7, 8, 9, 10, 11, 12 and 13, correlations between the actual concentration
in soil and leached concentrations for Mn, Pb, Fe, Ni, Na, Ca and K are presented.
There were good correlations for Mn (0.85), Pb (0.94) and Ni (0.8). Moderate
correlation was observed for Fe (0.69) and Ca (0.74); whereas K (0.46) and Na
(0.54) had poor correlations.

4.3 Leaching Behaviour of Soils of Games Village watershed

In Table 5, the percentage leaching and average leaching in Games Village
watershed are presented. For the samples of this site, the percentage leaching is in
the following order:

$$Na > Pb > Ni > Ca > K > Mn$$

Table 3 Correlation coefficient of all soil and leachate samples

	pH (S)	EC (S)	Na (S)	K (S)	Ca (S)	Pb (S)	Ni (S)	Fe (S)	Mn (S)	pH (L)	EC (L)	Na (L)	K (L)	Ca (L)	Pb (L)	Ni (L)	Fe (L)	Mn (L)
pH (S)	1.00																	
EC(S)	0.42	1.00																
Na (S)	0.12	0.02	1.00															
K (S)	0.34	0.06	0.13	1.00														
Ca (S)	0.57	0.24	0.28	**0.77**	1.00													
Pb (S)	0.37	0.16	**0.69**	0.46	**0.72**	1.00												
Ni (S)	0.23	0.33	−0.01	0.57	0.66	0.48	1.00											
Fe (S)	0.25	0.28	−0.03	**0.67**	**0.72**	0.40	**0.76**	1.00										
Mn (S)	−0.11	0.03	0.09	0.31	0.24	0.30	0.49	0.63	1.00									
pH(L)	**0.72**	0.35	0.14	0.64	**0.81**	0.51	0.50	0.65	0.16	1.00								
EC(L)	0.31	**0.69**	−0.06	0.02	0.12	0.16	0.48	0.17	0.12	0.17	1.00							
Na (L)	0.20	0.06	**0.97**	0.23	0.41	**0.74**	0.06	0.04	0.08	0.26	−0.08	1.00						
K (L)	0.40	0.06	**0.78**	0.25	0.38	0.64	0.19	0.07	0.07	0.37	0.14	**0.76**	1.00					
Ca (L)	0.49	0.26	0.49	0.40	0.68	**0.78**	0.59	0.28	0.01	0.57	0.32	0.54	0.69	1.00				
Pb (L)	0.52	0.27	0.03	0.61	**0.77**	0.43	0.58	**0.75**	0.33	**0.84**	0.12	0.14	0.18	0.41	1.00			
Ni (L)	0.62	0.37	0.12	0.62	**0.84**	0.51	**0.71**	**0.74**	0.22	**0.83**	0.33	0.20	0.37	0.63	**0.83**	1.00		
Fe (L)	−0.55	−0.35	−0.10	−0.60	**−0.72**	−0.46	−0.64	−0.65	−0.25	**−0.73**	−0.26	−0.18	−0.31	−0.54	−0.64	−0.67	1.00	
Mn (L)	−0.63	−0.35	−0.09	−0.62	**−0.71**	−0.40	−0.52	−0.59	0.02	**−0.79**	−0.23	−0.18	−0.31	−0.50	−0.63	**−0.72**	**0.83**	1.00

Table 4 Percentage leaching and average leaching in the samples of experimental watershed of IIT Guwahati

	Na	K	Ca	Pb	Ni	Fe	Mn
Range of percentage leaching from the initial concentration in soil (%)	13–38	0.1–2	1.7–6.9	2.5–10	4.8–9.7	0–1.4	0.5–14.5
Average leaching (μgm/gm)	20	15	22	0.89	0.44	139	42

Fig. 7 Correlation between concentration of Mn in soil (S) and leachate (L) in the samples of experimental watershed of IIT Guwahati

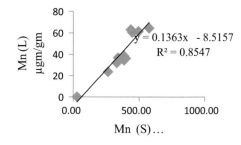

Fig. 8 Correlation between concentration of Pb in soil (S) and leachate (L) in the samples of experimental watershed of IIT Guwahati

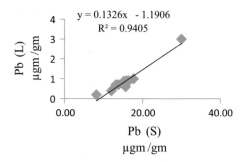

Fig. 9 Correlation between concentration of Fe in soil (S) and leachate (L) in the samples of experimental watershed of IIT Guwahati

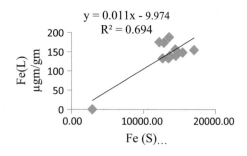

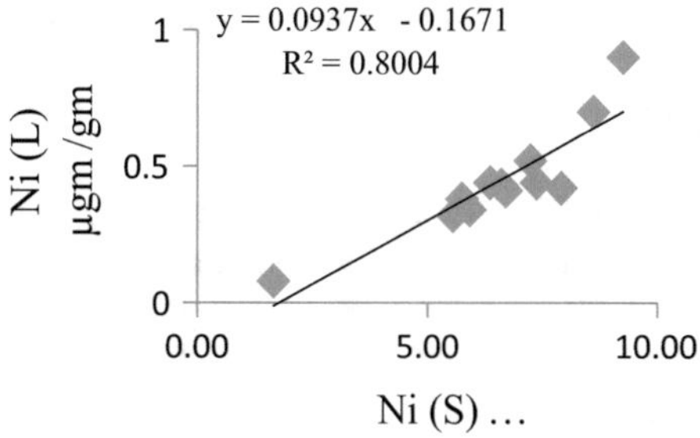

Fig. 10 Correlation between concentration of Ni in soil (S) and leachate (L) in the samples of experimental watershed of IIT Guwahati

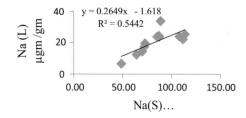

Fig. 11 Correlation between concentration of Na in soil (S) and leachate (L) in the samples of experimental watershed of IIT Guwahati

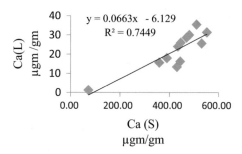

Fig. 12 Correlation between concentration of Ca in soil (S) and leachate (L) in the samples of experimental watershed of IIT Guwahati

Fig. 13 Correlation between concentration of K in soil (S) and leachate (L) in the samples of experimental watershed of IIT Guwahati

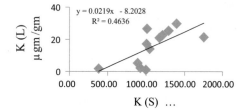

Table 5 Percentage leaching and average leaching in the samples of Games Village watershed

	Na	K	Ca	Pb	Ni	Mn	Fe
Percentage of leaching (%)	4–87%	0.35–0.89%	1.2–4.2%	6.3–7.1%	4.4–7.1%	0.01–0.03	BDL
Average leaching (µg/gm)	26	21	73	3	1.4	5	BDL

The average leaching is observed to be in the order as follows:

$$Ca > Na > K > Mn > Pb > Ni$$

For this site, Fe leaching was not detected. Also, good correlation between the actual concentration in soil and leachate concentration was observed for Na (0.91), K (0.89) and Ca (0.86). In Figs. 14, 15, 16, 17 and 18, correlations between the actual concentration in soil and leachate concentrations for Na, K, Ca, Ni and Mn are presented.

Fig. 14 Correlation between concentration of Na in soil (S) and leachate (L) in the samples of Games Village watershed

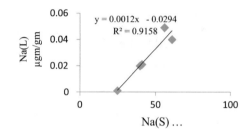

Fig. 15 Correlation between concentration of K in soil (S) and leachate (L) in the samples of Games Village

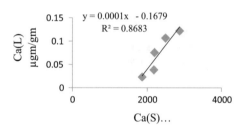

Fig. 16 Correlation between concentration of Ca in soil (S) and leachate (L) in the samples of Games Village watershed

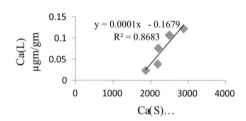

Fig. 17 Correlation between concentration of Ni in soil (S) and leachate (L) in the samples of Games Village watershed

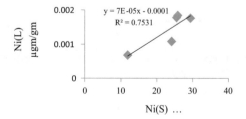

Fig. 18 Correlation between concentration of Mn in soil (S) and leachate (L) in the samples of Games Village watershed

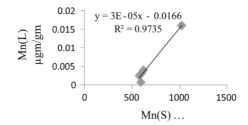

$y = 3E-05x - 0.0166$
$R^2 = 0.9735$

4.4 Leaching Behaviour of Bed Sediments

Samples from four sites (sample code B1-B4) were collected, of which two sites were moderately polluted (sample code B1-B2), one was unpolluted (Sample code B3) and one was highly polluted (sample code B4). Leaching from bed sediment was similar to that of surface sediment. However, the leaching behaviour varied from site to site. The concentration of leachates and their initial concentrations for Na, K, Ca, Pb, Ni, Fe and Mn are presented in Figs. 19, 20, 21, 22, 23, 24 and 25. The samples containing higher Na concentration also had higher leaching concentration. The K leaching was significantly low, except for the sample from the highly polluted sites, i.e. at Bharalu River at Bharalumukh (sample code B4). The Ca leaching was also not significant, except for the sample from the Bharalu river bed sediment at Bharalumukh (sample code B4) and Basistha river bed sediment at Kotabnari, Lokhra (sample code B1). The leaching concentrations of Pb and Ni

Fig. 19 Na concentration in the bed sediment and leachate samples

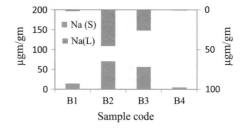

Fig. 20 K concentration in the bed sediment and leachate samples

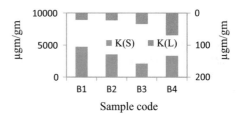

Fig. 21 Ca concentration in the bed sediment and leachate samples

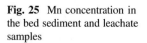

Fig. 22 Pb concentration in the bed sediment and leachate samples

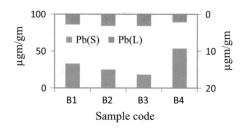

Fig. 23 Ni concentration in the bed sediment and leachate samples

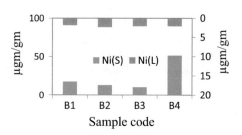

Fig. 24 Fe concentration in the bed sediment and leachate samples

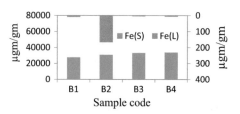

Fig. 25 Mn concentration in the bed sediment and leachate samples

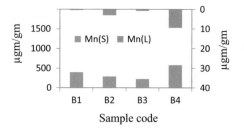

were almost similar in all the samples and their initial concentration did not seem to affect their leaching. However, Fe and Mn did not follow any trend in their leaching behaviour.

4.5   Estimation of Nutrients and Metal Flux from a Small Part of the Games Village watershed

With the understanding of the leaching behaviour of soil from the Games Village watershed, the amount of nutrients and metal loss from a small part this watershed has been estimated. The study area covered an area of about 0.17 km^2 and the RUSLE model was applied to determine the annual soil loss from the area under natural as well as disturbed conditions (Table 6). The average ionic concentration of various nutrients and metals as obtained in the soil of the Games Village watershed, the amount of annual nutrient/metal loss from the area that goes as sediment loss was computed (Table 7). Then by taking the average value of leaching concentration, the loss of leachable amount from the area in a year was computed. The value clearly showed that a significant amount of nutrient and metal loss occurs annually (Table 8).

Table 6 Annual soil loss in a small part of the Games Village watershed

Sediment yield in natural condition	Sediment yield in disturbed condition
2,116,970 kg	9,738,063 kg

Table 7 Computed nutrient/metal loss from the area

Computed nutrient/metal loss	Na	K	Ca	Pb	Ni	Fe	Mn
Natural (kg)	94	9647	4916	81	49	80,778	1445
Disturbed (kg)	434	44,377	22,615	371	228	371,577	6647

Table 8 Loss of leachable amount from the area

Loss of leachable amount	Na	K	Ca	Pb	Ni	Mn
Natural (kg)	55	45	155	5	3	11
Disturbed (kg)	254	205	713	25	14	51

5 Conclusions

The study of leaching behaviour of different soil types has indicated that the leaching of dissolved nutrients and trace metals varies with site condition and background concentration as well as level of saturation of the water. For example, the leaching of Na, K and Ca is higher for the soil samples from the urban areas representing comparatively polluted sites than the samples that are from the hilly watershed areas, i.e. in Games Village watershed and the experimental watershed of IIT Guwahati. The leaching of Pb and Ni was more in the soil samples from the urban areas compared to those samples from the hilly watershed areas. Similarly, the leaching of Fe and Mn was found to be higher for the soil samples of the experimental watershed of IIT Guwahati, which is a site covered with natural vegetation. Thus, it remains a challenge to suggest a generalized leaching behaviour for these soils. Overall analysis of all the samples from the 11 sites has however revealed that the soil samples with high initial concentration of ions do leach out more ions. Though a more or less linear trend was observed between the ionic concentration in soil and their subsequent amount of leaching, this need not perhaps be true for all ions and all samples. For samples of experimental watershed of IIT Guwahati, Mn, Pb, Ni and Ca have shown good correlation between their initial concentration in soil and the amounts of leaching. On the other hand, for samples of Games Village watershed, while Na, K, Ni and Mn have shown good correlation between their initial concentrations in soil and amount of leaching, while Fe leaching was not detected. In the study of bed sediment leaching, the samples from the polluted sites showed higher leaching of K, Ca and Mn.

The results of determination of the major cations and trace metals in soil and their amount of leaching into water are used to estimate the annual loss of nutrients and trace metals that occur due to soil erosion from the area. The area of the study watershed is a small part of Games Village watershed, with an area of 0.17 km^2. As obtained from RUSLE model, yearly soil loss is 2,116,970 kg (natural condition) and 9,738,063 kg (disturbed condition with residential development). Surface soil samples collected from this site were studied for their ionic content and leaching behaviour. Using these results, for the two scenarios—natural and disturbed—the computed nutrient and metal loss along with the leachable amount present in the soil were estimated and it was observed that watershed area loses significant amount of nutrients and trace metals annually and a part of it can go to water in the form of dissolved ions.

References

APHA, WEF, AWWA (1998) Standard method for the examination of Water and wastewater, 20th edn. APHA, Washington, DC

ASTM-D 3974-81 (2007) Standard soil digestion method for determining traces metal concentration in soil. ASTM International Standard, US

Butler BA (2009) Effect of pH, ionic strength, dissolved organic carbon, time, and particle size on metals release from mine drainage impacted streambed sediments. Water Res 43:1392–1402

Hansen EM, Djurhuu J (1997) Nitrate leaching as influenced by soil tillage and catch crop. Soil Tillage Res 41(3–4):203–219

IS 14767 (2000) Determination of the specific electrical conductivity of soils—method of test. Bureau of Indian Standards

Jain CK, Ram D (1997) Adsorption of lead and zinc on bed sediments of the river Kali. Water Resour 31(1):154–162

Jiang X, Jin X, Yao Y, Li L, Wu F (2008) Effects of biological activity, light, temperature and oxygen on phosphorus release processes at the sediment and water interface of Taihu Lake, China. Water Res 42:2251–2259

Schipper PNM, Bonten LTC, Plette ACC, Moolenaar SW (2008) Measures to diminish leaching of heavy metals to surface waters from agricultural soils. Desalination 226:89–96

Townsend T, Jang YC, Tolaymat T (2003) A guide to the use of leaching tests in solid waste management decision making. Report #03-01(A), prepared for The Florida Center for Solid and Hazardous Waste Management, University of Florida

Environmental Sustainability: Stabilized Mud Block

Atanu Kumar Dutta, Deheswar Deka, Arup Hatibaruah, Dipankar Goswami, Shashanka Sekhar Rai and Tridib Basumatary

Abstract The urban ecosystem is considerably dependent on its surrounding areas; green enriches it and pollution degrades the same. Brick kilns are a common sight in any urban periphery. In the advent of rapid urbanization, the demand for brick as filler material of buildings has been increasing exponentially. Conventional burnt brick puts tremendous pressure on the environment, as the burning process results in lot of greenhouse gases. An alternative to burnt brick is the call of the day. However, alternatives such as hollow concrete/fly ash blocks are costly. The paper examines the case of stabilized mud blocks as filler material. These blocks do not need burning and thus save the ecosystem from burden of burning fossil/ conventional fuel. The paper discusses the engineering viability of such stabilized mud blocks. The engineering properties of such blocks have been presented. The paper concludes that stabilized mud blocks are a viable alternative to conventional bricks with lot more intangible benefits attached to it.

Keywords Stabilized mud block · Stabilizer · Burnt brick

1 Introduction

Earth or mud is one of the few traditional materials that have been used by man ever since he started building his shelter. Earth is also among the very few low-cost, widely available economical building materials. Mud is a material which has the

A. K. Dutta (✉)
Department of Civil Engineering, Jorhat Engineering College,
JEC, JORHAT 785007, Assam, India
e-mail: emailatanu.dutta@gmail.com

D. Deka
Department of Civil Engineering, Girijananda Institute of Science and Technology,
Tezpur, India

A. Hatibaruah · D. Goswami · S. S. Rai · T. Basumatary
Department of Civil Engineering, Jorhat Engineering College, Jorhat, India

© Springer International Publishing AG 2018
A. K. Sarma et al. (eds.), *Urban Ecology, Water Quality and Climate Change*,
Water Science and Technology Library 84,
https://doi.org/10.1007/978-3-319-74494-0_5

flexibility for use in variety of structural elements such as walls, fireplaces, cooking *chulhas*, chimneys, furnaces and also in forms of ornamentations. However, the traditional burnt brick consumes a lot of wood/fossil fuels resulting in pollution as well as greenhouse gases. As the brick kilns are located at the periphery of the urban conglomerate, these contribute a lot to degradation of urban ecology. Therefore, a method for using mud, without degrading the environment poses a real challenge to the environmental planners. Here, the stabilized mud blocks offer a viable alternative.

Stabilized mud blocks (SMB) are made by mechanically pressing stabilized soil in a mould. The soil is stabilized either by cement or lime, and mixed with water of suitable percentage. A manually operated mechanical device is sufficient to produce high-density blocks of stabilized soil.

Walker (2004) tested cement stabilized pressed earth blocks using different blended soils and compacted using constant volume manual press. Effect of specimen geometry on experimental compressive strength was determined and aspect ratio correction factors for unconfined unit strength outlined. Proposal for unified compressive strength testing was outlined. Venkatarama Reddy and Gupta (2005) studied strength and durability issues of soil-cement blocks using highly sandy soils. Venkatarama Reddy and Latha (2014) studied the influence of soil grading on the characteristics of cement stabilized soil compacts.

In this present work, the SMB is attempted with local material particular to North-Eastern India's alluvial plain. Also, the locally available natural fibre straw is incorporated as additional stabilizer to see its efficacy.

The scope of this work includes looking at different soils available in the north-eastern part of India and also to incorporate different locally available natural fibre, for which the region is famous for.

2 Salient Feature of SMB

SMB are not the same as burnt bricks, rather they are a comparable alternative. This is an energy efficient, interesting and aesthetic alternative to burnt brick. The interesting characteristics of SMB are listed below:

(a) Lower cost
(b) Substantial energy saving
(c) Comparable strength with conventional burnt brick
(d) More functional efficiency
(e) Better appearance/aesthetic features
(f) Labour-intensive production
(g) Use of local resources and materials

3 Production Method

3.1 Selection of Soil

Main constituents of soil are sand, silt and clay. Sand and silt are inert material, where clay is an active material. There are several types of clay. Montmorillonite, Kaolinite and Illite are the most commonly occurring clay minerals. The behaviour of any soil is mainly controlled by the composition of sand, silt and the type of clay mineral.

Soil containing 60–70% sand and 10–20% clay are ideally suited for cement stabilized mud blocks. If a soil is deficient or excess in either clay or sand, it can be reconstituted by manipulating its sand and clay content. Soil containing too much silt and fine sand can lead to poor handling strength of the block in green condition. Such soil can be utilized by adding suitable quantity of coarse sand and clayey soil.

Soil containing predominantly non-expansive clay minerals such as Kaolinite can be used for mud block production. Cement stabilization is suited for such soil.

Soil containing considerable amount of Montmorillonite clay is often not suitable for cement stabilization. Such soil is characterized by very hard lump in dry condition. Such soil can be stabilized by adding sand and lime only. Detailed investigation is to be carried out to use such soil for mud block preparation (Terzaghi and Peck 1962, p. 13).

It is preferable to avoid soil with high organic matter (Terzaghi et al. 1996, p. 16). Soil containing more than 0.5% organic matter can be used for soil-cement block only if its suitability is established with laboratory investigation. Acidic soil adversely affects cement stabilization. Acidic soil adversely affects cement stabilization (Neville and Brooks 2010, pp. 10, 74). Addition of lime can remove the acidity of soil. It has been reported that presence of sulphates in excess of 0.5% in the soil will completely inhibit stabilization with lime or cement. Stabilized mud block with such soil tends to disintegrate on exposure to moisture.

Proper testing of soil is mandatory for soil to be used in mud block. It is preferable to have a grain size analysis to understand its characteristics. However, certain field tests can roughly indicate the nature of the soil. The procedure, observations and interpretations of the tests are given as follows:

3.1.1 Visual Analysis

Colour of soil can be good indicator of its properties.

(a) Deep yellow, orange and red, ranging from deep rich brown indicate iron content. This is a good soil.
(b) Soil rich in clay have a colour greyish to dirty white.
(c) Dull brown with slightly greyish colour indicates organic presence.

3.1.2   Dry Strength Test

The procedure follows from standard literature (Terzaghi et al. 1996, p. 24).

Procedure:

(a) Prepare two or three pots of soil
(b) Place the pots in sun or in an oven, until they have completely dried
(c) Break a soil pot and attempt to pulverize it between thumb and index finger
(d) Estimate the strength of the pot.

Observation and Inference:

(a) Difficult to break, when it does it breaks with a snap like a dry biscuit. Cannot be crushed between thumb and forefinger, it can be merely crumbled though without reducing to dust: **Almost pure clay**
(b) Not too difficult to break. Can be crushed to powder between thumb and forefinger with a little effort: **Silty or sandy clay**
(c) Easily broken and can easily be reduced to powder: **Silt or fine sand, low clay content**.

3.1.3   Water Retention Test

Procedure:

(a) Prepare a ball of 'fine mortar' 2 or 3 cm in diameter
(b) Moisten the ball so that it sticks together but does not stick to the fingers
(c) Slightly flatten the ball so that in palm of extended hand. Hit vigorously the palm of hand so that the water runs out. The appearance of the ball may be shiny or greasy.

Observation and Inference:

(a) Five to six blows are enough to bring the water to the surface. When pressed, the water disappears and the ball crumbles: **Very fine sand or coarse silt**
(b) Twenty to Thirty blows required. When pressed does not show any cracking nor does it crumble of flattens: **Slightly plastic silt or silty clay**
(c) No water appears. When pressed appears shiny: **Clayey soil**.

3.1.4   Consistency Test

The general procedure can be formulated as follows:

Procedure:

(a) Prepare a ball from the soil sample. Diameter of the ball is to be 2 to 3 cm
(b) Moisten the ball so that it can be moulded without being sticky

(c) Roll the ball on a flat clean surface until a thread is slowly formed
(d) If the thread breaks before its diameter is reduced to 3 mm, the soil is too dry, add water
(e) The thread should break when its diameter is 3 mm
(f) When the thread breaks, remould it into a ball again and crush it between thumb and index finger.

Observation and Inference:

(a) The reconstituted ball is difficult to crush, does not crack or crumble: **High clay content**
(b) The ball cracks or crumbles: **Low clay content**
(c) Impossible to reconstitute the ball: **High sand or silty clay, very little clay**
(d) Thread has a soft or spongy feel: **Organic soil**.

3.1.5   Cohesion Test

Procedure:

(a) Make a roll of soil about size of a sausage with diameter of 12 mm
(b) The soil should not be sticky and should be capable of being shaped so that it makes a continuous thread 3 mm in diameter
(c) Place the thread in the palm of hand. Starting at one end, carefully flatten it between index finger and thumb to form a ribbon of between 3 and 6 mm width as long as possible
(d) Measure the length obtained before the ribbon breaks.

Observation and Inference:

(a) Long ribbon (25–30 cm): **High silt content**
(b) Short ribbon (5–10 cm): **Low clay content**
(c) No ribbon at all: **Very low clay content**.

3.2   Stabilization of Soil

Stabilization is a process of altering the strength and durability characteristics of soil (Das 2004) Cement is generally recommended for red sandy loams. The percentage can vary from 2.5% by weight to 10% by weight depending on the strength requirements. If clay content of the soil is in the higher side, it is advisable to use a combination of lime and cement. If the black cotton soil or high clay soil is to be used, addition of sand and lime (5–10%) is preferable. Lime should be hydrated and powdered before mixing, but should be used before 3–4 weeks. If Lump formation occurs in hydrated lime, it is not desirable for use.

Some other easily available stabilizers are listed below:

(a) Straw: Non-chemical. Minimizes cracking and makes the green block easy to handle
(b) Cow dung: Contains lot of fibres and results in good mud
(c) Urine: Urea acts as a binder
(d) Gum Arabic: Good binder.

3.3 Block Making Process

3.3.1 Soil Preparation

(a) Soil sieved through 5 mm sieve
(b) 20–30 scoops of dry soil are spread over level ground in 10–15 cm thickness
(c) Appropriate quantity of stabilizer is spread over the top layer
(d) Mixed thoroughly. May be done in concrete mixer
(e) Water is mixed with dry soil and the soil is hand-mixed till it can be made into a ball and does the stick to the palm in the process so that when dropped from waist height it will break into four to six pieces. If water is more, it does not break.

3.3.2 Soil Block Preparation in Machine

The machine: Hand pressing machine 'TARA Balram™' has been used for the compaction of the mud blocks. It produces two size of blocks: $23 \times 10.9 \times 7.6$ cm and $23 \times 23 \times 7.6$ cm. The compaction capability of the machine is 160 cycles per hour: In this way in an 8 h day, it can produce 1,000–1,500 for Type A units; 500–750 for Type B units, number of blocks including time for soil preparation, block stacking and curing.

Six to ten persons are required to operate a manual machine. One skilled worker is required, the rest are unskilled. Typically, the workers can be trained to operate any machine within 10–12 days.

The process: The process comprises of mixing the raw materials using the pan mixer for 4–5 min and then transferring the material to the TARA Balram™ Mechanized machines. The compressed blocks are then transferred to a curing yard with the help of wooden pallets on which the blocks are placed and moved using a hydraulic pallet trolley. The blocks are then water cured for minimum 14 days. After 14 days of curing, bricks are kept for 7 days in dried place, after that these are ready for sale.

3.4   Laboratory Test for Compression

Test is conducted in a compression testing machine. The test is conducted as per IS 3495 (Part 1): 1992.

3.5   Laboratory Test for Durability

The blocks should possess good rain erosion capability. It is checked with spray test by simulation of rain (IS: 1725 1982).

3.5.1   Procedure

The diameter of each shower is 10 cm with 36 holes of 2 mm diameter. A facility for providing a device pump to create a constant pressure of 1.5 ± 0.2 kg f/cm should be available for this test. The block to be tested is to be mounted on a test rig, such that only one face is exposed to shower and discharged water should find an exit without wetting the other faces or getting collected such that blocks get immersed. These showers are placed at a distance of 18 cm from the block and are arranged by the side, such that the complete face gets exposed. The period of exposure is limited to 2 h and then the exposed surfaces are examined for possible pitting. The tests are carried out on at least 3 blocks. The limiting diameter of the pit formed is to be within 1 cm for passing this weathering test.

3.6   Waterproofing Coat

Erosion of soil is hastened by water absorptivity of soil. Also continuous exposure to water leads to fungal growth which can accelerate the disruption of soil.

Bituminous coat is recommended to protect mud blocks. However, this coating is disadvantageous as far as aesthetics is concerned. The following alternative coatings can be used:

3.6.1   In situ Calcium Soap Coating

Steps:

(a) 10% solution of laundry soap is prepared in warm water
(b) 3% solution of sodium chloride is prepared in a different container
(c) The soap solution is applied with a fine brush on the dry surface of wall of mud block and the painted surface is allowed to dry for few minutes. The calcium chloride solution is applied immediately.

3.6.2 Varnish Coating

Touchwood varnish can be applied after diluting with 1:0.5 ratio of turpentine. It improves appearance in addition to increasing the erosion resistance.

4 Test Results and Discussion

4.1 Physical Tests

4.1.1 Classification

The soil sample selected has the following Atterberg limits: Liquid Limit: 37, Plastic limit: 23.48 and Plasticity Index: 13.52. Based on this, the soil has been classified as Clay of Intermediate Compressibility. Detail hydrometer test carried out on the sample shows the composition as

Sand: 8%
Silt: 75%
Clay: 17%.

4.1.2 Other Properties

(a) Optimum Moisture Content is found to be 17.1%
(b) Maximum dry bulk density is: 1.685 gm/cc.

4.2 Samples

4.2.1 Percentage of Sand

Three blocks were prepared with 50% sand by weight of soil, three blocks with 60% sand and three with 70% sand. Cement of 5% of total wt has been added and the blocks were tested for 7 days for compressive strength.

The results have been presented in Table 1. It is clear from the table that 60% sand will be suitable for making stabilized mud block.

Table 1 Comparison of compressive strength of stabilized mud block with varying percentage of sand

S. No.	Sand by wt of soil (%)	Cement by wt of (sand + soil) (%)	Average water absorption	Average 7 day strength (N/mm^2)
1	50	5	26.12	4.87
2	60	5	26.57	7.51
3	70	5	28.34	7.40

Table 2 Effect of straw as additional stabilizer straw on strength and water absorption

S. No.	Straw by wt% of soil	Water absorption (%)	Average water absorption (%)	7-day compressive strength (N/mm^2)	Average 7-day compressive strength (N/mm^2)
1	0.2	21.92	22.92	11.46	11.33
2	0.2	22.91		10.67	
3	0.2	23.94		11.85	

4.2.2 Additional Stabilizer

Additional stabilizer of straw is added to the block and the result is presented in Table 2. It has been observed that addition of straw increases the average compressive strength by 50% while reducing its water absorption by 15%.

5 Conclusion

The paper concludes the use of stabilized mud block as a viable construction material for the low-cost housing. If used in mass scale, it will prove to be highly cost-effective and will contribute to environmental protection at the same time.

References

Das BM (2004) Principles of foundation engineering 5E. Thomson, pp 704–708

Indian Standard Code IS: 1725 (1982) Reaffirmed 2002

Neville AM, Brooks JJ (2010) Concrete technology, 2nd edn. Pearson, London, p 10

Product Brochure, TARA, Development Alternatives, B-32, TARA Crescent, Qutab Institutional Area, New Delhi—110 016. www.devalt.org

Terzaghi K, Peck RB (1962) Soil mechanics in engineering practices, Asia Publishing House, p 13

Terzaghi K, Peck RB, Mesri G (1996) Soil mechanics in engineering practices. Wiley, New York, pp 16–24

Venkatarama Reddy BV, Gupta A (2005) Characteristics of soil-cement blocks using highly sandy soils. Mater Struct 38(280):651–658

Venkatarama Reddy BV, Latha MS (2014) Influence of soil grading on the characteristics of cement stabilized soil compacts. Mater Struct 47(10):1633–1645

Walker P (2004) Strength and erosion characteristics of earth blocks and earth block masonry. J Mater Civ Eng 16(5):497–506

Contribution of Urbanization to Emissions: Case of Guwahati City, India

Rachna Yadav and Anamika Barua

Abstract The focus of research over the years has been mostly on industrial greenhouse gas emissions. While there has been an extensive analysis of the drivers of aggregate CO_2 emissions from fossil fuel combustion and cement production, analysis of the drivers of greenhouse gases emissions from agriculture, forestry, and other land uses which are also known as non-industrial emissions are limited (Sanchez and Stern in Ecol Econ 124:17–24, 2016). Agriculture, forestry and other land use (AFOLU) represents 20–24% of the global GHG emissions, the largest emitting sector next to energy. In Asia, the AFOLU sector is important and accounts for the largest proportion of global AFOLU emissions. India is the world's fourth largest economy and fifth largest global GHG emitter. The net AFOLU emissions in India were 146.7 million tCO_2e, accounting for about 11% of its net national emissions. The agricultural emissions were 355.6 million tCO_2e, accounting for 23% of gross national emissions and 96% of gross AFOLU emissions in the same year. AFOLU is not the largest emitter in India. The forestry and other land use (FOLU) is, on the other hand, an important sink with net removals of 236 million tCO_2e as reported in 2000. Cities and towns have been found to be settled after clearing large areas under forest cover in Assam. The present study examines the non-Industrial (AFOLU) emissions in the city of Guwahati. This study analyses 100 years trend (1911–2015) of deforestation and conversion of forests to settlements, wetlands and agricultural land and fuelwood burning within the city limits. The area values of AFOLU sectors were computed from maps and satellite images. Emission factor (EF) values were obtained from available literature to study the AFOLU emissions in the city. The findings indicate that the share of deforestation in CO_2 emission increased from 49% in 1911 to 85% in 2011, and contributed almost 0.91 tCO_2 per capita to the total emissions. Past 100 years average AFOLU emission per capita for Guwahati was found to be 1.81 tCO_2

R. Yadav (✉) · A. Barua
Department of Humanities and Social Sciences, Indian Institute of Technology Guwahati, Guwahati, India
e-mail: rachnayadav@yahoo.com

A. Barua
e-mail: anamika.barua@gmail.com

© Springer International Publishing AG 2018
A. K. Sarma et al. (eds.), *Urban Ecology, Water Quality and Climate Change*,
Water Science and Technology Library 84,
https://doi.org/10.1007/978-3-319-74494-0_6

against 1.03 tCO$_2$ for the last 50-year average. The results would be useful for policymakers given the fact that the city of Guwahati is one of the 100 cities in India that has been taken up for the smart city project presently underway in the country. Moreover, the results of the study would also be useful for further research and decision-making for achieving the SDG 11.

Keywords Ecological economics · Carbon emission · AFOLU
Urbanization · Source and sink · Ecosystem services

1 Introduction

The biophysical and ecological limits are posing a great challenge to the economic growth and the development process, especially rapid urbanization. Georgescu-Roegen (1971) proposed the biophysical limits to growth and pointed out to irreversibility of energy and matter used in the economic process. It was only later that other natural resources, especially exhaustible resources began to be seen as factors of production (Ayres 2001). Economic growth is constrained by the planetary boundaries (Rockström et al. 2009). The process of urbanization is largely dependent on natural resources extracted from within the city as well as spaces much beyond. While proving to be propellers of economic growth, on the one hand, these urban areas display a parasitic character, on the other hand. Although cities cover only 2% of the earth's surface, they consume 75% of its resources (García et al. 2008). Quality infrastructure in the form of transportation networks, power supply, telecommunication networks, housing infrastructure, modernized medical facilities, industrial centres and educational centres are a prerequisite for economic growth in cities. Urban dwellers play only a minor functional role within many 'in city' urban ecosystems, but they are virtually the sole macro-consumers in vast areas of cropland, pasture, and forest outside the city, scattered all over the world. Similarly, many wastes generated by people in the city are injected into the global commons—the atmosphere, rivers, and ultimately the oceans—for processing and possible recycling (Rees 2003). Urbanization in India has been closely following this global trend (Sridhar 2010). Economic growth in urban areas constitutes close to half of India's gross domestic product. While there has been an extensive analysis of the drivers of aggregate CO$_2$ emissions from fossil fuel combustion and cement production, analysis of the drivers of greenhouse gases emissions from agriculture, forestry, and other land uses which are also known as non-industrial emissions are limited (Sanchez and Stern 2016). With the objective of understanding the relationship between urbanization and natural ecosystems with focus on AFOLU components such as forest, agriculture and wetland ecosystems, the present study was undertaken for Guwahati city, located in north-eastern India. The study is based on the principles of ecological economics.

1.1 Urbanization and Deforestation

Deforestation in the tropics is seen as being majorly driven by urge to urbanize. Biello (2009) in an article in Scientific American quoted Pedro Sanchez, Director of the Tropical Agriculture and Rural Environment Program at The Earth Institute at Columbia University saying that by 2050, the world will host nine billion people, while, at least one billion people are chronically malnourished or starving today, and in order to simply to maintain this sad state of affairs, it would require deforestation of 900 million additional hectares of land. According to Sanchez, there is not that much land available, and at most, we might be able to add 100 million hectares to the 4.3 billion already under cultivation worldwide. DeFries et al. (2010) used satellite-based estimates of forest loss for 2000–2005 to assess economic, agricultural and demographic correlates across 41 countries in the tropics. They correlated with urban population growth and forest loss during the time period. Their findings indicate that rural population growth is not associated with forest loss, indicating the importance of urban growth as a driver of deforestation. The strong trend in movement of people to cities in the tropics is likely to be associated with greater pressures for clearing tropical forests, which needs strategic afforestation interventions. Using the satellite images from Landsat and Aqua (MODIS), it found that the typical 'fish-bone' signature of deforestation in the tropical forests indicative of small-scale operations has given way to removal of large chunky blocks of cleared land indicating the changed circumstances of large enterprises feeding urban demand. Another study by Hasse and Lathrop (2010) showed that the Garden State of New Jersey lost 4300 acres forest per year from 1986 to 1995, thereafter 5901 acres per year from 1995 to 2002, and 8490 acres per year from 2002 to 2007 in the process urbanization and new settlements. WWF-India carried out a report on urbanization and biodiversity in two cities of India, namely, Kolkata and Coimbatore (WWF 2011). According to this report, urbanization in India is occurring at a rate that is faster compared to many other parts of the developing world. It is estimated by the Planning Commission of the Government of India that about 40% of the country's population will be residing in urban areas by 2030. The report suggests that the rapid urbanization of both Coimbatore and Kolkata has led to drastic changes in land use, destruction of natural ecosystems and increase in the demand for natural resources. For instance, expansion of Coimbatore city in recent decades has led to the degradation of the Noyyal River that served the city's water needs. Similarly, the spatial growth of Kolkata has led to changes in the biodiversity of the East Kolkata wetlands in the city as well it negatively impacted the Sundarbans. These cases indicate that proper environment planning is needed for sustainable urban development.

1.2 Agriculture, Forestry and Other Land Use (AFOLU), GHG Emissions and Climate Change

According to Biello (2009), agriculture is responsible for one-third of global greenhouse gas emissions from human activity caused by deforestation, nitrous oxide emissions from fields, methane from cattle and rice paddies. The quantum of contribution of these emissions is so high that, quoting the ecologist Jonathan Foley, Director of the Institute on the Environment (IonE) at the University of Minnesota, Biello mentions that the emissions from transporting food, also known as food miles, get relegated to decimal places as rounding errors. Agriculture, forestry and other land use (AFOLU) represents 20–24% of the global GHG emissions, the largest emitting sector next to energy (Smith et al. 2014). The AFOLU sector is particularly important in Asia, which accounts for the largest proportion of global AFOLU emissions. India is the world's fourth largest economy and fifth largest global GHG emitter. In 2000, India's net AFOLU emissions were 146.7 million tCO_2e, accounting for about 11% of the total net national emissions. The agricultural emissions were 355.6 million tCO_2e, accounting for 23% of gross national emissions and 96% of gross AFOLU emissions in the same year (AFOLU Working Group 2016). Although AFOLU is not the largest emitting sector in India, forestry and other land use (FOLU or previously LULUCF) is an important sink with net removals of 236 million tCO_2e in 2000 (AFOLU Working Group 2016). Forests are the largest terrestrial store of carbon, but deforestation is the largest source of carbon dioxide emissions after fossil fuel burning, causing 15% of global greenhouse gas emissions, with a range from 8 to 20% (van der Werf et al. 2009).

1.3 Objective of Study and Research Questions

The main objective of the study is to arrive at linkages between urbanization and its impact on various ecosystems within the city limits, and also to examine the urban growth in this perspective over a time period. The following research questions have been attempted to be answered in the study:

1. Is there any non-industrial (AFOLU) contribution to GHG emissions in Guwahati city?
2. What is the contribution of AFOLU to emissions in Guwahati city?
3. How the different AFOLU sectors have contributed to the emissions in the city?

In order to address the above questions, a case study of Guwahati city was conducted. Section 2 describes the study area and its topography, forests, wetlands and agriculture. Section 3 describes the methodology used in the study and data sources. Section 4 gives the results and analysis, and Sect. 5 contains conclusions and limitations of the study.

2 Study Area

The study area comprises the Guwahati city metropolitan area (GMA) under the administrative control of the Guwahati Metropolitan Development Authority (GMDA). The hills and forests within this boundary have been taken into consideration for the present study.

2.1 About Guwahati City

The study area comprises Guwahati city which is located in the north-eastern region of India and situated between 26° 5'–26° 13'N latitude and 91° 35'–91° 52'E longitude, on the banks of the river Brahmaputra. For the study, the area under the Guwahati Metropolitan Development Authority (GMDA) was considered. GMDA's jurisdiction extends over an area of 262 km^2 covering the entire Guwahati Municipal Corporation (GMC) area, entire North Guwahati Town Committee area and some revenue villages of Silasundari Ghopa Mouza, Pub Barsar Mouza, Dakhin Rani Mouza, Ramcharani Mouza and Beltola Mouza. The city falls within the civil jurisdiction of Kamrup (metro) district, which was a part of the erstwhile Kamrup district (GMDA 2009; Gogoi 2011).

2.2 Topography of Guwahati

The topography of the city is undulating varying in elevation from 49.5 to 55.5 m above mean sea level (MSL). The land is interspersed with a large number of hills. The central part of the city has small hillocks, namely, Sarania hill (193 m), Nabagrah hill (217 m), Nilanchal hill (193 m) and Chunsali hill (293 m) (GMDA Plan 2009). The Agiathuri hill and Maliata hill lie on the northern and western boundary of the city limits. The Buragosain Parbat in the east and the hills of Rani and Garbhanga in the south form the major hill formations of the city. These hills make contiguous formations with the hills of Meghalaya. There are a total of 18 hills in the city. The total reported area covered by hills in GMDA area is 6881 ha (Anon 2010). The hill tracts are mostly covered with forests. The hills within the GMDA area are shown in Map 1.

2.3 The Forests of Guwahati

The hills are mostly covered, barring the rocky outcrops, with **forests** of various formations ranging from Sal forests, mixed moist deciduous forests, evergreen

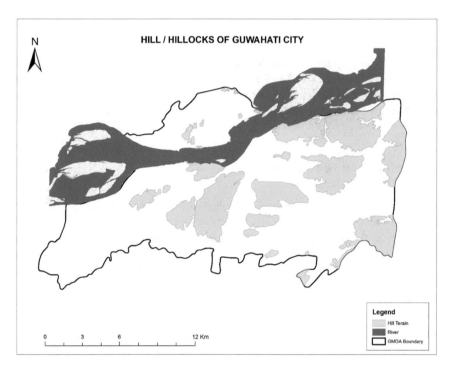

Map 1 Hills of Guwahati. *Source* Yadav and Barua (2016)

forest, bamboo brakes and secondary scrub forests. The forests in and around the city fall under the jurisdiction of the Kamrup (East) Forest Division. The management of the forest tracts is carried out as per prescriptions of the working plans. As per the working plans, there are a total of 14 reserved forests (RF) within and on the immediate periphery of the city area. The total RF area comes to 33,342.55 ha comprising of Rani RF (1882, 4361.584 ha), Maliata RF (1915, 324.776 ha), Agiathuri hill RF (1917, 363.196 ha), Garbhanga RF (1926, 11,441.28 ha), Garbhanga 1st Addition (1990, 7395 ha), Khanapara RF (1953, 994 ha), Fatasil RF (1966, 669.02 ha), Amchang RF (1972, 5318 ha), South Amchang RF (1990, 1550 ha), Hengrabari RF (1972, 579 ha), Gotanagar RF (1984, 171 ha), Sarania RF (1989, 7.99 ha), South Kalapahar RF (1989, 70 ha) and Jalukbari RF (1990, 97.70 ha) (Jacob 1938; Das 1973; Swargowary 2002). The first figure in the brackets is the year of notification of the RF, while the second figure is area of the RF in hectares. The forests on the southern periphery of the city have Sal formations mixed with patches of evergreen and bamboo formations. The forests in the city show moist mixed deciduous forest formations. Where soil is shallow and poor, stunted growth of bamboo and scrub occurs. The working plan records over the years show that the density of the forests has progressively declined. To quote Jacob (1940), 'Existing Unclassed State Forests are being jhumed extensively, have been and being rapidly taken up for cultivation by immigrants from Bengal as well as the indigenous people and are deteriorating rapidly under uncontrolled exploitation of forest produce given free to settlement holders and by grazing. It is, therefore, only a question of time before this type of forest is wiped out'. Increase in population is one of the most important parameters leading to forest depletion.

2.4 Wetlands of Guwahati

The Guwahati city is drained on the north by the Brahmaputra River. The other major rivers and streams are Amchang nadi, Bashsitha nadi (also known as Barpani), Bharalu nadi, Bonda nadi (also nala), Bukat nadi, Kalmani nadi, Kana nadi and Mora nadi. There are several other unnamed streams and rivulets. The Deepar Beel is the largest waterbody and a Ramsar site. Part of its also declared as wildlife sanctuary. The other main waterbodies are Borsola (Borhola) Beel, Bordal Beel, Chunchuki Beel, Damal Beel, Hachora Beel, Khalkhowa Beel, Rangagora Beel, Silsako Beel, Tepal Beel, Thengbhanga Beel and Thupdhara Beel. There are several other unnamed waterbodies, swamps and mud pools in the city limits. In addition, the largest man-made water tank is the Dighalipukhuri. The other major water tanks are Silpukhuri and Jurpukhuri (two ponds together). There are several other artificial water tanks in the city. Some of these names have been variously extracted by the author from the topographic sheets pertaining to Guwahati.

2.5 Agriculture in Guwahati

Most of the low lying areas in the city limits, in the earlier days, have been subject to cropping. As the city grew, the agricultural areas have steadily shrunk. Agriculture is mostly confined now in the outskirts of the city, especially in areas surrounding the Deepar Beel and North Guwahati. Since most of the areas are low lands, and agriculture is largely rainfed, the cropping system can be described as lowland rainfed floodplain (RFFP) agricultural regime.

3 Materials and Methods

The study is confined to the administrative boundary of the Guwahati Metropolitan Development Authority, and the hills and forest ecosystem existing within these boundaries. This also includes, incidentally, the Guwahati Municipal Corporation (GMC) areas. In the instant study, the following components of AFOLU have been selected:

1. Deforestation and conversion of forest to settlement (in short FS),
2. Burning of biomass (in short FW),[1]
3. Methane emission from cropland (in short CL) and
4. Net CO_2 emission from wetlands (in short WL).

According to Asner (2009), two kinds of measurement are needed to estimate GHG emissions from deforestation, namely, the rate of change in the forest cover which is termed as deforestation rate and secondly the amount of carbon stored in the forest which is termed as carbon stock. The forest activities that release GHGs have been divided into two categories: first, deforestation which is the clear-cutting and often burning of entire biomass, and second degradation which includes selective logging, thinning, burning and other disturbances that do not completely remove the forest canopy but lower the carbon-storing capacity of the forests. The loss of forest is best studied using satellite images and other imaging technologies such as Lidar. For carbon stock assessment, IPCC prescribes, as a good practice, in Tier-I approach, which is the most general approach, based on generic estimates of forest carbon density values (e.g. tonnes of carbon per hectare). Tier-II approach is more detailed and uses forest maps and forest carbon inventories that are more accurate than Tier-I default values. Tier-III is the most rigorous approach and is based on very detailed landscape-specific or even species-specific carbon stock estimates with regular, ongoing reassessments (IPCC 2000a). In the present study, guidelines laid down in the IPCC Guidelines for National Greenhouse Gas Inventories, Vol 4, Agriculture, Forestry and Other Land Use, 2006 (IPCC 2006) have been used within the constraints and limitations of the study.

3.1 Key Questions

1. What is the relationship between deforestation due to settlements and carbon emission in Guwahati city?
2. What is the contribution of other components of AFOLU to the city CO_2 emissions?

3.2 Methodology

The contribution of AFOLU (agriculture, forests and other land uses) to carbon emission is given by (IPCC 2006; WRI 2015)

$$\Delta C_{AFOLU} = \Delta C_{FL} + \Delta C_{CL} + \Delta C_{GL} + \Delta C_{WL} + \Delta C_{SL} + \Delta C_{OL}, \qquad (1)$$

---

[1]Under biomass, fuelwood burning only has been considered in this study. Though this is contributory to energy sector emission, it has been considered here as it is one of the main drivers of deforestation.

where

ΔC   Carbon stock change
FL   Forest land
CL   Cropland
GL   Grassland
WL   Wetland
SL   Settlements
OL   Other land

3.2.1 Estimating Carbon Emissions from Deforestation and Fuelwood Consumption

This section is divided into two subsections, namely, carbon dioxide emissions from deforestation and secondly carbon dioxide emissions from fuelwood burning.

CO_2 Emission from Deforestation

The deforestation in the Guwahati city is primarily from conversion of forest land to settlements by the people. Here it can be assumed that the total carbon is lost, as the trees are cut, felled and burnt as firewood, and also the soil is dug up to make space for housing, etc. Though there are no direct estimates of carbon content of the forests of the Guwahati city, data published by the Forest Survey of India was used (FSI 2011). The FSI has assessed carbon stocks in various states of the country, and according to its latest available report (2011), the forests of Assam have been found to have an average carbon pool of 61.10 t C ha^{-1}. Therefore, if 100 ha (=1 km^2) of forests are lost to settlements, the corresponding loss is 6110 T carbon per km^2 which translates to 22,403.33 tCO$_2$ ($E_{tCO_2} = E_{tC} * 44/12$). Therefore, an emission factor of about 22,400 tCO$_2$ per km^2 has been used in event of 1 km^2 of forested area changing to settlements.

CO_2 Emission from Fuelwood Burning

It has been estimated that 2 metric tonnes of firewood is required per family annually in the north-eastern region (Mathew 1987). In 1991, a study on domestic energy of Kamrup district in two blocks, namely, Dimoria and Hajo, was carried out (Borah 1995). According to Borah J, the per capita urban fuelwood consumption was 334.6 kg annually. A nationwide review of fuelwood sampling studies was carried out by Pandey (2002). According to this study, the share of fuelwood in Assam domestic energy consumption was 44.1% in 1993 and came down to 34.1% in 1999/2000. According to the study, the per capita consumption of firewood in the forested districts of Assam stood at 435 kg per annum. As per 2011 census details, the percentage of households in urban area using fuelwood as fuel for cooking comes to 22% in the Kamrup metropolitan district (Census 2011).

If

f Per capita fuelwood consumption in kg per annum
fr Percentage of households using firewood
P Population
F Total quantity of fuelwood in metric tonnes

Then

$$F = P \cdot fr \cdot f \cdot 10^{-3} \text{ tonnes year}^{-1}, \tag{2}$$

$$E_{CO_2} = F \cdot NCV_{\text{fuelwood}} \cdot EF_{\text{fuelwood}}, \tag{3}$$

where

E_{CO_2} Emission of CO_2 in metric tonnes
NCV_{fuelwood} Net calorific value of fuelwood (=0.015 TJ/tonne for fuelwood as default value by IPCC 1996)
EF_{fuelwood} Emission factor for fuelwood (=109.6 tCO_2/TJ for fuelwood as default value by IPCC 1996)

3.2.2   Estimating Methane Emission from Agricultural Fields of Assam

Methane emissions from agricultural fields across the world vary due to a variety of factors such as soil type and texture, organic content and kind of fertilizers applied and the water regime. Therefore, the methane fluxes are largely site specific. However, methane flux studies are not many in the state. Mitra et al. (2012) have elaborately dwelt upon the emission factors for methane emission from crop fields of Assam, and reported values from literature vary from 6.92 to 46 g m^{-2}. According to Gupta et al. (2009), estimated methane emissions from paddy fields of Assam as seasonally integrated flux (SIF) are shown to be in two studies at 9.18 and 7.14 g m^{-2} at the Assam Agricultural University, Titabor farm under National Campaign 2002 (NC-2002). However, the authors reported an average emission factor of 19.0 ± 6.0 g m^{-2} for rainfed floodplain agricultural systems. Another study by Gupta et al. (2002) showed that in Jorhat fields, the seasonal integrated flux was 46 g m^{-2}. A study by Gogoi et al. (2008) showed wide variation in Ahu and Sali crop methane fluxes. Whereas the Ahu paddy (pre-monsoon crop) methane SIF was found to be 7.51 g m^{-2}, while for Sali paddy (monsoon crop) was found to be 16.39 g m^{-2}. Further as reported by Jain et al. (2004) the values for lowland rainfed flood-prone cropping system, the average SIF was found to be 19.0 ± 6.0 g m^{-2}. There being such a wide variation, the value for the lowland rainfed flood-prone (RFFP) cropping system under the Methane Asia Campaign 1998 as reported by Jain et al. (2004) and Gupta et al. (2009) was used in this study. Therefore, the E_f for methane from rice fields was assumed at 19 g m^{-2} for the city fields. Assuming global warming potential (GWP) of methane to be 21, the formula for estimation of CO_2 emission from agriculture is

$$CO_{2_{eq}} = E_{CH_4} \cdot GWP_{CH_4}$$
$$= 21 \cdot E_{CH_4} \tag{4}$$

$$= 21 * E_{CH4}, \tag{5}$$

where E_{CH_4} is given by

$$E_{CH4} = A * EF_{CH4} \text{ Tonnes year}^{-1}, \tag{6}$$

where A is in km^2, and EH_{CH_4} is in g m^{-2}. These equations have been adopted from IPCC (1996, 2006) and RTI (2010).

3.2.3 Estimating Emissions from Waterbodies

The data on emission from freshwater wetland and marshes and their sequestration potential is lacking. As per IPCC (2000b) Special Report on LULUCF, the freshwater wetlands are responsible for 7–40 g m^{-2} year^{-1} CH$_4$ emissions, and sequestration rates vary from small negative values to 0.35 tC ha^{-1} year^{-1}. Assuming an emission rate of 7 g m^{-2} year^{-1} CH$_4$ and sequestration rate of 0.35 tC ha^{-1} year^{-1} for 100 ha (=1 km^2) of a wetland, it can be seen that it would emit 147 teCO$_2$ (=40.09 tC ycar^{-1}), while it would also sequester 35 tC year^{-1}, making it a net source of 5.09 tC year^{-1} (=18.66 tCO$_2$ year^{-1}). Therefore, for this study, the wetlands have been treated as net source with an emission of 18.66 tCO$_2$ year^{-1} per km^2.

3.2.4 Combined Equation for CO_2 Emission from AFOLU

By combining the various parameters and factors for deforestation, fuelwood burning, cropland and wetland, the AFOLU emissions can be estimated as below:

$$CO_2 = E_{FS} + E_{FW} + E_{CL} + E_{WL}, \tag{7}$$

$$CO_2 = A_{FS} * EF_{FS} + F * EF_{FW} + A_{CL} * EF_{CL} + A_{WL} * EF_{WL}, \tag{8}$$

$$= A_{FS} * 22,400 + F * 1.644 + A_{CL} * 399 + A_{WL} * 18.66, \tag{9}$$

where

CO_2 CO_2 emissions in tonnes per year
E_{FS} CO_2 emissions from conversion of forests to settlements
E_{FW} CO_2 emissions from burning of fuelwood
E_{CL} CO_2 eq emissions from lowland rainfed floodplain cropping system (methane emissions only)
E_{WL} Net CO_2 emissions from wetlands (after deducting sequestration)
A_{FS} Area of forest land converted to settlement (in km^2)
E_{FS} Emission factor for forest land converted to settlement (=22,400 tCO_2 per km^2)
F Mass of fuelwood consumed in Tonnes
EF_{FW} Emission factor for fuelwood (=$NCV_{fuelwood} \cdot EF_{fuelwood}$ = 0.015 * 109.6 = 1.644 tCO_2 per tonne fuelwood)
A_{CL} Area under cropland in km^2
EF_{CL} Emission factor for cropland in RFFP cropping system (=GWP * EF = 21 * 19 g m^{-2} = 399 in g m^{-2})
A_{WL} Area of wetland in km^2
EF_{WL} Emission factor for wetland (net emission =18.66 tCO_2 $year^{-1}$ per km^2 assumed)

3.2.5 Rates of Change with Baseline Year

Changes in various parameters such as loss of forest areas, increase in built-up areas, etc. over consecutive time periods were computed by simple ratio of value difference and time difference. However, to arrive at long-term perspective, the differences were computed from the base year 1911. Assuming values V_{t0}, V_{t1}, V_{t2} for a particular component in year t_0 (1911), t_1 and t_2, the rate of growth or decline were computed as below:

$$R = (V_{t2}, -V_{t2})/(t_2 - t_1), \tag{10}$$

$$R_0 = (V_{t2}, -V_{t0})/(t_2 - t_0). \tag{11}$$

3.2.6 Population Growth Rates and Projections

The following formulae were used to arrive at the population growth rates. For years for which data at 10-year interval was available, the decadal growth rate (DGR) was calculated as (Bruce Seymour 2004)

$$\mathrm{DGR} = \frac{P_n - P_0}{P_0} * 100. \tag{12}$$

Wherever the data was not 10 years apart, the following formula was used:

$$\mathrm{DGR} = \left[\left(\frac{P_n}{P_0} \right)^{\frac{10}{N}} - 1 \right] * 100. \tag{13}$$

The annual growth rate (AGR) was obtained using the formula

$$\mathrm{AGR} = \left[\left(\frac{P_n}{P_0} \right)^{\frac{1}{N}} - 1 \right] * 100. \tag{14}$$

The projected population in future was obtained using the formula

$$P_f = P_n \cdot (1 + R)^N, \tag{15}$$

where

DGR   Decadal growth rate in %
AGR   Annual growth rate in %
P_n Population in the year n
P_0 Population in the beginning
P_f Projected population in future time f
R   Annual growth rate in decimal ($R = \mathrm{AGR}/100$)
N   Time interval between two populations P_n and P_0, and P_n and P_f

3.3   Data Sources

The secondary data sets used in the study were obtained from the related departments of the Government of Assam. Inputs from the Forest Department included working plans for the forest areas of Kamrup district since the year 1938–39 till the year 2011–12, stock maps of the forests, data and maps concerning the hills of Guwahati. The master plan, GMDA boundary and other secondary data pertaining to the city and urban sprawl were obtained from GMDA.

The mapset studied included the Survey of India topographic sheets 78N12, 78N16 first on 1 = 1 Mile scale and survey year 1911–12, and second on 1:50,000 scale and survey year 1967–68, 78N12 (NE, NW, SE and SW) and 78N16 (NE, NW, SE and SW) on 1:25,000 scale and survey year 1986–87. The satellite imagery for the study area was obtained from USGS for the year 2010 (Landsat TM 5 P137 R42 DoP 30.01.2010) and 2015 (TM 8 P137 R42 DoP 28.01.2015). The results were refined by cross-checking from available Google Earth satellite data in the public domain. The land use and land cover map of Guwahati city available in the Microzonation study of Guwahati city by AMTRON and other agencies under the aegis of the Department of Science and Technology, Government of India, based on satellite data of 2003, was also used.

The population figures for the Guwahati city were taken from various sources such as Census 2001; 2011, GMDA Master Plan and the Statistical Handbook (2014), Government of Assam.

3.3.1 GIS and Remote Sensing

The mapsets and satellite data were brought on a single coordinate reference system (CRS) using the EPSG:32646-WGS 84/UTM Zone 46N projection system on the QGIS platform. The required features of built-up area, hills and forests were extracted digitally within the GMDA boundary vector. The forest cover was computed only within the hill vectors using unsupervised classification and quick reconnaissance type ground truthing. The agricultural areas and wetlands were computed for all the areas falling within the GMDA boundary. However, there would be some overlap between agricultural areas and wetlands as some of the shallow areas of the wetlands are used for rice cultivation.

4   Results and Analysis

Based on the primary and secondary data, the population growth trend, built-up area, value of forest ecosystem services, forest loss and carbon footprint estimation were computed. The results and findings are discussed below.

4.1   Guwahati City Population Growth

The Guwahati city population was estimated at 8394 in 1891 (GMDA 2005). The population of the city at different periods of time is given in Table 1 (Yadav and Barua 2016).

The decadal growth rates shown in column 3 of Table 1 are based on the year-wise entries in the said table and may differ from the official decadal growth rates published based on the census. Since the population census unit for the city was different at different times as municipality, Guwahati Municipal Corporation (GMC) and Guwahati metropolitan area (GMA) under the GMDA, the figures may not be exactly comparable. Based on the above growth rates, the following population estimates have been arrived at for the Guwahati city for further analysis, as given in Table 2, adopted and modified from Yadav and Barua (2016).

Table 1 Population and decadal growth of Guwahati city

Year	Population	Decadal growth rate (%)
1891	8394	
1921	16,480	25.22
1961	199,482	86.52
1971	293,219	46.99
1991	646,169	48.45
2001	890,773	37.85
2011	968,549	8.73

Table 2 Population of Guwahati city in selected years

Year	Population
1911	13,785
1967	255,724
1986	557,932
2003	906,394
2010	963,255
2015	1,097,751

4.2 Guwahati City Built-Up Growth (1911–2015)

Built-up area of the Guwahati city was calculated for the years 1911, 1967, 1986 and 2010. The year-wise built-up area, the growth in built-up from 1911–1967, 1967–1986, 1986–2010 and 2010–2015 and the corresponding growth rates are tabulated in Table 3, adopted and modified from Yadav and Barua (2016). The growth rate of the built area of the city from the 1911 baseline, along with the year to year growth rate is pictorially presented in Chart 1. The built-up area in the Guwahati city has grown within the span of 100 years starting from 1911 to 2015 from modest 8.59–176.19 km^2 at a rate of about 1.61 km^2 per annum. The map of the built-up and growth of the city during the period is shown in Map 2.

Table 3 Guwahati city built-up area and its growth rate

Year	Built-up area (km^2)	Growth (km^2)	Growth rate (km^2 per year)	Growth rate 1911 baseline (km^2 per year)
1911	8.59	0	0	0
1967	54.48	45.89	0.82	0.82
1986	90.65	36.17	1.90	1.09
2003	116.99	26.34	1.55	1.18
2010	142.75	25.76	3.68	1.36
2015	176.19	33.44	6.69	1.61

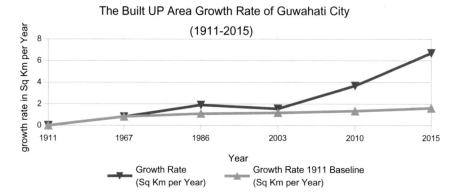

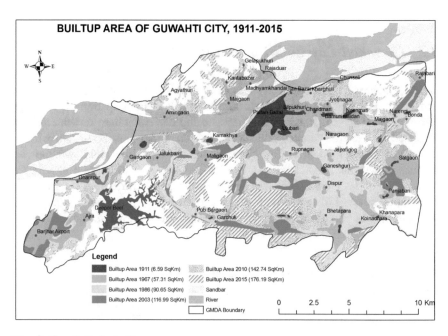

Chart 1 The built-up area growth rate of Guwahati city

Map 2 Guwahati city built-up area

4.3 Guwahati City Decline in Forest Ecosystem (1911–2015)

4.3.1 Loss of Forests to Settlements

A detailed account of loss of forest ecosystem and ecosystem services values has been presented at length by Yadav and Barua (2016) for the Guwahati city. Table 4, adopted and modified from Yadav and Barua (2016), depicts the land use change pattern of the hills/forests of the city since 1911. The forest areas from 1911 to 2015 have also been shown in Map 3.

Table 4 Year-wise statement of forest area, degraded forest and rate of forest loss

Year	Dense forest (ha)	Degraded forest (ha)	Total forest area (ha)	Forest converted to settlement (ha)	Rate of forest loss (ha year^{-1})	Cumulative forest loss (ha year^{-1})
1911	6708.63	0	6708.63	172.63	0	0.00
1967	6158.44	0	6158.44	722.82	9.82	9.82
1986	5619.44	0	5619.44	1261.82	28.37	14.52
2003	2334.21	2562.31	4896.52	1984.74	42.52	19.70
2010	1722.84	1500.62	3223.46	3657.80	99.83	35.20
2015	1438.49	983.27	2421.76	4459.50	160.34	41.22

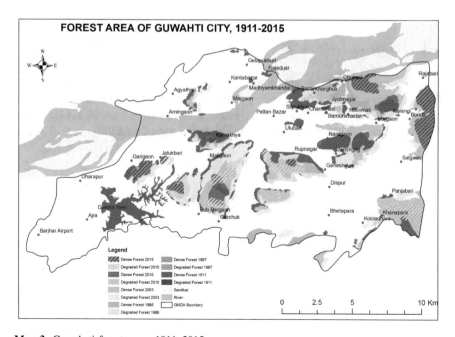

Map 3 Guwahati forest areas, 1911–2015

Table 5 Fuelwood consumption of Guwahati city

Year	Population [P]	Fraction of population using fuelwood [fr]	Per capita per year fuelwood consumption (kg Year^{-1}) [f]	Fuelwood consumption (metric tonnes) [F = P * fr * f * 10^{-3}]	Per capita fuelwood consumption (metric tonnes) [F/P]
1	2	3	4	5	6
1911	13,785	1[a]	435[b]	5,996.48	0.435
1967	255,724	1[a]	435	111,239.94	0.435
1986	557,932	0.46[c]	435	111,642.19	0.200
2003	906,394	0.34[c]	435	134,055.67	0.148
2010	963,255	0.22	435	92,183.50	0.096
2015	1,097,751	0.22[d]	435	105,054.77	0.096

[a]Assumed to be 100% dependent upon fuelwood
[b]As per Pandey (2002)
[c]As per Pandey (2002). 1983/1984 NSSO value for all India used in 1986
[d]As per 2011 household census data for Kamrup metropolitan district

4.3.2   Fuelwood Consumption in Guwahati City

The fuelwood consumption from 1911 to 2015 has been estimated based on the methodology suggested earlier in this study. The values are tabulated in Table 5.

4.4   Status of Croplands in Guwahati City (1911–2015)

The areas of cropland (agricultural areas) were estimated within the city limits of GMDA from 1911 to 2015 based on the available topographic sheets, satellite images and land use and land cover maps available from various sources. The finding is presented in Table 6. Map 4 depicts the agricultural areas from 1911 to 2015 pictorially.

Table 6 Agricultural area of Guwahati city

Year	Agricultural area (km^2)
1911	74.57
1967	58.32
1986	52.49
2003	42.26
2010	16.51
2015	14.41

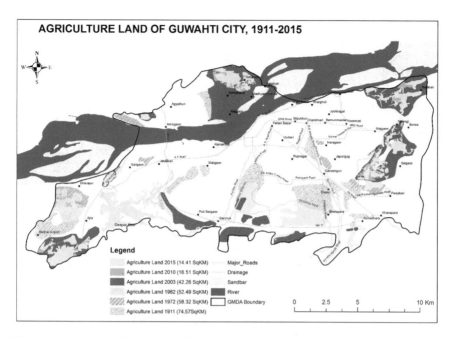

Map 4 Guwahati agricultural areas, 1911–2015

4.5   Status of Wetlands in Guwahati City (1911–2015)

The wetlands of the Guwahati city were classified into three categories namely:

1. Marshy lands and swamps (dry waterbodies during pinch period),
2. Waterbodies (retaining water even during drier periods) and
3. Important man-made ponds and tanks.

These were delineated within the city limits of GMDA from 1911 to 2015 based on the available topographic sheets, satellite images and land use and land cover maps available from various sources. The finding is presented in Table 7. The wetland areas from 1911 to 2015 have been depicted in Map 5, and the wetland status of Guwahati city was shown in Chart 2.

4.6   AFOLU Contributions to CO_2 Emissions in Guwahati City (1911–2015)

The CO_2 emissions arising out of the various components of the AFOLU, namely, forest area lost to settlement, fuelwood burning, emissions from agricultural activities (only cropland CH_4 emission for a lowland rainfed floodplain system),

Table 7 Year-wise area of wetlands in Guwahati city

Year	Marshy land (km^2)	Waterbodies (km^2)	Ponds/ tanks (km^2)	Total wetland area (km^2) [W]	Loss of wetlands (km^2 Year^{-1}) [$\Delta W/\Delta T$]	Loss of wetlands with 1911 baseline (km^2 Year^{-1}) [$\Delta W/\Delta T_0$]
1911	17.71	9.25	0.18	27.14	0	0
1967	9.88	7.83	0.16	17.87	0.17	0.17
1986	6.90	7.88	0.16	14.94	0.15	0.16
2003	6.65	6.51	0.16	13.32	0.10	0.15
2010	6.61	6.02	0.16	12.79	0.08	0.14
2015	6.01	5.73	0.16	11.90	0.18	0.15

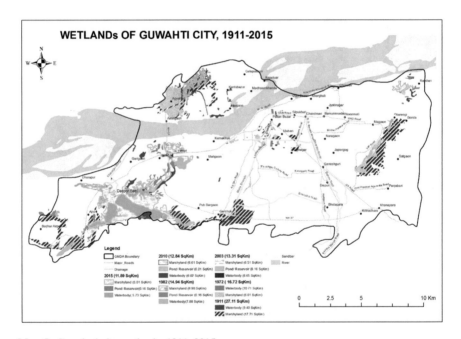

Map 5 Guwahati city wetlands, 1911–2015

and net emissions from wetlands, were computed from Eq. (9). The total AFOLU emissions, thus arrived, along with per capita emissions are given in Table 8. To understand individual contributions of the various components of the study, component-wise emissions along with per capita emissions have been computed as well based on Eq. (9) and presented in Table 9.

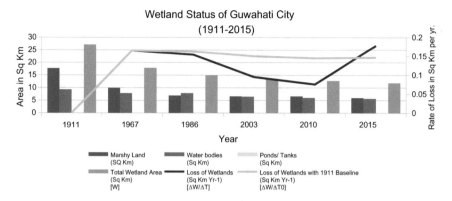

Chart 2 Wetland status of Guwahati city

4.7   Discussion and Analysis

The population of the city has grown from a modest 13,785 in 1911 to 1,097,751 in 2015, indicating a growth of about 80 times over a century. The built-up area has grown from 8.59 km^2 in 1911 to 176.19 km^2 in 2015, indicating a growth of 20.5 times during the period. Taking 1911 as baseline, the built-up growth rate has been steadily on the rise and is currently at 1.61 km^2 per year (see Table 3). This rise in built-up and population growth appears to have come from the areas covered by hills and forests, wetlands and agricultural open fields. Almost a quarter of the city built-up is on forest areas (Yadav and Barua 2016). The fuelwood scenario of the city is such that as of 2011 (Census 2011), only 22% of the population (in terms of households) depended upon fuelwood as a means of cooking energy. This amounts to about 0.1 MT of fuelwood being consumed annually in the city. Assuming higher levels of fuelwood use in the past among the city households, collating from scanty data available in the literature, it is seen that fuelwood consumption has remained almost more than 0.1 MT per annum since 1967. At this rate in the past 50 years, the city households have consumed almost 5 MT of fuelwood. The changes in AFOLU components over time have been depicted in Chart 3.

Carbon emissions from agriculture were very significant and contributed almost 2.16 tCO$_2$ per capita in 1911 towards the total AFOLU emissions of 5.72 tCO$_2$ per capita. As the area of the agricultural fields dwindled rapidly by five times in the last 100 years, the cropland emissions also dropped proportionately. In 2015, the per capita cropland contribution dropped to just 0.01 tCO$_2$ per capita showing more than 200 times drop. The CO$_2$ emissions from wetlands have been very low, amounting to just 500 tCO$_2$ in 1911 which dropped to 222 tCO$_2$ in 2015. The CO$_2$ emissions from fuelwood have been of the order of 0.1–0.2 MT since 1911. The per capita emission from fuelwood, however, dropped from 0.72 tCO$_2$ per capita in

Table 8 Year-wise area under AFOLU emissions

Year	Population	Forests to settlements (km^2)	Fuelwood consumption (metric tonne)	Area under cropland (km^2)	Area under wetlands (km^2)	AFOLU emissions (tCO$_2$)	Per capita AFOLU emissions (tCO$_2$)
1911	13,785	1.7263	5,996.48	74.57	27.14	78,787.2	5.72
1967	255,724	7.2282	111,239.94	58.32	17.87	368,393.28	1.44
1986	557,932	12.6182	111,642.19	52.49	14.94	487,409.73	0.87
2003	906,394	19.8474	134,055.67	42.26	13.32	682,079.57	0.75
2010	963,255	36.5780	92,183.50	16.51	12.79	977,723.03	1.02
2015	1,097,751	44.5950	105,054.77	14.41	11.90	1,177,609.69	1.07

Table 9 Year-wise AFOLU emissions and per capita emissions

Year	Population	E_{FS}	E_{FW}	E_{CL}	E_{WL}	E_{FS}	E_{FW}	E_{CL}	E_{WL}
1911	13,785	38,669.12	9858.21	29,753.43	506.43	2.81	0.72	2.16	0.037
1967	255,724	161,911.68	182,878.46	23,269.68	333.45	0.63	0.72	0.09	0.001
1986	557,932	282,647.68	183,539.76	20,943.51	278.78	0.51	0.33	0.04	0.000
2003	906,394	444,581.76	220,387.52	16,861.74	248.55	0.49	0.24	0.02	0.000
2010	963,255	819,347.20	151,549.67	6587.49	238.66	0.85	0.16	0.01	0.000
2015	1,097,751	998,928.00	172,710.04	5749.59	222.05	0.91	0.16	0.01	0.000

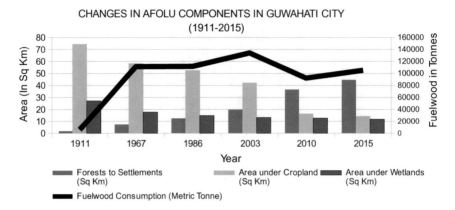

Chart 3 Changes in AFOLU components in Guwahati city

1911 to 0.16 tCO$_2$ per capita in 2015. Fuelwood burning is the second largest contributor to AFOLU emissions in Guwahati, after deforestation. Deforestation and conversion of forest land to settlement is the largest contributor to AFOLU emissions in the city which contributed to almost 50% in 1911, but its share increased to 85% in 2015. Due to increased urbanization activities, the per capita contribution of deforestation and loss of forests to settlements stands at 0.91 tCO$_2$ per capita in 2015. The contributions of various components of AFOLU to CO$_2$ emission have been depicted in Chart 4. The vertical bars indicate total emissions, while the lines show the per capita emission from various components.

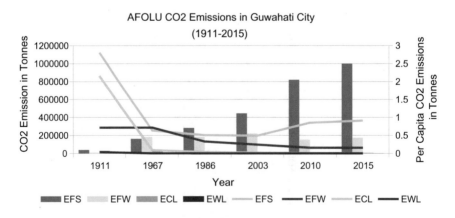

Chart 4 AFOLU CO$_2$ emissions in Guwahati city

5 Conclusions and Further Research

Yadav and Barua (2016) in the study on carbon footprint of the Guwahati city had assumed an average of 1.80 tCO_2 per capita emission within the city for 2015. The present study reveals that almost 1 tCO_2 per capita is contributed from the AFOLU sector alone. Further, almost 85% of this contribution comes from the loss of forest areas to settlements. Guwahati is a very fast growing city and has already crossed one million population mark, and also is selected as one of the 100 smart cities under the SMART city mission programme. The planners would have to ensure that the rapid decline of forests and wetlands is put to halt. A suggested programme, as it emerges from this study, is enlisted below:

1. Total halt on fragmentation of forest areas and total arrest of deforestation. This would completely wipe out the emission related to forests, and the forests would become net sinks for the city.
2. Wetland areas should be freed from encroachment, and improvement programme should be taken up so that wetlands become sinks.
3. Further conversion of agricultural areas within the existing city limits should be stalled, and these areas should be converted into green spaces, so that the remaining 10–12 km^2 of croplands could become sinks of the future, giving the much needed relief to the city dwellers.
4. The number of households using firewood is 22% now, which could be brought to 8–10% level in next 5 years by adopting to schemes such as Ujwala Yojana of the Hon'ble Prime Minister of India and other schemes.
5. The city authorities need to adopt massive afforestation drive planting at least a million plants a year, and removing encroachment from the forest areas.

Thus, the AFOLU emissions could be reduced from 1 tCO_2 per capita to almost 0.1 tCO_2 per capita. The forests and wetlands and the croplands converted to forestlands would become sinks sequestering the CO_2 for the citizens. These steps would greatly offset the other sources of emissions such as use of electricity and other fossil fuels. The city administrators should take up the above-mentioned recommendations at policy level and device suitable management strategies and institutional mechanisms to expedite reduction in CO_2 emissions emanating due to AFOLU. This would help in achieving the SDG 11 '**Make cities and human settlements inclusive, safe, resilient and sustainable**', especially the Targets 11.5, 11.6, 11.7 and 11.b. Afforestation on hill slopes would help in building resilience by future-proofing and reducing soil erosion and landslides which aggravate urban flooding.

5.1 Some Limitations of the Study

The study has the following limitations:

1. The study does not take into accounts the enteric fermentation of livestock, emissions from landfills and N_2O release from synthetic fertilizers.
2. It has been assumed that forest areas lost to settlements contribute entirely to carbon emission inclusive of above ground and below ground biomass. However, some trees may be either standing originally or planted afresh by residents. These have not been taken into account.
3. The emission factors have been taken from published literature. However, there is no specific publication on emission factors pertaining to Guwahati city per se.
4. The estimates of AFOLU, including fuelwood, in area and value may be considered as upper limits, as satellite data of coarse resolution has been used prior to 2010.
5. The carbon sequestration for the city forests also required to be worked out in order to arrive at the actual mitigation potential.

Acknowledgements The authors wish to acknowledge the Forest Department, the Assam Electronics Development Corporation Ltd. (AMTRON) and the Guwahati Metropolitan Development Authority for providing secondary data including maps and reports. The authors also wish to thank the GIS consultants and experts who helped with the maps and satellite image analysis. The authors are also grateful to the anonymous referees. The authors are also very thankful to the ENSURE 2012 organizers who provided an opportunity for presenting the findings in the Workshop, as well as allowing the authors to update the results for publication.

References

AFOLU Working Group (2016) Role of Agriculture, Forestry and Other land Use Mitigation in INDCs and National Policy in Asia, February, 2016, USAID LEAF Program and Winrock International

Anonymous (2010) Seuji Prakalp Project, Central Assam Circle, Guwahati, Assam Forest Department

Asner GP (2009) Measuring carbon emissions from tropical deforestation: an overview. Department of Global Ecology Carnegie Institution for Science, Stanford University, May, 2009

Ayres Robert U (2001) Resources, scarcity, growth and the environment. INSEAD, France

Biello D (2009) Another inconvenient truth: the world's growing population poses a Malthusian Dilemma, Scientific American, 2 October 2009

Borah J (1995) Domestic energy use in Kamrup District, Assam—a geographical analysis of selected areas, Ph.D. Thesis, Gauhati University

Bruce Seymour D (2004) Calculating decadal growth rate, March, 2004, OC research, OC International. http://ocresearch.info

Census (2011) Census of India, District Census Handbook, Kamrup Metropolitan, Assam, Series-19, Part-XIIB, 2011. http://censusindia.gov.in

Das (1973) Working plan for the Reserved Forests of South Kamrup Division, 1973–74 to 1982–83, Forest Department, Govt. of Assam

DeFries RS, Rudel T, Uriarte M, Hansen M (2010) Deforestation driven by urban population growth and agricultural trade in the twenty-first century. Nat Geosci 3:178–181 Published online: 7 February 2010. https://doi.org/10.1038/ngeo756

FSI (2011) Carbon Stock in India's Forests. Forest Survey of India, Dehradun, p 2011

García G, Abajo B, Olazabal M, Herranz K, Proy R, García I, Izaola B, Santa Coloma O (2008) A step forward in the evaluation of urban metabolism: definition of urban typologies. In: Havranek M (ed) Con Account 2008, Urban Metabolism, Measuring the Ecological City, Book of Proceedings, Charles University Environment Center, Prague

Georgescu-Roegen N (1971) The entropy law and the economic process. Harvard University Press, Cambridge, Massachusetts, London, England

GMDA (2009) Master Plan of Guwahati Metropolitan Area, 2025, Guwahati Metropolitan Development Authority, Govt. of Assam

Gogoi D (2011) Issues of environment degradation in the context of urbanisation and development, Northeast Town Planners Newsletters, Institute of Town Planners India Northeast Regional Chapter Dec 2011

Gogoi N, Baruah KK, Gogoi B, Gupta PK (2008) Methane emission from two different rice ecosystems and lower Brahmaputra valley zone of North East India. Appl Ecol Environ Res 6 (3):99–112, 2008 change, National Physical laboratory, 2002

Gupta PK, Sharma C, Mitra AP (2002) Measurements from rice fields in India, New Delhi, Centre on Global Change, National Physical Laboratory, 2002

Gupta PK, Gupta V, Sharma C, Das SN, Purkai N, Adhya TK, Pathak H, Ramesh R, Baruah KK, Venkatratnam L, Singh G, Iyer CSP (2009) Development of methane emission factors for Indian paddy fields and estimation of national methane budget, Chemosphere, 2009, January, 74(4):590–598

Hasse JE, Lathrop RG, 2010 Changing landscape in the Garden State: Urban growth and open space loss in NJ 1986 thru 2007, Rowan University Rutgers, 2010

IPCC (2000a) Good Practice Guidance and Uncertainty Management in National Greenhouse Gas Inventories, IPCC, 2000

IPCC (2000b) IPCC Special Report on LULUCF, 2000

IPCC (2006) IPCC Guidelines for National Greenhouse Gas Inventories, vol 4, Agriculture, Forestry and Other Land Use, 2006

Jacob MC (1938) A revised working plan for the Kamrup sal forests, Assam, 1938–39 to 1947–48, Forest Department, Govt. of Assam

Jacob MC (1940) Forest Resources of Assam, Forest Department, Govt. of Assam

Jain N, Patha H, Mitra S, Bhatia A (2004) Emission from rice fields—a review. J Sci Ind Res 63:101–115 (February, 2004)

Mathew P (1987) North-eastern region status paper in rural energy planning for the Indian Himalaya. In: Vinod Kumar (ed), Wiley Eastern Ltd., New Delhi, 1987, pp 236–239

Mita SN, Mitra S, Rangan L, Dutta S, Singh P (2012) Exploration of 'hot spots' of methane and nitrous oxide emission from the agricultural fields of Assam, India. Agric Food Secur 1:16

Pandey D (2002) Fuelwood studies in India—Myth and Reality, CIFOR, 2002, Jakarta

Rees WE (2003) Understanding urban ecosystems: an ecological economics perspective, Chapter, II-8. In: Understanding Urban Ecosystems, Springer-Verlag, New York

Rockström J, Steffen W, Noone K, Persson Å, Chapin III, FS, Lambin E, Lenton TM, Scheffer M, Folke C, Schellnhuber H, Nykvist B, De Wit CA, Hughes T, van der Leeuw S, Rodhe H, Sörlin S, Snyder PK, Costanza R, Svedin U, Falkenmark M, Karlberg L, Corell RW, Fabry VJ, Hansen J, Walker B, Liverman D, Richardson K, Crutzen P, Foley J (2009) Planetary boundaries: exploring the safe operating space for humanity. Ecol Soc 14(2):32. http://www.ecologyandsociety.org/vol14/iss2/art32/

RTI (2010) GHG emissions estimation methodology for selected biogenic source categories, RTI International Draft Report, December, 2010 for US EPA

Sanchez LF, Stern DI (2016) Drivers of industrial and non industrial greenhouse gas emissions. Ecol Econ 124(2016):17–24

Smith P, Bustamante M, Ahammad H, Clark H, Dong H, Elsiddig EA, Haberl H, Harper R, House J, Jafari M, Masera O, Mbow C, Ravindranath NH, Rice CW, Robledo Abad C, Romanovskaya A, Sperling F, Tubiello F (2014) Agriculture, forestry and other land use (AFOLU). In: Edenhofer O, Pichs-Madruga R, Sokona Y, Farahani E, Kadner S, Seyboth K, Adler A, Baum I, Brunner S, Eickemeier P, Kriemann B, Savolainen J, Schlömer S, von Stechow C, Zwickel T, Minx JC (eds) Climate change 2014: mitigation of climate change. Contribution of Working Group III to the Fifth Assessment Report of the Intergovernmental Panel on Climate Change. Cambridge University Press, Cambridge, United Kingdom and New York, NY, USA

Sridhar KS (2010) Carbon emissions, climate change, and impacts in India's cities. India Infrastructure Report 2010, 345–354

Swargowary A (2002) Working plan for Kamrup East Forest Division, Part I & II, 2002–03 to 2011–12, Forest Department, Govt. of Assam

van der Werf GR, Morton DC, DeFries RS, Olivier JG, Kasibhatla PS, Jackson RB, Collatz GJ, Randerson JT (2009) CO_2 emissions from forest loss, Nat Geosci, 2:737–738 (November 2009). www.nature.com/naturegeoscience

WRI (2015) Stephen Roe (lead), Mashen Brander, Jason Funk, Apurba Mitra, Stephen Russell, Marion Vieweg Policy and Action Standard, Agriculture, Forestry and Other land Use (AFOLU) Sector Guidance, Draft, May, 2015, World Resource Institute

WWF (2011) Impact of urbanization on biodiversity—case studies from India, WWF-India, 2011

Xepapadeas A (2008) Ecological economics. The New Palgrave Dictionary of Economics 2nd edn. Palgrave, MacMillan

Yadav R, Barua A (2016) A study of urbanization and ecosystem services of Guwahati city from forest footprint perspective. J Ecosys Ecograph S5:004. https://doi.org/10.4172/2157-7625.S5-004

User Preferences and Traffic Impact Evaluation of Bus Rapid Transit System for a Medium-Sized City

K. Krishna Kumar and K. V. R. Ravi Shankar

Abstract Recognizing the problem of urban transport, a number of cities are coming up with mass transit proposals. Amongst the domain of high capacity public transport systems available, Bus Rapid Transit System (BRTS) with dedicated bus lanes is one of the viable options. BRTS will create a high-quality public transport to enhance the mobility pattern and demonstrate that protection of environmental conditions, energy savings, reduction in road accidents, etc. This study examines applicability of BRTS for a medium-sized city like Visakhapatnam. The ridership improvement upon introduction of BRTSis collected through a stated preference survey. The modal shift improvement from personalized vehicles is predicted. The present study analyzes the travel behavior of the people in the selected study area. The results identified a qualitative difference in the choice of people in preferring the mode based on their income levels, travel cost, and purpose of the trip. The end results will be applicable in determining the fare structure and enhancing the existing system.

Keywords Sustainable transport modes · Bus rapid transit system
Stated preference survey · Modal shift

K. Krishna Kumar
Civil Engineering Department, Woldiya Univeristy, North Wollo Zone,
400 Woldiya, Ethiopia
e-mail: krishnakumarkorada@gmail.com

K. V. R. Ravi Shankar (✉)
Department of Civil Engineering, National Institute of Technology Warangal,
Warangal 506004, Telangana, India
e-mail: kvrrshankar@gmail.com

© Springer International Publishing AG 2018
A. K. Sarma et al. (eds.), *Urban Ecology, Water Quality and Climate Change*,
Water Science and Technology Library 84,
https://doi.org/10.1007/978-3-319-74494-0_7

97

1 Introduction

Mass transit is movement of people within urban areas using group travel technologies such as buses and trains. The main objective of mass transportation is that many people can travel in a single vehicle or attached vehicles (trains). This makes it possible to move people in the same travel corridor with greater efficiency, leading to lowered travel costs and also seeking to reduce travel demand by encouraging a better integration of land use and transport planning (Chandra 2006; Erik et al. 2007). Various mass transport options available are Bus Rapid Transit System (BRTS), Metro, Light Rail Transit (LRT), Mono Rail, etc. Wright (2005) defines BRT as a "bus-based mass transit system that delivers fast, comfortable, and cost effective urban mobility." BRTS can be applied to a variety of transportation systems using buses to provide faster, more efficient service than an ordinary bus line. BRT is regarded as sustainable, environmental-friendly transport mode and is being implemented in many cities of the world (Vidwans 2006). Often this is achieved by making improvements to existing infrastructure, vehicles, and scheduling. The goal of these systems is to approach the service quality of rail transit while still enjoying the cost savings and flexibility of bus transit. The first BRTS in the world was the Rede Integrada de Transporte (RIT, translated as "Integrated Transportation Network"), implemented in Curitiba, Brazil in 1974 (Caulfield and O'Mahony 2009). Some of the major BRTS projects in service/under implementation in India are Ahmedabad (88.50 km), Delhi (310 km), Pune (101.77 km), Pimpri Chinchwad (Maharashtra, 42.22 km), Indore (11.45 km), Bhopal (21.71 km), Jaipur (39.45 km), Vijayawada (15.50 km), Vizag (42.80 km), Rajkot (29.00 km), and Surat (29.90 km) (Vyas et al. 2011).

The objective of this study is to assess the possible modal shift of personal vehicle users toward the proposed BRTS using an appropriate mode choice modeling: To describe the current urban transport system and identify travel behavior of the people in the study area. The ridership improvement upon introduction of BRTS is analyzed through stated preference survey. Details of the study area are described in the next section.

2 Study Area

Visakhapatnam is one of the fastest growing cities and is the second largest city of Andhra Pradesh. It is primarily an industrial city, apart from being a port city with an estimated area of 550 km^2. The city doubled its population from 1990 to 2000 owing to a large migrant population from surrounding areas and other parts of the country coming to the city to work in its heavy industries (Vyas et al. 2011).

Gopala Raju (2011) observed that about 4.5 lakh registered vehicles ply on the city roads in Visakhapatnam, 90% of which are cars and motorized two-wheeled vehicles. National Institute of Urban Affairs (NIUA) reported the present travel demand in the city is about 12 lakh trips per day; a significant 65% of these trips are catered by private modes. The current modal split studies indicate that only 20% favor public transport. According to the National Urban Transport Policy (NUTP), cities with one million-plus population must target a minimum public transport mode split of 50%. Therefore, a higher modal split of above 50% has been recommended for Visakhapatnam to be achieved in phases. This can be attained by giving impetus to public transport to arrest the trend of personal modes.

To overcome the current problems and with the objective to promote and enhance regional mobility and serve the public by providing quality transit services and solutions that improve the overall quality of life of the residents of Visakhapatnam, building of a BRTS service is one of the solutions. Two corridors have been identified by the authorities (Fig. 1) on which opportunities for developing BRTS were explored (Vyas et al. 2011). These would enhance the connectivity of the core of the city with all other areas through an integrated transportation network of bus corridors. It was proposed to conduct the stated preference survey on corridor 1 (Pendurthi to RTC Complex). This stretch was selected for the reason that it is one of the most preferred routes for travel towards the city center.

Fig. 1 BRTS corridors in Visakhapatnam. *Source* Vyas et al. (2011)

Table 1 Characteristics of the TAZ

TAZ	Number of HH	HH size	Income	Number of PV	Number of Trips
1	70	2.32	12,550	40	225
2	64	2.14	17,000	44	197
3	62	2.31	17,000	42	105
4	76	1.89	11,400	38	132
5	73	2.48	14,500	58	170
6	77	1.38	14,700	51	166
7	63	2.35	12,800	42	103
8	70	2.34	13,400	72	139

TAZ Traffic Analysis Zones; *HH* Household; *PV* Personal Vehicles

3   Data Collection

A stated preference household survey was conducted in BRTS corridor 1 and the area around this corridor was divided into eight zones. A household interview survey was carried out to collect household characteristics of the people and their travel behavior in commuting their daily trips (work or educational trips).

The zones located for data collection fall within 1 km from the main route, i.e., corridor 1. In total, 554 household samples were collected. The survey questionnaire consisted of queries pertaining to household information, trip characteristics, and opinions on their willingness to shift to proposed BRTS. The data collected here is the socioeconomic information and travel characteristics such as household size, monthly household income, occupation, personal vehicles, number of trips, etc. The characteristics of each traffic analysis zone are shown in Table 1. The lowest average income was reported for zone 4 and highest numbers of trips were observed in zone 1 (225). The average household size was low at 1.89 (zone 4) and high at 2.35 (zone 7). These results show that the zones selected and the samples collected have a wide variation in the personal and travel characteristics.

4   Trip Characteristics

4.1   Trip Purpose

The trip purpose was collected as a part of the travel characteristics of the users and this was segregated based on work trip, academic, shopping and entertainment, etc. The trip purpose composition is as shown in Fig. 2. Out of the total number of trips (1237), most of the trips are for work and academic based followed by shopping trips.

Fig. 2 Purpose of trip

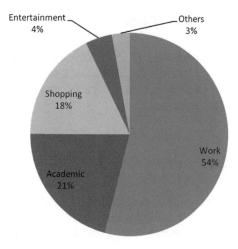

4.2 Trip Length

The trip lengths are collected from the household survey, and the data obtained was presented in Fig. 3. Among all trip lengths, it has been observed that less than 5 km trip lengths were high in number, and more than 40 km trip lengths are low in number.

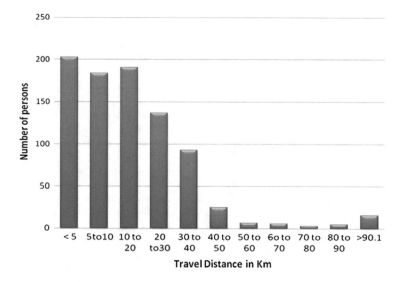

Fig. 3 Distribution of trip lengths of households

4.3   Travel Cost

The travel cost of each individual is categorized from the household interview survey, and the distribution of travel cost among households is as shown in Fig. 4. On an average, the travel cost was observed to range between 10 and 50 Indian rupees per day.

4.4   Income Range

The income of each individual is categorized from the household interview survey, and Fig. 5 indicates the income ranges of households. More than 50% of the data has the income in the range between 4 and 15 thousand Indian rupees.

4.5   Distance to the Nearest Bus Stop

The time required for each individual to reach the nearest bus stop is collected from home interview survey, and Table 2 shows the frequency of walking times. It can be observed that most of the bus stops are positioned within a reach of 6–10 min walking distance.

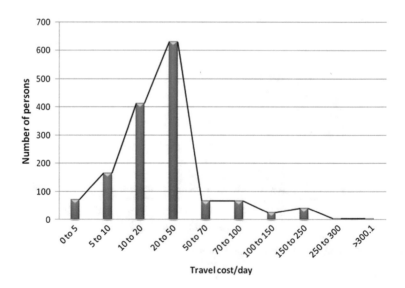

Fig. 4 Travel cost

Fig. 5 Income range

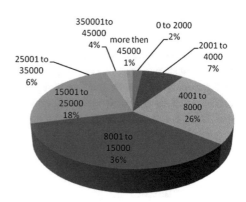

Table 2 Walking time to nearest bus stop

T	<2	2–4	4–6	6–8	8–10	>10
F	124	17	279	12	224	88

T Walking time to the nearest bus stop in minutes
F Number of persons (frequency)

5 Modal Choice Analysis

Modal split has considerable implications for transportation policy, particularly in large metropolitan areas. For instance, the decision whether to implement the new BRTS should be informed by the predictions of the number of people who will switch to the BRT bus in their trips. The development and application of the individual choice model such as Multiple Linear Regression (MLR) models were developed (Bajracharya 2008; Singh 2011; Tischer and Dobson 1979). Mode choice models have always formed a critical part in analyzing the travel demand the choice is for the mode used for traveling between a particular origin zone and a particular destination zone. The probability of individuals choosing a given option is a function of their socioeconomic characteristics and the relative attractiveness of the option. The basic structure of MLR equation derived is of the following form (Golob 1970; Meyer et al. 1978):

$$P_k = a + b * I + c * PV + d * TL + e * NBS, \tag{1}$$

where

P_k	Proportion of commuters choosing mode k,
I	Family income for group in region,
PV	Personal vehicle,
TL	Trip length in km,
NBS	Distance to nearest bus stop,
a	Constant, and
b, c, d, e	model coefficients to be calibrated.

The estimation also considers the possibility of examining the effect the variables influencing the choice of mode. In this study, the stated preference (willingness or otherwise to shift) of the respondent is the dependent variable; and income, personal vehicle, distance to the nearest bus stop, trip purpose, trip length, and trip cost difference are the independent variables considered for model estimation. The model form presented above has been validated for the collected data, and it was estimated that the proportion of trips increase by 1.8 times on the proposed BRTS.

6 Conclusion

This research work quantified the travel characteristics of people and identified the choice of mode of commuters in Visakhapatnam for the proposed BRTS. Further, user preferences in selecting the proposed transit system were analyzed. The choice parameters were formulated, and multiple regression analysis models were used for estimating the proportion of commuters choosing the BRTS. The average household size and trip length were observed as 2.15 and 29.2 km, respectively. The user preferences suggest that the proposed BRTS has a higher potential for attracting the commuters from personal mode to public transportation and a well-planned system is in much need for a city like Visakhapatnam.

References

Bajracharya AR (2008) The impact of modal shift on the transport ecological footprint: a case study of proposed BRTS in Ahmedabad, India. Thesis international institute for geo-information science and earth observation enschede, the Netherlands, VREF (SP-2006-09)

Caulfield B, O'Mahony M (2009) A stated preference analysis of real-time public transit stop information. J Publ Transp 12(3):1

Chandra R (2006) National urban transport policy and its implication (Journal), vol 26. National Institute of Urban Affairs, New Delhi, July–December—2

Erik A, Vigina L, Hamed B, Ban, Liaun C (2007) An integrated methodology for corridor management planning. TRB J 8:27

Golob TF (1970) The survey of user choice of alternate transportation modes. High Speed Ground Transp J 4:103–116

Gopala Raju SSSV, Balaji KVGD, Durga Rani K (2011) Vehicular growth and its management: Visakhapatnam city in India—a case study. J Indian Sci Technol 4(8). ISSN: 0974-6846

Meyer RJ, Levin IP, Louviere JJ (1978) Functional analysis of mode choice. Transp Res J TRB 673:1–6 (Washington, D.C.)

Singh I (2011) Attitudinal and trade-off studies on mode choice in a multi-modal environment. Indian Highways 39(12):33–47

Tischer ML, Dobson R (1979) An empirical analysis of behavioral intentions of single occupant auto drivers to shift to high occupancy vehicles. Transp Res A 13:143–158

Vidwans N (2006) Ahmedabad takes the bus, India Together Magazine, 17 May. URL: http://www.indiatogether.org/2006/may/eco-brts.htm. Accessed on Dec 2011

Vyas SS, Roy S, Sharma P (2011) Urban transport initiatives in India: best practices in PPP. Report on "Visakhapatnam BRTS", National Institute of Urban Affairs

Wright L (2005) Module 3a mass transit options. Institute of Transportation development policy Division 44 Environmental and infra structure project-Transport policy Advice

Green Road Approach in Indian Road Construction for the Sustainable Development

Rajiv Kumar, G. Appa Roa and Teiborlang Lyngdoh Ryntathiang

Abstract Growing public awareness of climate change requires transportation professionals to integrate green concepts into the transportation planning, design, construction, and operation processes. Green highways are a relatively new concept although the implementation of technologies involved in green highway design has been encouraged for many years. A green road may not look like a normal road at first glance, but with closer inspection, a driver will notice the subtle difference. More plant life grows along the shoulder; trees are planted along the road and act as wildlife buffers, thus giving a pleasant riding experience. In towns, highways become more aesthetically pleasing, and in rural areas, highways become a more natural part of the environment. Green highway can be defined by five broad topics such as conservation and ecosystem management; watershed-driven strong water management; life cycle energy and emissions reduction; recycle, reuse, and renewable; and overall societal benefits. This paper describes the factor affecting green road and the construction process of microsurfacing as a preventive wearing course of a pavement that can reduce both direct and indirect costs. Microsurfacing is environmental friendly which reduces the greenhouse gas and fuel consumption.

1 Introduction

Road transport is considered to be one of the cost-effective and preferred modes of transport for both freight and passengers. India has an extensive road network of 5.4 million km—the second largest in the world. The national highways have a total length of 97,991 km and serve as the arterial road network in the country (MORTH 2013). It is estimated that more than 70% of freight and 85% of passenger traffic in the country are being handled by roads. While highways/expressways constitute only about 2% of the total road length, the rest are state highway, major district

R. Kumar · G. Appa Roa · T. L. Ryntathiang (✉)
Civil Engineering Department, IIT Guwahati, Guwahati 781039, India
e-mail: lyngdoh@iitg.ernet.in

© Springer International Publishing AG 2018   107
A. K. Sarma et al. (eds.), *Urban Ecology, Water Quality and Climate Change*,
Water Science and Technology Library 84,
https://doi.org/10.1007/978-3-319-74494-0_8

road, district roads, and rural and other road which is considered as low-volume road (Kumar and Ryntathiang 2012).

Growing public awareness of climate change requires transportation professionals to integrate green concepts into the transportation planning, design, construction, and operation processes. Green highways are a relatively new concept although the implementation of technologies involved in green highway design has been encouraged for many years. More plant life grows along the shoulder; trees are planted along the road and act as wildlife buffers thus giving a pleasant riding experience. In towns, highways become more aesthetically pleasing, and in a rural area, highways become a more natural part of the environment a such as conservation and ecosystem management, watershed-driven strong water management, life cycle energy and emission reduction, recycle reuse and renewable, and overall societal benefit (Bryce 2008).

Around 90% of roads in India are flexible pavement which generally comes in hot mix asphalt categories (Jain et al. 2014). As we know that hot mix production involves the use of fossil fuel both as a raw material and energy source and hot mix plants vary according to their age and efficiency. On average, the mixing temperature of the hot mix asphalt is more than 140 °C and another new technology is warm mix asphalt which reduces the emission around 20–40% (Kumar and Chandra 2016). Using cold mix and cold mix recycling, the carbon emission could reduce by 100% that comes during mixing and compaction of hot mix asphalt or warm mix asphalt as shown in Fig. 1. Warm mix asphalt approximately saves 1.5–2.0 L fuel per ton of materials, which reduces to 20–35% of CO_2, and it is equivalent to 4.1–5.5 kg of CO_2 (Croteau and Tessier 2008). For the production of aggregate which includes quarrying, hauling, crushing, and screening, the greenhouse gas (GHG) emission range is 2.5–10 kg of CO_2/ton and for asphalt, it is about 221 kg of CO_2/ton (Chehovits and Galehouse 2010). Therefore, conservation of nonrenewable resources and energy, together with reduced environmental pollution and working conditions, is global issues that are becoming increasingly important to civil engineers. As a result, authorities in various countries are creating

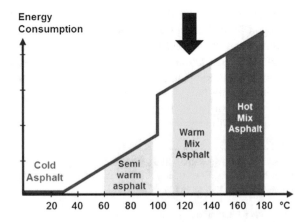

Fig. 1 Production temperature of bituminous mixtures (Kumar and Chandra 2016)

legislation to reduce energy consumption by using cold mix construction and recycling of pavement (European Commission 2010). These incentives are bound to shape the development of new processes in the road construction especially in the section where the highest consumption occurs, i.e., in the hot mixes asphalt production.

Microsurfacing, a cold mix technology, has been used in Germany, Spain, and France since 1976 and was introduced to the United States in 1980. The technology has been in use in the world over for a very long time as a routine form of maintenance in preference to conventional hot mix overlays (Kumar and Ryntathiang 2012). This technology was introduced in India in the year 1999–2000 under the brand name of Macroseal by Yala construction and Elsamex SA, Spain (Kumar and Ryntathiang 2016) as a pilot study. The initiation by the Delhi government to maintain the roads of Delhi was to use bitumen emulsion instead of hot mix in the light of ban on hot mix plants imposed by the Supreme Court. The study was conducted approximately 10 years back, by Elsamex S.A, Spain with the active involvement of Central Road Research Institute, Delhi (Kumar and Ryntathiang 2012). In the study report, the specifications for the use of cold emulsion for maintaining various types of roads were standardized with the latest state-of-the-art technology. The study report was utilized by Delhi public works department, New Delhi municipal council and Ministry of road transport and highways (MORTH 2013). Subsequently, the technology was given a go-ahead for maintaining the roads of Delhi. It has also been included in the MORTH specification in clause 516 as a slurry seal. IRC subsequently brought out SP: 81 (IRC 2008) "Tentative Specification for Microsurfacing". NHAI has also issued a circular recommending microsurfacing for renewal of wearing course for maintenance of national highways. Now, microsurfacing is used in the many states of India for low-volume road, highway, and expressway maintenance.

Cold in-place recycling is an eco-friendly paving process for any road structure, which is in irreparable condition. Recycling and reusing the existing pavement layer does away with the need for fresh aggregate and therefore the costs of purchasing and transporting the aggregate. During cold in-place recycling, two to five inches of the current road surface are pulverized to a specific aggregate size, mixed with a rejuvenating asphalt emulsion, and then reused then and there; and this reduces the overall labor costs. Since no heating is needed to heat the aggregate or bitumen, cold in-place recycling reduces the noxious fumes and pollution associated with many other processes which are safe for the environment and also safe for the road construction workers.

The objective of this paper is to discuss the materials used for microsurfacing and step involved in microsurfacing (cold mix technology) and to discuss the environmental impact of microsurfacing compare with hot mix asphalt overlay and advantages of cold mix recycling.

2 Microsurfacing

Microsurfacing is a mixture of polymer-modified asphalt emulsion, mineral aggregate, mineral filler, water, and other additives, properly proportioned, mixed, and spread on a paved surface in variable thick cross section which resists compaction (Gransberg 2010).

2.1 Materials

The microsurfacing shall consist of a mixture of an approved emulsified asphalt, mineral aggregate, water, and specified additives which are described below.

2.1.1 Emulsified Asphalt

An emulsion is a thermodynamically unstable system consisting of at least two immiscible liquid phases one of which is dispersed as globules in other liquid phase stabilized by the third substance called emulsifying agent. In general, microsurfacing suppliers supply the emulsion to the contractor along with a mix design. Typically, in the manufacturing of emulsion, the polymer is milled together with the bitumen and emulsifier and the produced emulsion is called as polymer-modified emulsion. Different polymers or a combination of the polymer can be added to the emulsion, and these tend to be proprietary. Each polymer has its own unique properties that will enhance the performance characteristics of the emulsion. These performance characteristics could be stiffness of the emulsion at high temperatures, resistance to flushing, and elasticity of the emulsion at low temperatures. The amount and type of the emulsifier will affect the setting characteristics and compatibility of the emulsion.

The emulsified asphalt shall be a quick-set polymer-modified cationic-type CSS-1H emulsion and shall conform to the requirements specified in ASTM 2397 2012 (Standard Specification for Cationic Emulsified Asphalt). It shall pass applicable storage and settlement tests. The polymer material shall be milled into the emulsion or blended into the asphalt cement prior to the emulsification process. The cement mixing test shall be waived for this emulsion. The residue of the emulsion shall have a minimum ring and ball softening point of 60 °C. According to MORTH (2013), the emulsified bitumen shall be a cationic rapid setting type as approved by the Engineer, conforming to the requirements as per IS: 8887 (2004).

2.1.2 Aggregate

The mineral aggregate used shall be the type specified in the particular application requirements of the slurry seal. The aggregate shall be crushed stone such as granite, slag, limestone, chat, or another high-quality aggregate, or combination thereof. To assure the material is 100% crushed, the parent aggregate will be larger than the largest stone in the gradation to be used. The smooth textured crusher fines shall have less than 1.25% water absorption. The aggregate shall be gray in color and clean and free from organic matter, other deleterious substances, and clay balls. Oversized granular material and/or presence of clay balls will require the project to be stopped and shall meet the following requirements:

Quality Tests

The aggregate should meet agency specified polishing values and these minimum requirements are given in Table 1.

Gradation

The mix design aggregate gradation shall be within one of the following bands (or one recognized by the local paving authority) which are described below:

Type II. This aggregate gradation is used to fill surface voids, address more severe surface distresses, seal, and provide a durable wearing surface and is given in Table 2.

Type III. This aggregate gradation provides maximum skid resistance and an improved wearing surface. This type of microsurfacing surface is appropriate for heavily traveled pavements, rut filling, or for placement on highly textured surfaces requiring larger size aggregate to fill voids and is given in Table 3.

Table 1 Quality test on aggregate

Test	Test method		Specifications
Sand equivalent value of soil and fine aggregate	T 176	D 2419	45 minimum
Soundness of aggregate by use of sodium sulfate or magnesium sulfate	T 104	C 88	15% maximum w/Na$_2$SO$_4$ 25% maximum w/MgSO$_4$
Resistance to degradation of small size coarse aggregate by abrasion and impact in Los Angeles machine	T 96	C 131	35% maximum

Table 2 Microsurfacing gradation of type II

Sieve size	9.5	4.75	2.36	1.18	0.6	0.3	0.15	0.075
% passing	100	90–100	65–90	45–70	30–50	18–30	7–16	5–15

Table 3 Microsurfacing gradation of type III

Sieve size	9.5	4.75	2.36	1.18	0.6	0.3	0.15	0.075
% passing	100	70–90	45–70	26–50	19–34	12–25	7–16	5–15

According to MORTH (2013), the aggregate shall be crushed rock or slag and may be blended, if required, with clean, sharp, naturally occurring sand free from silt pieces and organic and other deleterious substances to produce a grading as given in Table 500–33. The aggregates shall meet the requirements of the film stripping test as given in Indian Standard (IS 2013), and a suitable amount and type of anti-stripping agent added, as may be needed.

2.1.3 Mineral Filler

Mineral filler may be used to improve mixture consistency and to adjust mixture breaking and curing properties. Portland cement, hydrated lime, limestone dust, fly ash, or other approved filler meet the requirements of ASTM protocol and shall be used if required by the mix design (ASTM 2001). Typical use of filler levels are normally 5–15% and considered as part of the aggregate gradation. According to MORTH (2013), it is usual to use ordinary Portland cement, hydrated lime, or other additives to control consistency, mix segregation, and setting rate. The proportion of the additive should not normally exceed 2% by weight of dry aggregates.

2.1.4 Water

The water shall be free of harmful salts and contaminants. If the quality of the water is in question, it should be submitted to the laboratory with the other raw materials for the mix design.

According to MORTH (2013), the quality of water should be such that its quality will not separate from the emulsion before placing the microsurfacing on the surface of a pavement. The pH of the water must lie in the range 4–7, and if the total dissolved solids in the water amount to more than 500 ppm, the engineer may reject it, or order the contractor to conduct a trial emulsion mix to demonstrate that it does not cause early separation.

2.1.5 Additives

Additives may be used to accelerate or retard the break/set of the microsurfacing and its range should be appropriately approved by the laboratory as part of the mix design and they are aluminum sulfate crystal, ammonium sulfate, inorganic salts, liquid aluminum sulfate, amines, and anti-stripping agents (Broughton and Lee 2012).

2.2 Mixing Process of Microsurface

The mix of microsurface can be prepared at site, and it depends mainly on the availability of material and transportation facility. The mixing procedure should follow the mix design obtained from laboratory. The following type of mixing can be batch mixing, continuous mixing, and hand mixing.

2.3 Mixing Procedure

The required aggregate quantity is mixed with water according to the surface area of aggregate—fine and course aggregate separately. After 10 min, the prewetted aggregate is placed in concrete mixer in batch-wise manner. Thereafter, emulsion in required proportion is poured in the mixer followed by cement and mix till aggregate; cement and emulsion is proper and ready to be poured out (Kumar and Ryntathiang 2013).

2.4 Laying Procedure

The procedure for laying of microsurfacing can be summarized as follows:

- Use rope to ensure straight edges and cover thickness,
- Use squeegee to spread,
- Drag hessian burlap for smooth finish, and
- Only open to traffic once dry (±4 h).

The breaking time of emulsion depends on the emulsion temperature which is shown in Fig. 2; so the mix starts braking after this time (ISSA 2010).

2.5 Environmental Impact

Pavement preservation is inherently green owing to its focus on conserving energy and raw materials, and reducing greenhouse gasses by keeping good roads good (Kumar and Ryntathiang 2012). Microsurfacing's environmental footprint is lower than most common pavement preservation and maintenance treatments (Takamura et al. 2001). Figure 3 shows a study on the environmental impact of several

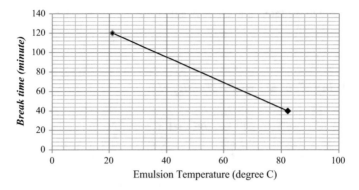

Fig. 2 Relationship between temperature and emulsion breaking time (ISSA 2010)

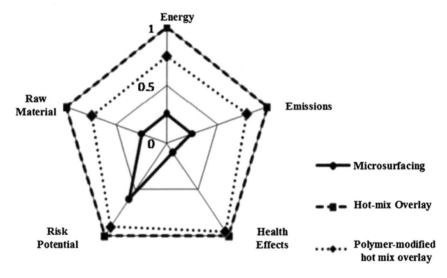

Fig. 3 Microsurfacing environmental footprint compared with two types of pavement preservation overlays (Takamura et al. 2001)

commonly used pavement preservation and maintenance treatments. The study developed "eco-efficiency" indices for the five categories shown in the figure and found that microsurfacing had a substantially lower environmental footprint than the other options (Takamura et al. 2001). This study does not include the reduced greenhouse gas emissions resulting from microsurfacing's ability to greatly reduce traffic delays in work zones. Additionally, the "risk potential" and "health effects" categories did not include the reduction in work zone accident risk inherent to microsurfacing (Gransberg 2010). Therefore, microsurfacing's "true" footprint may be even smaller in relation to hot mix asphalt options for pavement preservation and maintenance programs. When looking at options to address pavement preservation

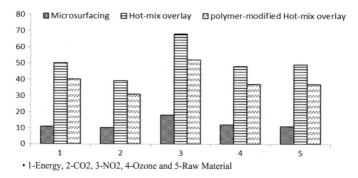

Fig. 4 Comparative environmental impacts of three pavement preservation and maintenance treatments (Takamura et al. 2001)

and maintenance issues, the engineer can use the environmental and safety benefits of microsurfacing as a possible justification to offset any marginal increase in construction cost versus other alternatives.

Figure 4 (Takamura et al. 2001) illustrates the output from that study and provides greater detail with respect to the greenhouse gas emissions, as well as information on raw material consumption. Both studies merely constructed a simplified snapshot of the comparative environmental impact of microsurfacing. It neither included the impact of work zone delays nor the life safety benefits accrued from microsurfacing owing to its ability to minimize the duration of work zone delays nor increased congestion during pavement maintenance operations.

Chehovits and Galehouse (2010) performed several comparisons for different structure pavement sections and determined that for different structures, yielding the same structure performance, energy, and generate GHG emission can vary as much as 80%.

3   Recycling of Pavements

Recycling of pavements is one such methodology where many countries are using for over 30 years (Willis 2015). In the United States of America, products such as aluminum, plastic, paper, glass, etc. are recycled but asphalt pavement is the product which is 2nd widely recycled (at about 18%) after cement concrete pavement (EPA 2015). Various products which are recycled in the United States of America and their percentages for the year 2010 are shown in Fig. 5. It can be easily concluded that clear winner in recycling is asphalt pavement.

The Federal Highway Administration reports that out of 91 million metric tons of asphalt pavement, 73 million metric tons are recycled, i.e., an amount of 80% of asphalt pavements are recycled in the USA. The recycled material is used in rehabilitation projects, road widening projects, construction of embankments and

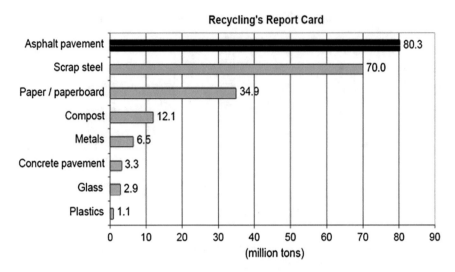

Fig. 5 Products recycled in USA (Han et al. 2011)

shoulders, etc. Performance aspects of the recycled pavement also indicate that recycled pavements perform equal to or better than the conventional mixes (Kandhal and Mallick 1997). All these statistics show a successful implementation of the recycling of pavements in the USA.

The main reason behind the successful implementation of recycling is its benefits over the conventional methods. Two of the most highlighted benefits are the cost savings and preservation of the environment. These benefits prove to be significant because the Government wants to conserve budgets in every possible situation and presently the whole world is giving special attention to the environment by focusing on the "Green Technology". Adoption of recycling of pavements results in the savings of around $35–$80 per ton. So, if an amount of 15 ton is recycled per week over a year, then cost savings of around $30,000–$60,000 can be achieved (Asphalt Recycling 2012). This is a significant amount of saving and if such a saving is achieved in India, much stress on the economy can be relieved by reducing the budgets allotted to the road development projects.

Hence, it can be concluded that recycling of pavements is an efficient solution for construction and maintenance of roads which offers enormous benefits like cost savings, preservation of virgin materials, preservation of environment, etc. Hence, it is an open alternative for implementation on Indian roads to save the construction costs as well as sustain the sources of raw materials.

Recycling of pavements offers various advantages over the conventional methods of construction. Some of the advantages are mentioned below (FHA 2017):

- **Reduction in cost of construction**: The construction cost of a bituminous pavement can be significantly reduced if recycling is adopted. As this methodology uses RAP to produce new materials, the requirement of new material gets reduced. Thus, material costs can be reduced.

- **Preservation of virgin aggregates and binder**: Road construction projects have increased the material requirement, and nowadays it has become difficult to find good quality of aggregates for road construction. Recycling of pavements offers use of existing material, thus sustaining the virgin aggregate sources. Also, virgin bitumen can be saved using recycling which is more advantageous as it is a petroleum product.
- **Preservation of pavement geometrics**: The conventional rehabilitation technique is to overlay an existing pavement. This overlay increases the pavement thickness extensively causing a problem for pavement geometry and especially pavement drainage. Also in the case of urban areas, the increased pavement thicknesses cause problems to the sidewalks and also to the adjacent properties like houses, buildings, shops, etc. Pavement recycling, if adopted, can sustain the thickness of a pavement to avoid such problems.
- **Preservation of the environment**: Pavement recycling can be implemented at ambient temperatures with the use of bitumen emulsion reducing the emissions and fuel consumption. Also, it can be implemented on the site itself which eliminates the transportation of materials and in turn reduces the fuel consumption. Hence, recycling methodology preserves the environment and conserves energy.

4 Conclusion

The development of a new process or product requires substantial research and development before a sufficient level of reliability is achieved to enable implementation on a large scale. Microsurfacing is a pavement preservation and maintenance tool with very few technical or operational limitations. One ton of HMA resurfacing produces about 5.25 kg of carbon which can stop by implementing the cold mix technology like microsurfacing. For the production of aggregate, GHG emission range is 2.5–10 kg CO_2/t and for asphalt, it is 221 kg CO_2/t. Microsurfacing environmental footprint compared with hot mix overlay and polymer-modified hot mix overlay is less. Yielding of greenhouse gasses for the same structure performance, energy, and generate greenhouse gasses (GHG) emission can vary as much as 80%. Recycling of pavements offers various advantages over the conventional methods of construction (Hot mix overlay), and it can be implemented in Indian road construction. To minimize energy use and GHG over the life of a pavement, all preservation treatments can be done by cold mix technology, like microsurfacing.

References

Asphalt Recycling (2012) Asphalt Recycling. https://asphaltrecycling.com/news-details.php?news=29. Accessed 22nd January 2018

ASTM (2001) Standard specification for mineral filler for bituminous paving mixtures. In: ASTM D242, ASTM International, West Conshohocken, Pennsylvania, United States, pp 1–2

ASTM (2012) Standard specification for cationic emulsified asphalt. In: ASTM D2397, ASTM International, West Conshohocken, Pennsylvania, United States, pp 1–3

Broughton B, Lee SJ (2012) Microsurfacing in Texas. Texas Department of Transportation. Report No. FHWA/TX-12/0-6668-1

Bryce J (2008) Developing sustainable transportation infrastructure. In: American Standards and Testing Materials (ASTM). West Conshohocken, Pennsylvania, United States

Chehovits J, Galehouse L (2010) Energy usage and greenhouse gas emissions of pavement preservation processes for asphalt concrete pavements. In: First international conference on pavement preservation, pp 27–42

Croteau J, Tessier B (2008) Warm mix asphalt paving technologies: a road builder's perspective. In: Proceedings, annual conference of the transportation association of Canada, pp 1–12

EPA (2015) Advancing sustainable materials management: facts and figures 2013, USA Environmental Protection Agency. https://archive.epa.gov/epawaste/nonhaz/municipal/web/pdf/2013_advncng_smm_rpt.pdf. Accessed 22nd January 2018

European Commission (2010) Green public procurement—road construction and traffic signs background report. In: European Commission, D G Environment-G2, B-1049, Brussels

FHA (2017) Selection of pavement for recycling and recycling strategies. https://www.fhwa.dot.gov/pavement/recycling/98042/03.cfm. Accessed 22nd January 2018

Gransberg DD (2010) NCHRP synthesis 411 microsurfacing. Washington, D.C.

Han J, Thakur SC, Chong O, Parsons RL (2011) Laboratory evaluation of characteristics of recycled Asohalt Pavement in Kansas

Indian Roads Congress (IRC) (2008) Tentative specification for slurry seal and microsurfacing. Indian Roads Congress, New Delhi

Indian Standard (IS) (2003) Method of test for determination of stripping value of road aggregates. IS-6241:2003, Bureau of Indian Standards, New Delhi

Indian Standard (IS) (2004) Bitumen emulsion for roads (Cationic Type)—Specification. IS-8887:2004, Bureau of Indian Standards, New Delhi

ISSA (2010) Recommended performance guideline for emulsified Asphalt Slurry Seal. ISSA A105

Jain PK, Rongali UD, Chourasiya A, Ramizrazam M (2014) Laboratory performance of polymer modified warm mix asphalt. J Indian Roads Congress 24–32

Kandhal SP, Mallick RB (1997) Pavement recycling guidelines for state and local Governments Participant's reference book. National Centre for Asphalt Technology, Federal Highway Administration, Washington, D.C. 20590

Kumar R, Chandra S (2016) Warm mix asphalt investigation on public roads—a review. Civil Eng Urban Plan Int J (CiVEJ) 3(2):75–86

Kumar R, Ryntathiang LT (2012) Rural road preventive maintenance with microsurfacing. Int J Comput Appl (IJCA) 4–8

Kumar R, Ryntathiang LT (2013) IIT Guwahati: department of transportation experience with microsurfacing. In: Proceedings, 1st annual international conference on architecture and civil engineering (ACE 2013), Singapore, pp 437–442

Kumar R, Ryntathiang LT (2016) New laboratory mix methodology of microsurfacing and mix design. Transp Res Procedia 17:488–497

MORTH (2013) Basic road statistics of India 2012. Ministry of Road Transport and Highways, New Delhi

Takamura K, Lok KP, Wittlinger R (2001) Microsurfacing for preventive maintenance: eco-efficient strategy. In: International slurry seal association meeting, Maui, Hawaii, pp 1–8

Willis PX (2015) Analysis of the use of Reclaimed Asphalt Pavement (RAP) in Europe. MS Thesis in Civil Engineering, Politecnico Di Milano, Italy

Part II
Climate Change Issues in Urban Sector

Climate Change Consideration in Planning and Development of Semi-Urban Area

R. Vinnarasi and Arup K. Sarma

Abstract The present scenario clearly indicates that urbanization is taking place in a fast rate. Due to the rapid increase in population and constraint in the temporal space, urbanization has become an ideal choice for satisfying the needs of the growing population. The present trend shows that there are a lot of chances for the semi-urban area to be urbanized in the near future. Urbanization has lot of advantages and disadvantages too. Urbanization leads to lot of problems in drainage, water distribution system, sediment transport, etc. Urbanization has lot of impacts on climate, one of which is greenhouse gas emission. Urbanization leads to climate change, and climate change in turn increases the risk of environmental hazards like floods, hurricanes, cyclones, etc. To tackle the problem, proper planning and development should be made to avoid and mitigate the effects of climate change on urbanization of semi-urban area. The design of hydrological systems like city drain, water distribution system, etc. needs to be done considering future development due to urbanization. In this study, climate change study for Bokakhat Township, a semi-urban area, has been carried out. The precipitation and number of dry days have been forecasted using statistical downscaling of climatic variable obtained from large-scale global climate model (GCM). NCEP reanalysis data have been used to model the precipitation, and the model for dry days has been created by comparing two GCMs, without using NCEP reanalysis data. The downscaling has been carried out using CGCM3 and HadCM3 models with A2 scenario. HadCM3 model has been chosen for precipitation and CGCM3 model for number of dry days. The results obtained from downscaling have been used to predict the rainfall intensity in the future. The result shows 27% increase in the average annual precipitation and 31% increase in the average rainfall intensity.

R. Vinnarasi (✉) · A. K. Sarma
Department of Civil Engineering, Indian Institute of Technology, Guwahati, India
e-mail: vinnarasi.ruban@gmail.com

A. K. Sarma
e-mail: aks@iitg.ernet.in

© Springer International Publishing AG 2018
A. K. Sarma et al. (eds.), *Urban Ecology, Water Quality and Climate Change*,
Water Science and Technology Library 84,
https://doi.org/10.1007/978-3-319-74494-0_9

Keywords Climate change · Global climate model · Statistical downscaling
Multiple regression analysis · Urbanization

1 Introduction

Urbanization includes three inter-related dimensions, namely, change in the size
distribution of cities, growth in individual city population sizes, and growth in
number of cities (Henderson and Wang 2007). Due to the increasing trend of
population, the number of cities is increasing in the developing countries. It clearly
shows that the semi-urban areas are becoming urbanized areas. Urbanization is
always related to the climate, i.e. urbanization induces climate change and the
climate change causes more problems in the urbanized area. Therefore, the
knowledge about both climate and planning is very important (Eliasson 2000).
Climate should be the fundamental consideration for proper urban planning and
development (Zhao et al. 2011).

In recent years, the climate change and its consequences have affected the
precipitation pattern. It causes flood as well as drought. These effects are intensified
in urban areas because of the anthropogenic effects they have on the water cycle,
such as reducing the infiltration capacity of basins. Disposal of sediment and solid
waste into channels will decrease the channels' safe carrying capacity (Karamouz
et al. 2011). Therefore, proper design of hydrological system like city drain, water
distribution system, etc., in semi-urban area, requires the knowledge about precipitation variation in future.

Global Climate Models (GCMs) have been used extensively to describe the
climate behaviours like weather forecasting, understanding climate, and projecting
climate change. GCMs are generally coarse in size, which are used to simulate
climatic variables such as wind speed, sea level pressure, etc., (Ghosh and
Mujumdar 2007) but these GCMs are poor in performance while predicting precipitation because it is inherently nonlinear and extremely sensitive to physical
processes (Stockdale et al. 1998). Therefore, to bridge the gap between climatic
variables to local hydrological variables and to account for the inaccuracies in
describing precipitation extremes, downscaling methods are commonly used in
practice (Willems et al. 2012).

Downscaling technique is classified into two types: they are dynamic downscaling and statistical downscaling. Dynamic downscaling is a technique, which is
used to bring output from GCM without using any equations (statistical relationship). It is complex in operation and requires large demand of computer resources.
Statistical downscaling makes statistical relation between large-scale climate variable and local quantities. This relation is used to project the future (Ghosh et al.
2010). Many researches have been done using statistical downscaling such as
prediction of future variation in river flow and reservoir flow using SDSM and
LARG-WG (Dibike and Coulibaly 2004), prediction of soil erosion and crop
production by WEPP model (Zhang 2005), determination of spatial–temporal water

availability in river system (Gosain et al. 2006), forecast of the station scale precipitation in the Philippines and Thailand (Kang et al. 2007) and modelling the stream flow behaviour using SVM and RVM (Ghosh and Mujumdar 2007).

Basically, the northeast region of India is a hilly and heavy rainfall area. Nowadays, urbanization is rapidly taking place in this region. Therefore, the study about the impact of climate change is very important. It will help to design the hydrological structure in proper way to reduce the vulnerable situation. In this study, climate change study for Bokakhat Township, a semi-urban area, has been carried out.

2 Data Collection

Observed monthly precipitation and number of dry days data used in this study have been taken from Bokakhat tea garden, a member of NETA.

The predictor data have been taken from two GCM model outputs. These GCM data have been downloaded from Intergovernmental Panel on climate change and Canadian centre for climate modelling and analysis. The details of the model have been given in Table 1.

NCEP reanalysis data are also used in this study. NCEP data are used to choose the best GCM model. The spatial resolution of reanalysis data is 2.5° of latitude and longitude.

3 Methodology

The strategy for prediction of precipitation and number of dry days is described as the following three steps.

3.1 Standardization

Basically, the difference between observed and climate variables occurs due to parameterization. The difference between observed and climatic variables is called bias. Standardization is used to reduce the biases in the mean and variance of GCM

Table 1 Selected models for downscaling

Model	Centre	Resolution
HadCM3	UK Met. Office, UK	3.75° × 2.5°
CGCM3	Canadian Centre for Climate Modelling and Analysis, Canada	T47 (3.75°)

predictors relative to those of the observed or NCEP data. The baseline period for number of dry days is from 2001 to 2010, which is a very short time. The baseline period for precipitation is from 1965 to 2010. In this process of standardization, the mean is subtracted from the baseline period and divided by standard deviation for both NCEP and GCM outputs. It is performed as given in Eq. 1, given by Ghosh and Mujumdar (2007):

$$v_{s\tan,t}(k) = \frac{v_t(k) - \mu_v(k)}{\sigma_v(k)}, \tag{1}$$

where

$v_t(k)$ Original value of the kth predictor variable at time t,
$\mu_t(k)$ The mean value of the kth predictor variable and
$\sigma_v(k)$ Standard deviation of the kth predictor variable.

3.2 Selection of Predictors

The predictors are selected by stepwise regression method for the numerical model given in Eq. 2. Stepwise regression consists of two main approaches, namely forward selection and backward elimination. In this study, the combination of two approaches is used. The predictor selection carried out by automatic procedure includes simple correlation and partial correlation. The process continues until it reaches the best f test, t test, adjusted R^2 (coefficient of determination), Akaike information criterion, Bayesian information criterion and Mallows'Cp.

$$y_i = \beta_o + \beta_1 x_{i1} + \beta_2 x_{i2} + \cdots \beta_p x_{ip}, \tag{2}$$

where y_i is the predictand, x_i is the predictor and β is the constant. Simple correlation is the correlation between two variables without the influence of other variables. When the correlation coefficient between y and x is computed by first eliminating the effect of all other variables, it is called partial correlation.

3.3 Statistical Downscaling

In this study, statistical downscaling has been done by Multiple Linear Regression (MLR) analysis. The calibration and validation of the models have been done by three approaches with the following relations:

(1) Multiple linear regressions without additive residual

$$y_i = \beta_1 x_{i1} + \beta_2 x_{i2} + \cdots \beta_p x_{ip} \tag{3}$$

(2) Multiple linear regressions with residuals

$$y_i = \beta_1 x_{i1} + \beta_2 x_{i2} + \cdots \beta_p x_{ip} + r_i \tag{4}$$

(3) Multiple linear regressions with a multiplying factor.

$$y_i = \left(\beta_1 x_{i1} + \beta_2 x_{i2} + \cdots \beta_p x_{ip} + r_i \right) m, \tag{5}$$

where y_i is the precipitation, x_i is the predictor, β is the coefficient, r_i is the residual and m is the multiplying factor. The calibration and validation for each station have been explained below with the graphical plots.

3.4 Intensity of Rainfall

The average intensity of rainfall (i_{avg}) is calculated by dividing the monthly average daily precipitation (P) to the number of rainy days as given in Eq. 4. Number of rainy days has been computed through the subtraction of number of dry days (n_d) from the total number of days (n_t) in a month.

$$i_{avg} = \frac{P \times 30}{(n_t - n_d)} \tag{6}$$

4 Results and Discussion

The study has been conducted in a place called Bokakhat. It is a semi-urban area in Jorkhat district. The future forecast has been made for precipitation and number of dry days. The outputs from downscaling have been used to predict the rainfall intensity per day.

4.1 Model for Precipitation

In precipitation model, the NCEP reanalysis data have been used for calibration and validation, and this model has been used to project the future precipitation using HadCM3 model in IPCC. The predictor selection is based on stepwise regression as

Table 2 Predictor selection using third assessment data

Model	Predictors
NCEP	Sea level pressure Air temperature @ 500 hpa
HadCM3	Sea level pressure Air temperature @ 850 hpa

Fig. 1 Calibration using NCEP data

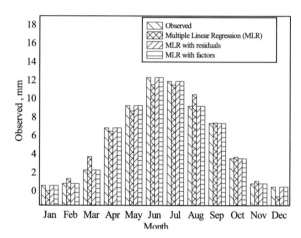

mentioned above. The predictand is precipitation, and the selected predictors for each model are listed in Table 2.

Model calibration and validation for two models have been done using three approaches as mentioned above in Sect. 3.3. The model has been developed using NCEP data, which is observed climatic data and it is available for a long time period (1965–2010); therefore, first 18 years have been chosen for calibration and remaining years for validation. The NCEP model has been used to find the best GCM model. The results are given in Figs. 1, 2, 3 and 4.

Among these three approaches, MLR with residuals has given good correlation between observed precipitation and predicted precipitation and has been selected for use. The correlations obtained for NCEP and HadCM3 are 0.966 and 0.984, respectively (Figs. 2 and 4).

4.2 Model for Dry Days

This model has been developed using IPCC data. Observed daily precipitation data was not available for Bokakhat station. Therefore, the precipitation data of two nearby stations within a distance of 25 km have been compared and data from Jorhat station have been chosen. Due to less availability of historical data, two GCM models CGCM3 and HadCM3 have been compared. The predictor selection is based on stepwise regression as mentioned above in Sect. 3.2. The predictand is

Fig. 2 Validation using
NCEP data

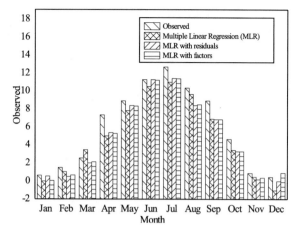

Fig. 3 Calibration using
HadCM3 data

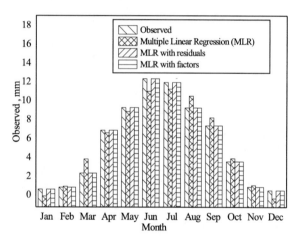

Fig. 4 Validation using
HadCM3 data

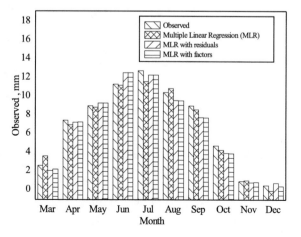

Table 3 Predictor selection using fourth assessment data

Model	Predictors
CGCM3	Sea level pressure
HadCM3	Sea level pressure

Table 4 Deviation of each model from the observed data

Month	Monthly series (S.D)	Validation (D) with residuals	
		CGCM3	HadCM3
Jan	0.8	−0.737	0.033
Feb	5.293	0.932	1.023
Mar	1.923	0.177	−0.010
Apr	9.487	−1.319	−1.516
May	8.343	−2.160	−1.992
Jun	10.887	−3.916	−4.464
Jul	12.473	−0.662	−1.160
Aug	9.533	−0.888	−1.329
Sep	8.445	1.612	1.724
Oct	5.228	−0.842	−0.316
Nov	1.516	−0.675	−0.501
Dec	0.678	0.184	0.271

Fig. 5 Calibration using CGCM3 data

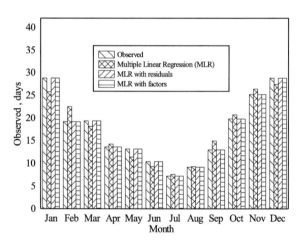

number of dry days, and the selected predictors for each model are listed in Table 3 and Table 4 .

Model calibration and validation for two different GCMs have been done using three approaches as mentioned above in Sect. 3.3. The observed historical data are available only for a short time period (2001–2010); therefore, alternative years have been chosen for calibration and remaining years for validation. The results are given in Figs. 5, 6, 7 and 8.

Fig. 6 Validation using
CGCM3 data

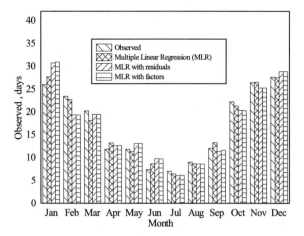

Fig. 7 Calibration using
HadCM3 data

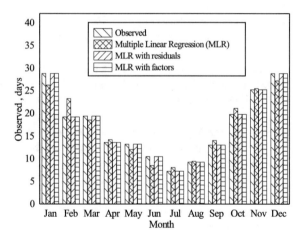

Fig. 8 Validation using
HadCM3 data

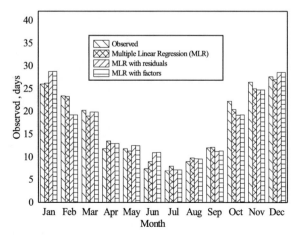

MLR with residuals has given good correlation between observed number of dry days and predicted number of dry days for this model also and has been selected for use.

The standard deviation and deviation of each model from the observed data have been compared. The model whose actual deviation is less than the standard deviation has been chosen for the study.

The correlations obtained for CGCM3 and HadCM3 are 0.951 and 0.924, respectively.

4.3   Future Prediction

Based on the correlation of observed predictand and predicted predictand, HadCM3 has been chosen to project the precipitation, and CGCM3 has been chosen for projecting the number of dry days. The future period has been divided into three sets of three decades. They are 2011–2040, 2041–2070 and 2071–2100. The variation of number of dry days and precipitation in future has been given in Figs. 9 and 10.

Figure 10 shows that the precipitation for Bokakhat increases for all the months. Highest precipitation occurs in the month of June with an increase of 18%. Figure 9 shows either the number of dry days increase or remain same in the future. It shows clearly that in the future, the total precipitation increases and at the same time number of dry days also increases. Due to this phenomenon, the variation of average rainfall intensity in future rainfall in the future has been computed for Bokakhat station.

Fig. 9 Future prediction for number of dry days

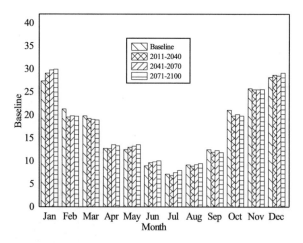

Fig. 10 Future prediction for precipitation

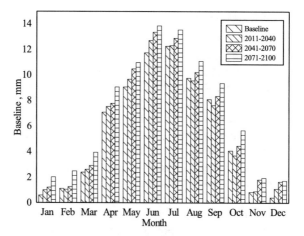

Fig. 11 Future rainfall intensity

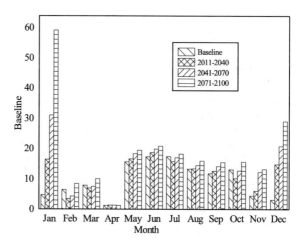

Figure 11 shows that the average intensity of rainfall increases throughout the year. Especially, the rainfall intensity is too high in November month. The average annual rainfall intensity increases by 31%. It indicates that the increasing trend of rainfall intensity increases runoff, therefore proper planning for city drains, water distribution system, etc., is needed.

5 Conclusion

In this study, the future precipitation, number of dry days and rainfall intensity have been predicted for Bokakhat Township considering climate change impact using GCM output. The predictor selection is based on the stepwise regression. The statistical downscaling has been done by multiple linear regression. The model calibration and validation have been done by three approaches. Among which MLR with residuals has been selected due to its better performance. Due to short period of historical data, the alternative year has been chosen for calibration and remaining years for validation. HadCM3 and CGCM3 data are used to project the future for precipitation and number of dry days, respectively. From this study, it can be concluded that in future, the precipitation as well as rainfall intensity increases. In future, this will lead to severe flood over some period and severe drought over other periods. The precipitation increases by 27% and the rainfall intensity increases by 31%. Considering this situation, proper planning for hydrological design is essential.

References

Dibike BY, Coulibaly P (2004) Hydrologic impact of climate change in the Saguenay watershed comparison of downscaling methods and hydrologic models. J Hydrol 307:145–163

Eliasson I (2000) The use climate knowledge in urban planning. Landsc Urban Plan 48(23):31–44

Ghosh S, Mujumdar PP (2007) Statistical downscaling of GCM simulations to streamflow using relevance vector machine. Adv Water Resour 31:132–146

Ghosh S, Raje D, Mujumdar PP (2010) Mahanadi stream flow: climate change impact assessment and adaptive strategies. Curr Sci 98:1084–1091

Gosain AK, Rao S, Basuray D (2006) Climate change impact assessment on hydrology of Indian river basins. Curr Sci 90:346–353

Henderson JV, Wang HG (2007) Urbanization and city growth: the role of institutions. Reg Sci Urban Econ 37(10):283–313

Kang H, An KH, Park CK, Solis ALS, Stitthichivapak K (2007) Multimodel output statistical downscaling prediction of precipitation in the Philippines and Thailand. Geophys Res Lett 34

Karamouz M, Hosseinpour A, Nazif S (2011) Improvement of urban drainage system performance under climate change impact: case study. J Hydrol Eng ASCE 16:395–412

Stockdale TN, Anderson DLT, Alves JOS, Baimaseda MA (1998) Global seasonal rainfall forecasts using a coupled ocean-atmosphere model. Nature 392:370–373

Willems P, Nielsen AK, Olsson J, Nguyen VTV (2012) Climate change impact assessment on urban rainfall extremes an urban drainage: methods and shortcomings. Atmos Res 103: 106–118

Zhang XC (2005) Spatial downscaling of global climate model output for site-specific assessment of crop production and soil erosion. Agric For Meteorol 135:215–229

Zhao C, Fu G, Liu X, Fu F (2011) Urban planning indicators, morphology and climate indicators: a case study for a north-south transect of Beijing, China. Build Environ 46:1174–1183

Effect of Climate Change Over Kashmir Valley

Manzoor Ahmad Ahanger and Mudasir Ahmad Lone

Abstract The modern and quite commonly used techniques for projection of climate change are the General Circulation Models or Global Climate Models (GCMs). However, these models predict climate at a much coarser spatial resolution. Often many studies involve the assessment of climate change impacts at smaller scales, viz., a catchment area of a river or even point scales, viz., a gauging station. In order to use the output of a GCM for conducting hydrological impact studies, downscaling is used. In the present study, the effect of climate change on meteorological parameters, viz., precipitation and temperature at four metrological stations, viz., Srinagar, Pahalgam, Qazigund, and Kupwara of Kashmir Valley, were examined. The data of mslpas (mean sea level pressure), tempas (mean temperature at 2 m), p500-as (500 hpa geopotential height), humas (specific humidity at 2 m), and p5_uas (500 hpa zonal velocity) obtained from Canadian third-generation Climate Model (CGCM3) were used as predictors along with National Center for Environmental Prediction/National Center for Atmospheric Research (NCEP/NCAR) reanalysis climatic data set. The locally observed temperature and precipitation were used as predictands. The methods of Multiple Linear Regression (MLR) and Statistical Downscaling Model (SDSM) were used as downscaling techniques. The large-scale GCM predictors were related to observed precipitation and temperature. It was found that the temperature of the Kashmir Valley is likely to increase in the coming decades while as the precipitation is going to decrease with each coming decade.

Keywords General circulation models · Downscaling · Multiple linear regression SDSM model

M. A. Ahanger · M. A. Lone (✉)
Department of Civil Engineering, N.I.T. Srinagar, Hazratbal, Srinagar, Jammu and Kashmir, India
e-mail: lonessite@gmail.com

M. A. Ahanger
e-mail: maahanger@gmail.com

© Springer International Publishing AG 2018
A. K. Sarma et al. (eds.), *Urban Ecology, Water Quality and Climate Change*,
Water Science and Technology Library 84,
https://doi.org/10.1007/978-3-319-74494-0_10

1 Introduction

Climate change is a long-term change in the statistical distribution of weather patterns and that persists for an extended period, typically decades or longer. It refers to any change in climate over time, whether due to natural variability or as a result of human activity (IPCC 2007a). The term sometimes is used to refer specifically to climate change caused by human activity, as opposed to changes in climate that may have resulted as part of Earths' natural processes. The Inter-governmental Panel on Climate Change (IPCC) declared that "warming of the climate system is unequivocal" (IPCC 2007b). The total temperature increase from 1850–1899 to 2001–2005 is 0.76 °C $\pm$ 0.19 °C, and the IPCC concluded that most of the observed increase in global average temperature since the mid-twentieth century is the result of human activities that are increasing Greenhouse Gas (GHG) concentrations in the atmosphere. The IPCC (2007c) predicts that the pace of climate change is likely to accelerate with continued GHG emissions, with globally averaged surface temperatures estimated to rise by 1.8–4.0 °C by the end of the twenty-first century.

Hydrologic impacts of climate change can be modeled using simulation results from General Circulation Models (GCMs). GCMs are the most credible tools designed to simulate time series of climate variables globally, accounting for the effects of greenhouse gases in the atmosphere. GCMs perform reasonably well in simulating climatic variables at larger spatial scale (3.75° lat. × 3.75° long.), but poorly at the smaller space and timescales relevant to regional impact analyses. Such poor performances of GCMs at local and regional scales necessitate the requirement of downscaling techniques. The methods used to convert GCM outputs into local meteorological variables required for reliable hydrologic modeling are usually referred to as "downscaling" techniques (Coulibaly et al. 2005).

Downscaling can be achieved in two ways such as dynamic downscaling and statistical downscaling.

Dynamic downscaling involves the nesting of a higher resolution Regional Climate Model (RCM) within a coarser resolution GCM. The noteworthy limitations of dynamic downscaling, which restricts its use in climate change impact studies, are its complicated design and being computationally as demanding as GCMs. RCMs are also inflexible in the sense that expanding the region or moving to a slightly different region requires redoing the entire experiment.

The statistical downscaling involves deriving empirical relationships that transform large-scale features of the GCM (Predictors) to regional or local-scale variables (Predictands) such as precipitation and temperature. There are three implicit assumptions involved in statistical downscaling (Hewitson and Crane 1992). First, the predictors are variables of relevance and are realistically modeled by the host GCM. Second, the empirical relationship is valid also under altered climatic conditions. Third, the predictors employed fully represent the climate change signal. Statistical Downscaling Methodologies (SDM) can be broadly classified into three categories (Murphy 1999): weather generators, weather typing, and transfer function.

Weather generators are statistical models of observed sequences of weather variables that replicate the statistical attributes of a local climate variable (such as the mean and variance) but not the observed sequence of events. There are two basic types of daily weather generators, based on the approach to model daily precipitation occurrence: the Markov chain approach and the spell-length approach (Ghosh and Mujumdar 2008).

Weather typing approaches (Bardossy and Plate 1992) involve grouping of local, meteorological variables in relation to prevailing patterns of atmospheric circulation. Future regional or local climate scenarios are constructed either by resampling from the observed data (variable) distribution (conditioned on the circulation pattern produced by a GCM) or by first generating synthetic sequences of weather pattern and then resampling from the generated data. The mean or frequency distribution of the local climate is then derived by weighting the local climate states with the relative frequencies of the weather groups or classes.

Transfer function (Wigley et al. 1990) downscaling methods rely on empirical relationships between local-scale climate variables (predictands) and the variables containing the large-scale climate information in the form of GCM outputs (predictors). Individual downscaling schemes differ according to the choice of mathematical transfer function, predictor variables, or statistical fitting procedure. To date, linear and nonlinear regressions, Artificial Neural Network (ANN), canonical correlation, etc. have been used to derive predictor–predictand relationship.

Ojha et al. (2010) used MLR and ANN models for downscaling of precipitation for lake catchment in arid region in India. The results of downscaling models show that precipitation is projected to increase in future for A2 and A1B scenarios, whereas it is least for B1 and COMMIT scenarios using predictors.

Khan et al. (2006) compared three downscaling models, namely SDSM, Long Ashton Research Station Weather Generator (LARS-WG) model, and ANN model in terms of various uncertainty assessments exhibited in their downscaled results of daily precipitation, daily maximum, and minimum temperatures. The uncertainty assessment results indicate that the SDSM is the most capable of reproducing various statistical characteristics of observed data in its downscaled results with 95% confidence level, the ANN is the least capable in this respect, and the LARS-WG is in between.

Chu et al. (2010) used SDSM method by simultaneously downscaling air temperature, evaporation, and precipitation in Haihe River basin, China. The data used were large-scale atmospheric data encompassing daily National Center for Environmental Prediction/National Center for Atmospheric Research (NCEP/ NCAR) reanalysis data and the daily mean climate model results for scenarios A2 and B2 of the HadCM3 model. It was concluded that in the next 30 years, climate would be warmer and drier, extreme events could be more intense, and autumn might be the most distinct season among all the changes.

Chaleeraktrakoon and Punlum (2010) analyzed the variability of observed temperature data for the Chi and Mun river basins and described the linkage between climate simulations given by a global circulation model, GCM, (HadCM3) and the local temperature data, based on SDSM model. Data for average, maximum, and minimum daily temperature records over a period of 30 years (1976–2005) for 14 stations in the river basins were considered. Analysis of results has shown that the minimum, average, and maximum temperatures tend to slightly increase during the first 15 years (1976–1990), while the trend of the other period (1991–2005) is moderate.

In the present study, the multiple linear regression technique was employed to relate the GCM predictors with the predictands such as the locally observed precipitation and temperature at four meteorological observatories, namely Srinagar, Pahalgam, Qazigund, and Kupwara of Kashmir Valley India (Fig. 1). The predictors as obtained from Canadian third-generation Climate model (CGCM3) were

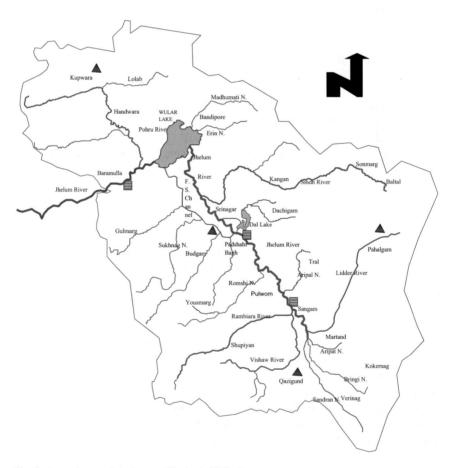

Fig. 1 Layout map of study area (Kashmir Valley)

mslpas (mean sea level pressure), tempas (mean temperature at 2 m), p500-as (500 hpa geopotential height), humas (specific humidity at 2 m), and p5_uas (500 hpa zonal velocity). SDSM was also used in the study for downscaling wherein in addition to GCM predictors the reanalysis data sets of NCEP/NCAR were also used.

2 Multiple Linear Regression (MLR)

The association of three or more variables can be investigated by the multiple regression and correlation analysis (Chow 1964). The multiple regression relation may be expressed in the form

$$x_1 = f(x_2, x_3, x_4 \ldots x_m), \tag{1}$$

where $x_1, x_2, x_3, x_4, \ldots, x_m$ are m variables. This equation gives the estimate of x_1 for given values of all other variables. If Eq. (1) is linear, the regression is referred to as MLR. The dependent variable x_1 may be expressed as a function of $m - 1$ independent variables by using the general form of MLR equation as

$$x_1 = B_1 + B_2 x_2 + \cdots + B_i x_i + \cdots + B_m x_m, \tag{2}$$

where B_1 is the intercept and B_i is the multiple regression coefficient of the dependent variable x_1 on the independent variable x_i with all other variables kept constant. The regression coefficients $B_1, B_2, \ldots, B_m$ are obtained from the equation set (3) as follows:

$$\left\{ \begin{array}{l} B_2 \sum (\Delta x_2)^2 + B_3 \sum (\Delta x_2 \Delta x_3) \cdots + B_m \sum (\Delta x_2 \Delta x_m) = \sum (\Delta x_1 \Delta x_2) \\ B_2 \sum (\Delta x_2 \Delta x_3) + B_3 \sum (\Delta x_3^2) \cdots + B_m \sum (\Delta x_3 \Delta x_m) = \sum (\Delta x_1 \Delta x_3) \\ \cdots \\ \cdots \\ B_2 \sum (\Delta x_2 \Delta x_m) + B_3 \sum (\Delta x_3 \Delta x_m) + \cdots + B_m \sum (\Delta x_m)^2 = \sum (\Delta x_1 \Delta x_m) \\ B_1 = \overline{x_1} - B_2 \overline{x_2} - B_3 \overline{x_3} - \cdots B_m \overline{x_m} + \cdots \end{array} \right\} \tag{3}$$

where $\bar{x}_i = \frac{\sum x_i}{N}$, $\Delta x_i = (x_i - \overline{x_i})$, and N is the sample size.

3 Statistical Downscaling Model (SDSM)

SDSM (Wilby et al. 2002; Wilby and Dawson 2007) is a user-friendly software package designed to implement statistical downscaling methods to produce high-resolution climate information from coarse-resolution climate model (GCM) simulations. The SDSM is a decision support tool, developed by Drs. Robert Wilby and Christian Dawson in the UK, for assessing local climate change impacts using a robust statistical downscaling technique. It is a hybrid of a stochastic weather generator and regression-based downscaling methods, and facilitates the rapid development of multiple, low-cost, single-site scenarios of daily surface weather variables under current and future climate forcing. SDSM is designed to help the user identify those large-scale climate variables (the *predictors*) which explain most of the variability in the climate (the *predictand*) at a particular site, and statistical models are then built based on this information. Statistical models are built using daily observed data—local climate data for a specific location for the predictand and larger scale NCEP data for the predictors—and these models are then used with GCM-derived predictors to obtain daily weather data at the site in question for a future time period. The version 4.2.2 of SDSM is selected for this research which generally reduces the task of downscaling daily climate from a global model into seven discrete processes, namely, quality control and data transformation; predictor variable(s) screening; model calibration; weather generation; statistical analyses; graphing model output; and scenario generation. To date, the downscaling algorithm of SDSM has been applied to a host of meteorological, hydrological, and environmental assessments, as well as a range of geographical contexts including Asia, Africa, North America, and Europe.

4 Study Area

The valley of Kashmir is called as the "paradise on earth". It has an approximate area of 15,948 km^2. The valley is a plain embedded in the midst of mountains lying in an oval shape, NW and SE between 33° 05′ and 34° 07′ north latitude and 74° and 75° 10′ east longitude. The valley is 150 km long from NW to SE and on width it varies between 32 and 100 km. In elevation, the valley varies from 2130 m above MSL (mean Sea level) down to 1585 m, with lowest portion along the north. Average height of the valley is 1850 m above the mean sea level, but the surrounding mountains which are perpetually snow-clad rise from 3000 to 4000 m above the mean sea level.

The valley of Kashmir is conveniently classed into a separate climatic region for its peculiarities in the variation of temperature, precipitation, and humidity compared to other regions of India.

Winter lasts up to March and is often severe. The mid-Mediterranean depressions called western disturbances blow over the Afghan frontier after passing over

Iran and precipitate in the valley and its surrounding mountains in the form of snow. The influence of the southwest monsoon is minimal over Kashmir Valley. The monsoon dies out of the south of the Pir Panjal ranges of the mountains and the summers have, therefore, very little in common with the general winds and pressures of India. In mid-summer, the temperature ranges from 32 to 35 °C and sometimes even 37 °C, and in winter it descends several degrees below freezing point. The mean temperature of the year is about 14 °C.

5 Data Analysis

The daily precipitation and temperature data at four meteorological stations of Kashmir Valley namely Srinagar, Qazigund, Pahalgam, and Kupwara for the period 1974–2004 were obtained from India Meteorological Department (IMD) Srinagar. The predictor data of mslpas (mean sea level pressure), tempas (mean temperature at 2 m), p500-as (500 hpa geopotential height), humas (specific humidity at 2 m), and p5_uas (500 hpa zonal velocity) were obtained from CGCM3 for A2 scenario for the grid location of 33.75°–37.25° (latitude) and 71.25°–75.75° (longitude). The above-mentioned predictor data along with NCEP/NCAR reanalysis climatic data set were downloaded for the period 1961–2100. For multiple linear regression (MLR) analysis, the data set for the period 1974–2000 was used for calibration and that of 2001–2004 were used for validation purposes. From "screen variable" feature of SDSM model and the parameter sensitivity analysis of MLR model, the most appropriate dependent predictors for modeling of temperature and precipitation were found. The GCM predictors of tempas, p500as, and p5_uas were found to be most appropriate dependent variables for temperature and those of mslpas, tempas, p500as, and humas for precipitation.

The average of the mean monthly precipitation and temperature recorded at the four meteorological stations were assumed to be the representative mean monthly precipitation and temperature of the Kashmir Valley. The MLR analysis was carried out to find the dependence relationship between temperature/precipitation and the appropriate GCM predictors. The relationship of temperature and precipitation with GCM predictors is expressed by Eqs. (4) and (5)

$$T = 12.29 + 11.57 \, \text{tempas} - 4.53 \, \text{p500as} + 0.64 \, \text{p5\_uas} \tag{4}$$

$$P = 88.91 - 34.2 \, \text{tempas} + 13.28 \, \text{humas} - 12.16 \, \text{mslpas} - 24.37 \, \text{p500as}, \tag{5}$$

where T and P are the mean monthly temperature and precipitation in °C and mm, respectively, over the Kashmir Valley.

The SDSM also used in this study was calibrated by NCEP/NCAR and observed temperature and precipitation data. The necessary operations of SDSM model were carried out to analyze the data and consequently, the results were obtained in the desired format.

Table 1 Regression statistics of MLR model

Variable	Std. error of estimate	Std. dev. of residuals	Mult. corr. coeff.	Coeff. of mult. det.
Temperature	2.11 °C	2.10 °C	0.96	0.91
Precipitation	44.06 mm	43.79 mm	0.61	0.36

6 Results and Discussions

The results of MLR analysis indicate that Eq. (4) replicates the future mean monthly temperatures over Kashmir Valley in an excellent manner, whereas Eq. (5) represents good correlation for mean monthly precipitation. The regression statistics for temperature and precipitation are given in Table 1. The regression models expressed by Eqs. (4) and (5) were validated on the observed data of mean monthly temperature and precipitation for the period 2001–2004. The results of validation are shown in Fig. 2. Furthermore, the future mean monthly temperatures of the Kashmir Valley for the period 2001–2099 were predicted using Eq. (4) and are shown in Fig. 3. It was observed that the mean monthly temperature of Kashmir Valley is expected to increase continuously over the twenty-first century, especially for the months of December to June. The average annual temperatures of Kashmir Valley during twenty-first century based on MLR predictions, which are shown in Fig. 5a, also depict an increasing trend.

Similarly, the future mean monthly precipitations of the Kashmir for the period 2001–2099 were predicted using Eq. (5) and are shown in Fig. 4. It was observed that the mean monthly precipitation of Kashmir Valley is expected to decrease continuously over the twenty-first century. This decrease in monthly total precipitation is more pronounced from December to April, the major precipitation season of the valley. The total annual precipitation of Kashmir Valley during the period 2001–2099 was also predicted using MLR model. Figure 5b represents the

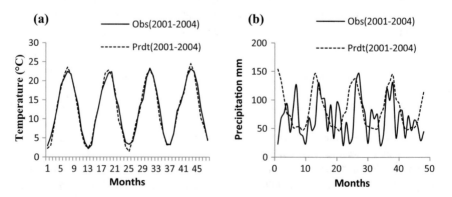

Fig. 2 Validation of **a** mean monthly temperature and **b** monthly total precipitation of Kashmir Valley for the period 2001–2004 using MLR

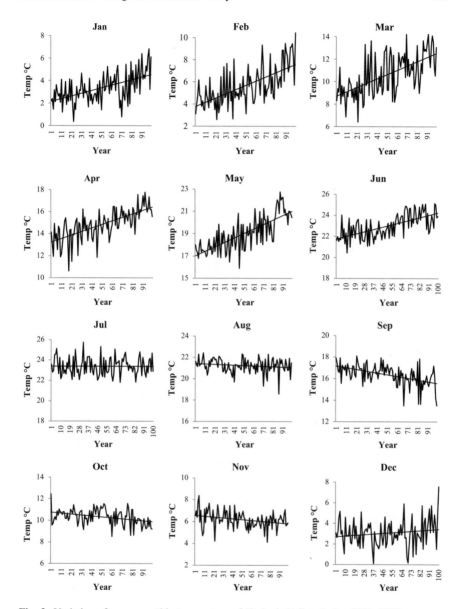

Fig. 3 Variation of mean monthly temperature of Kashmir Valley during 2001–2099

variation of this annual precipitation over a 100-year period of 2001–2099 and shows that annual precipitation is expected to decrease by about 38% by the end of twenty-first century.

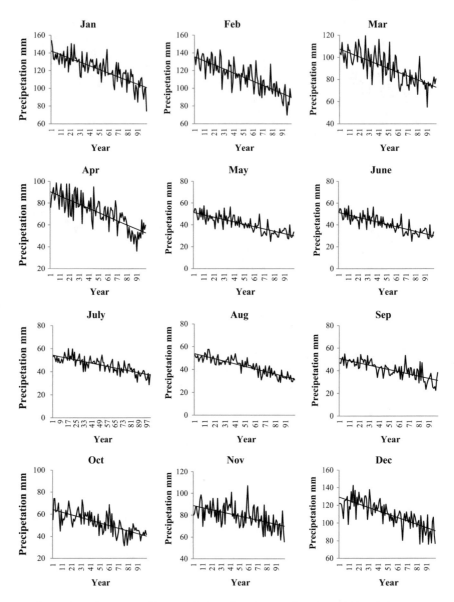

Fig. 4 Varition of mean monthly precipitation of Kashmir Valley during 2001–2099

The results of SDSM model analysis are shown in Figs. 6 and 7 which represent an increase in mean monthly temperature and decrease in monthly total precipitation in Kashmir Valley by the end of twenty-first century.

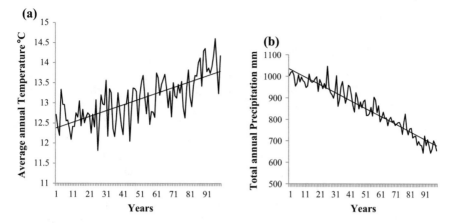

Fig. 5 Variation of MLR predicted **a** average annual temperature, **b** total annual precipitation of Kashmir Valley during twenty-first century

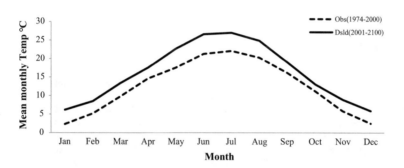

Fig. 6 Variation of mean monthly temperature of Kashmir Valley using SDSM model

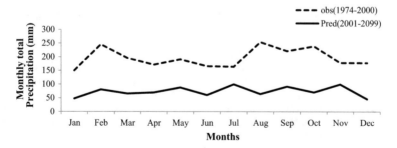

Fig. 7 Variation of monthly total precipitation of Kashmir Valley using SDSM model

7 Conclusion

From the results of the analysis, the following conclusions were drawn:

1. Both the MLR and SDSM models predict an increase in mean monthly temperature of Kashmir Valley over the twenty-first century. Although the MLR model predicts an increase in mean temperature during December to June months, the mean annual temperature of the valley is expected to increase by the end of twenty-first century.
2. The monthly total precipitation of Kashmir Valley was found to decrease by MLR as well as SDSM models. However, the MLR model predicted a pronounced decrease in the winter months of December to April (the major precipitation season of valley). The MLR model predicts that the total annual precipitation of the valley is expected to decrease substantially by about 38%.
3. As per the validation of data in the two methods applied, the MLR provides more accurate results as compared to SDSM.

References

Bardossy A, Plate EJ (1992) Space-time model for daily rainfall using atmospheric circulation patterns. Water Resour Res 28:1247–1259

Chaleeraktrakoon C, Punlum P (2010) Statistical analysis and downscaling for the minimum, average, and maximum daily-temperatures of the Chi and Mun River Basins. Thammasat Int J Sci Technol 15(4):64–81

Chow VT (1964) Handbook of applied hydrology. Tata Mc Graw-Hill, New York

Chu JT, Xia J, Xu C-Y, Singh VP (2010) Statistical downscaling of daily mean temperature, pan evaporation and precipitation for climate change scenarios in Haihe River, China. J Theor Appl Climatol 99(1–2):149–161

Coulibaly P, Dibike YB, Anctil F (2005) Downscaling precipitation and temperature with temporal neural networks. J Hydrometeorol

Ghosh S, Mujumdar PP (2008) Statistical downscaling of GCM simulations to streamflow using relevance vector machine. Adv Water Resour 31:132–146

Hewitson BC, Crane RG (1992) Large-scale atmospheric controls on local precipitation in tropical Mexico. Geophys Res Lett 19(18):1835–1838

IPCC (2007a) Climate change 2007: synthesis report. In: Pachauri RK, Reisinger A (eds) Contribution of Working Groups I, II and III to the fourth assessment report of the intergovernmental panel on climate change. Core Writing Team

IPCC (2007b) Climate change 2007: the physical science basis. In: Solomon S, Qin D, Manning M, Marquis M, Averyt K, Tignor MB, LeRoy Mil H (eds) Contribution of Working Group I to the fourth assessment report of the IPCC. Cambridge University Press

IPCC (2007c) Climate change 2007: impact adaptation and vulnerability. In: Parry ML, Canziani OF, Palutikof JP, van der Linden PJ, Hanson CE (eds) Contribution of Working Group II to the fourth assessment report of the IPCC. Cambridge University Press

Khan MS, Coulibaly P, Dibike Y (2006) Uncertainty analysis of statistical downscaling methods. J Hydrol 319:357–382

Murphy JM (1999) An evaluation of statistical and dynamical techniques for downscaling local climate. J Clim 12:2256–2284

Ojha CSP, Goyal MK, Adeloye AJ (2010) Downscaling of precipitation for lake catchment in the arid region in India using linear multiple regression and neural networks. Open Hydrol J 4:122–136

UNWTO (2007) Ministers summit on tourism and climate change. London Conference Pack

Wigley TML, Jones PD, Briffa KR, Smith G (1990) Obtaining sub-grid-scale information from coarse-resolution general circulation model output. J Geophys Res 95(D2):1943–1953

Wilby RL, Dawson CW (2007) Using SDSM Version 4.2—a decision support tool for the assessment of regional climate change impacts. User Manual, 90 p

Wilby RL, Dawson CW, Barrow EM (2002) SDSM—a decision support tool for the assessment of regional climate change impacts. Environ Model Softw 17:145–157

Assessment of Urban Heat Island Effect in Guwahati—A Remote Sensing Based Study

Juri Borbora, Apurba Kumar Das, Rajesh Kumar Sah
and Nabajit Hazarika

Abstract The manifestation of effects of urbanization on the thermal environment of a city is often related to Urban Heat Island (UHI) by various studies. The presence of abnormal warmth in urban areas as compared to rural surroundings is defined as UHI. Satellite images in thermal infrared can be used for assessing thermal environment as well as for defining Urban Heat Islands (UHIs) in urban areas. In this chapter, Guwahati city and its surrounding areas have been studied to observe the changes in the thermal environment that has occurred between the years 1989 and 2010 by means of satellite images provided by Landsat Thematic Mapper (TM) onboard Landsat 5. A range of influences have been investigated in the form of relationship between Normalized Difference Vegetation Index (NDVI) and Radiant Surface Temperature (Ts) to see the influence of sensible heat flux and urban density over the urban vegetation as an indicator of modified temperature regime over the years. The study tried to quantify the UHI phenomenon in the city by measuring Urban Heat Island Intensity (UHII) using remote sensing technique. The objective of the study was to quantify changes in the urban thermal environment of Guwahati, a rapidly growing city of Northeast India, which will help in studying various effects of urbanization in future.

1 Introduction

Urban Heat Island (UHI) is a phenomenon where surface and atmospheric modifications due to urbanization generally lead to a modified thermal climate that is warmer than the surrounding nonurbanized areas (Voogt and Oke 2003). UHI affects energy demand, human health, and environmental conditions related to pollution dispersion (Harlan et al. 2006; Crutzen 2004). Atmospheric heat island may be defined for the Urban Canopy Layer (UCL), the layer of the urban atmosphere extending upward from the surface to approximately mean building height.

J. Borbora · A. K. Das (✉) · R. K. Sah · N. Hazarika
Department of Environmental Science, Tezpur University, Tezpur 784028, Assam, India
e-mail: apurba@tezu.ernet.in; apurkdas@gmail.com

© Springer International Publishing AG 2018
A. K. Sarma et al. (eds.), *Urban Ecology, Water Quality and Climate Change*,
Water Science and Technology Library 84,
https://doi.org/10.1007/978-3-319-74494-0_11

The canopy-layer urban heat island is typically detected by in situ sensors at standard meteorological height or traverse of vehicle-mounted sensors. The Urban Boundary Layer (UBL) is the layer that extends above the UCL and is influenced by the underlying urban surface. The boundary-layer level urban heat island is detected by the specialized sensor platforms like tall towers, radiosonde, low flying planes, or tethered balloon flights. Thermal remote sensors observe the Surface Urban Heat Island (SUHI), or more specifically they "see" the spatial pattern of upwelling thermal radiance received by the remote sensors (Voogt and Oke 2003). Thus, thermal remote sensing of urban surface temperature is a special case of observing land surface temperature which varies in response to the surface energy balance. Consequently, the resultant surface temperature incorporates the effects of surface radiative and thermodynamic properties. It is known that heat islands of the UBL and the UCL are larger at night, while, SUHIs are larger during the day (Roth et al. 1989).

Three major applications of thermal remote-sensing have been put forward by Voogt and Oke (2003). The first approach includes examining the spatial structure of urban thermal patterns and urban surface characteristics, while, the second approach considers the relation between atmospheric heat islands (UHIs of UBL and UCL) and SUHI. The third approach is centered around studying urban surface energy balances by coupling urban climate models with remotely sensed data. In this chapter, the first approach of the thermal remote-sensing technique has been used to examine the relationship between the spatial structure of the urban thermal pattern of Guwahati and urban surface characteristics.

Normalized Difference Vegetation Index (NDVI) has long been used as an indicator of the urban climate. Higher values of NDVI typically indicate a larger fraction of vegetation in a pixel. The amount of vegetation determines LST by latent heat flux from the surface to the atmosphere via evapotranspiration. Lower LSTs are found in areas with high NDVI. The negative correlation between NDVI and LST is valuable for urban climate studies (Yuan and Bauer 2007). It has been used for assessing the influence of urban environment over observed minimum temperature (Gallo et al. 1993). Thermal responses of urban land cover types between day and night and the relation between land cover radiance and vegetation amount using NDVI have been studied using the data from Advanced Thermal and Land Application Sensor (ATLAS) (Lo et al. 1997). In this chapter, an attempt has been made to investigate the changes occurring in the NDVI values of the study area (Fig. 1) as compared to changes in the LST over a period of 21 years, i.e., 1989–2010.

1.1 Study Area

Guwahati, (26° 10′ 45″N, 92° 45′ 0″E) being a gateway to the northeastern region of the country has undergone a rapid urbanization change in the past decade. The urban area is around 262 km^2 and population of about 12 lakhs (2011 census), the population is projected to grow up to 21.74 lakhs by 2025 (GDMA 2009).

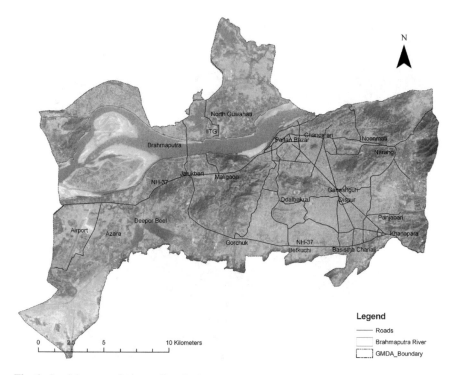

Fig. 1 Spatial extent of greater Guwahati

A considerable increase in the density of the population has occurred over the last decade that has resulted in expansion of urban areas into suburban extents. Replacement of natural vegetated areas with dry impervious surfaces, use of high heat capacity and low surface reflective building materials, reduced turbulent heat transfer due to street canyon geometry, and increased anthropogenic heat emission into the urban atmosphere are generally found to modify the thermal regime of a city. Presence of most of these factors in a city like Guwahati is believed to manifest itself in the form of observable "UHI" in coming years.

2   Materials and Method

A pilot study was done with the thermal maps prepared from Landsat images of 1991 (TM, 26/11/1991), 2002 (ETM+, 17/02/2002), and 2006 (ETM+, 26/10/2006) to get an impression of changes occurring to the thermal environment of Guwahati city (Fig. 2). NDVI values are subject to seasonal variations. Additionally, it is more convenient to study the thermal modification of atmosphere through LST if seasonal variations are kept to minimum. Both of these observations have motivated us to search for Landsat images of the study area of the same season. Finally,

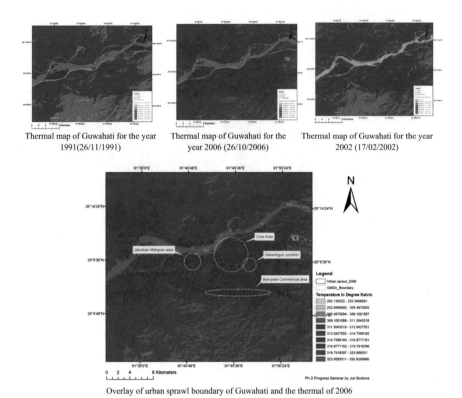

Thermal map of Guwahati for the year 1991(26/11/1991)

Thermal map of Guwahati for the year 2006 (26/10/2006)

Thermal map of Guwahati for the year 2002 (17/02/2002)

Overlay of urban sprawl boundary of Guwahati and the thermal of 2006

Fig. 2 LST retrieved images for the years of 1991, 2002, and 2006

Landsat TM images for the month of January for the years of 1989, 2000, and 2010 have been used for the study. The study area is covered under a 170×183 km scene (path 137, row 42). For all the years, band 6 was used for thermal analysis, while, bands 4 and 3 have been used for the calculation of NDVI.

1. Calculation of Normalized Difference Vegetation Index (NDVI):

NDVI was calculated by the following formula:

$$\text{NDVI} = \left(\frac{B4 - B3}{B4 + B3} + 1.0\right) \times 100.0,$$

where

$B4$ Reflectance measured in near-infrared wavelength.
$B3$ Reflectance measured in red wavelength.

2. Retrieval of Surface Temperature (Ts)

The thermal band (band 6) of TM was converted to Ts through the following steps (Chander and Markham 2003):

(a) **Conversion of Digital Number (DN) to Spectral Radiance (L_λ)**
Conversion of the image DN values to spectral radiance is carried out using the gain and offset values given in the image header file (Eq. 1). Thus,

$$L_\lambda = ((L_{MAX} - L_{MIN})/(QCAL_{MAX} - QCAL_{MIN})) \\ * (QCAL - QCAL_{MIN}) + L_{MIN}, \tag{1}$$

where $QCAL_{MIN} = 1$, $QCAL_{MAX} = 255$, and $QCAL$ = Digital Number L_{MIN} and L_{MAX} = spectral radiance for band 6 at DN = 0 and 255.

(b) **Conversion of Spectral Radiance to Brightness Temperature**
The TM thermal band data can be converted from spectral radiance to Brightness Temperature (BT) which assumes surface emissivity = 1 (Eq. 2)

$$BT = K2/\ln(K1/L + 1), \tag{2}$$

where

BT   Effective at-sensor brightness temperature in kelvin
$K1$ Calibration constant 1 [W/(m^2 sr μm)] (607.76)
$K2$ Calibration constant 2 in K (1260.56)
L Spectral radiance at-sensor [W/(m^2 sr μm)]

(c) **Emissivity Correction**
Corrections for emissivity differences were carried out by land cover type by ratioing the BT image with the classified image in which the pixel values for the land cover class were replaced with the corresponding emissivity value. Thus, the emissivity corrected surface temperature (Ts) is derived by Eq. (3).

$$Ts = BT/[1 + (\lambda BT/\rho) \ln \varepsilon], \tag{3}$$

where

λ Wavelength of emitted radiance (11.5 μm)
ρ hc/K (1.438×10^{-2} mK)
BT   Brightness temperature
ε Spectral surface emissivity

(d) **Generation of the Land Surface Temperature Image:**
By conversion of the Digital Number value to the spectral radiance values and further conversion to the BT values, a BT image in kelvin has been generated. After doing emissivity corrections, land surface temperature could be generated that directly depicted the temperature variations present over the urban areas as compared to the nonurbanized and the forest area.

3   Results and Discussion

1. Normalized Difference Vegetation Index:

NDVI values are generally expressed in the range of −1 to +1. A value of +1 generally depicts the highest value of NDVI indicating good health of vegetation, while −1 depicts the lowest value indicating the absence of any vegetation growth within the area. There is a considerable change observed in the NDVI values from 1989 to 2010 in and around Guwahati. In the NDVI image of 1989, we see that the lowest value of NDVI is in the range of −0.5 to −0.3. It was observed over the river Brahmaputra which could be due to the nonexistence of live vegetation in the freely flowing water. Riverine sand showed a low NDVI value in the range of −0.17 to 0. The highest values in the range of 0.33 to 0.5 were observed over parts of Khanapara, southeastern fringe of Deepor Beel, few areas within Azara and Plashbari showing the presence of thick green vegetation. The core city areas of Panbazar, Fancy Bazaar, Ganeshguri, and Chandmari areas showed relatively low NDVI values in between −0.67 and 0 depicting less greenery and more nonphotosynthesizing matter similar to sand which is most probably bare open land, roads, and concrete. Other areas showed moderate NDVI range of 0–0.67. In 2000, though the NDVI range for the city remained same, newer areas like Jalukbari, Maligaon, Noonmati, Naarengi, parts of Plasbari, Azara, Basistha Chariali, Betkuchi, Gorchuk, and North Guwahati were within the low NDVI range of −0.67 to 0, probably due to the removal of green areas for development activities. During 2010, low NDVI values in the range of −0.67 to 0 were not only present in the urban settlement areas but also had penetrated well into the suburban extents which were observed to be relatively greener in earlier two occasions. Thus, urban extension to accommodate rapid growth during the last decade could be well observed by decreased NDVI values over the settlement areas in 2010 as compared to 1989.

2. Change in the Land Surface Temperature

Like other tropical cities (Devadas and Rose 2009; Emmanuel 2003), careful comparison of thermal maps of 1991, 2002, and 2006 in the pilot study revealed that Guwahati too does not have a single UHI but has a collection of small UHIs separated by cooler areas. During the detailed study following the above, it was observed that almost uniform temperature regime prevailed within the study area during the year 1989 (Fig. 3). Except for a few places with riverine sand, no other place showed very high temperature as compared to rest of the surrounding. However, in the year 2000, it was observed that few pockets of land within the extent of the study area have shown warming up. These include areas like Azara, Polasbari, Betkuchi, Fatasil, Basisitha, Paltan bazaar, and North Guwahati. Urban Heat Island Intensity (UHII) of around 12 °C has been found in the year 2000. In the year 2010, it was observed that boundaries of the areas with high temperature have become more diffused and have brought more areas into its realm. However, UHII of around 5 °C could only be observed during this year. Warming up of the rural areas as relative to the previous years could be one of the reasons behind such a scenario.

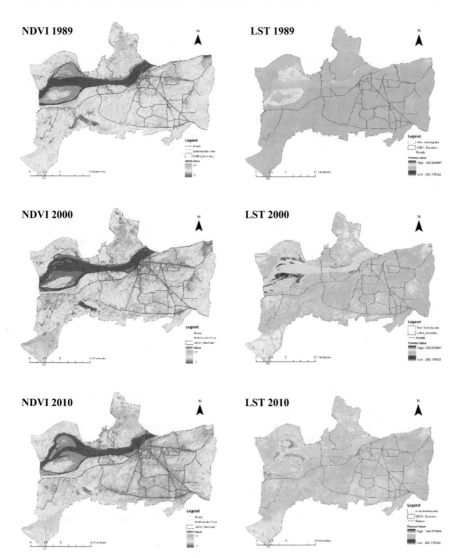

Fig. 3 NDVI and LST retrieved images for the years of 1989, 2000, and 2010

4 Conclusion

Relatively high temperature is found over Guwahati City as shown in the LST image of 2010 as compared to the one in 1989 (Fig. 3) where the average high temperature within the city was found mostly in certain urbanized pockets. The increased intensity of the surface layer temperature implies the presence of thermal anomaly in the form of UHI within its urban area. This can be attributed not only to the transition of urban landforms from vegetative cover to buildup areas but also as a cumulative effect of additional heat generation due to increased anthropogenic activities as a result of urbanization over the years. The urbanization has also caused an additional stress on the nearby vegetation cover as depicted by NDVI images which is often regarded as the representative of the vegetation vigor and expanse. The scenario within the urban area has much worsened during 2010 when the urbanization level had increased considerably. To improve the scenario, more plantation activities could be undertaken at relatively vulnerable core urban areas of Paltan bazaar–Fancy bazaar. Some stringent conservation measures are needed in Azara and near bypass area of Basistha Chariali to Gorchuk which are likely to witness more urbanization in future.

Acknowledgements The support of the Atmospheric Science Division, National Remote Sensing Centre, Hyderabad and Centre of Excellence, under the MoUD sponsored project for Integrated Land use Planning and Water Resource Management in the Department of Civil Engineering, Indian Institute of Technology, Guwahati is gratefully acknowledged.

References

Census of India (2011) Provisional population totals—2011—Assam. http://censusindia.gov.in/2011-prov-results/prov_data_products_assam.html

Chander G, Markham B (2003) Revised Landsat-5 TM radiometric calibration procedures and post calibration dynamic range. IEEE Trans Geosci Remote Sens 41(11):2674–2677

Crutzen PJ (2004) New directions: the growing urban heat and pollution "island" effect-impact on chemistry and climate. Atmos Environ 38:3539–3540

Devadas MD, Rose L (2009) Urban factors and the intensity of heat island in the city of Chennai. In: The seventh international conference on urban climate, Yokohama, Japan

Emmanuel R (2003) Assessment of impact of land cover changes on urban bioclimate: the case of Colombo, Sri Lanka. Archit Sci Rev 46(2):151–158

Gallo KP, MacNab AL, Karl TR, Brown JF, Hood JJ, Tarpley JD (1993) The use of NOAA AVHRR data for assessment of the urban heat island effect. J Appl Meteorol 32(5):898–908

GDMA (2009) Master plan for Guwahati Metropolitan area—2025. http://gmda.co.in/pdf/Part-I.pdf

Harlan SL, Brazel AJ, Prasad L, Stefanov L, Larsen L (2006) Neighbourhood microclimate and vulnerability to heat stress. Soc Sci Med 63:2847–2863

Lo CP, Quattrochi DA, Luvall JC (1997) Application of high-resolution thermal infrared remote sensing and GIS to assess the urban heat island effect. Int J Remote Sens 18:287–304

Roth M, Oke TR, Emery WJ (1989) Satellite derived urban heat islands from three costal cities and the utilization of such data in urban climatology. Int J Remote Sens 10:1699–1720

Voogt JA, Oke TR (2003) Thermal remote sensing of urban climates. Remote Sens Environ 86:370–384

Yuan F, Bauer ME (2007) Comparison of impervious surface area and normalized difference vegetation index as indicator of surface urban heat island effects in Landsat imagery. Remote Sens Environ 106:375–386

Climate Change and Its Effect on Hydrology and Threat to Biodiversity: A Study on Manas River (Upper Part) and the Manas National Park, Assam, India

Jyotishmoy Bora

Abstract Change has always been a powerful event of nature. Climate change is one of the most important environmental challenge faced by living organisms and non-living things. The nature, magnitude and consequences of this change are unprecedented especially in the present-day human society. The hydrology of rivers and biodiversity of wildlife sanctuaries and national parks have also badly affected by the change of climate all over the world. This chapter intends to study the changes both in the hydrology of Manas River (upper part) and a threat to the biodiversity of the Manas National Park occurred due to the climate change in the region. The chapter also attempts to suggest some measures for mitigation of the various problems associated with it. The study is carried out based on the data collected from the field and various secondary sources. The changes of hydrology and biodiversity have been assessed with the help of relevant data and information collected for different periods of time. It is found from the study that the changes have occurred in the hydrology of Manas River during the past few decades creating a lot of problems to its environment. Remarkable changes have also been noticed in the biodiversity of the Manas National Park as a result of climate change.

Keywords Climate change · Hydrology · Biodiversity

1 Introduction

Climate change is one of the most critical issues facing the present-day world. Climate change is a cross-cutting issue and it has always been a powerful force of nature. Significant changes in climate and their impacts are visible locally, regionally and globally. Magnitude and tempo of these changes are unprecedented

J. Bora (✉)
Department of Geography, Bajali College, Pathsala, Assam, India
e-mail: jyotishmoybora@gmail.com

© Springer International Publishing AG 2018   157
A. K. Sarma et al. (eds.), *Urban Ecology, Water Quality and Climate Change*,
Water Science and Technology Library 84,
https://doi.org/10.1007/978-3-319-74494-0_12

in human life. Apart from their causes, we must think what we can do to manage these impacts and adapt to the new circumstances they bring. Climate change is happening and people have begun to feel its impacts on their daily lives.

Over the past 100 years, the temperature of the earth surface rose by more than 1.4 °F (National Academy of Sciences 2012). Scientists predict the earth will continue to warm by about 2–6 °F over the next 100 years, whereas, during the last Ice Age, temperature was only 9 °F. Causes of climate change may come from both natural and human-induced factors.

Climate change has severe impacts and threatening on ecology, ecosystem, biodiversity, human life and economic activities of man. Intergovernmental Panel on Climate Change (IPCC) takes the global lead on assessing and summarizing information on past and future climate change. Climate change could upset the balance of various ecosystems causing lots of problems.

The change in climate has also a direct bearing on the hydrology of river and biodiversity of national parks or wildlife sanctuaries. Climate change is likely to lead an intensification of the global hydrological cycle and to have a major impact on regional water resources (Arnell 1999). Global climate change affects spatial and temporal patterns of precipitation and so has a major impact on surface and sub-surface water balances (Kunstmann et al. 2004). Therefore, the magnitude of change in climate and its consequences will be the matter of great concern for the scientists, environmentalists, planners, policymakers and others so as to regulate the balance in the ecosystem and to maintain it for future. Variations in climate, land use and water consumption can have profound effects on hydrology of river.

Manas River basin along with the Manas National Park has also been facing environmental challenges like some other areas in Assam. But to what extent the climate change has been affecting on the hydrology of Manas River and biodiversity of Manas National Park is yet unclear and studies on this aspect is lacking.

The study of climate change in this chapter is discussed by considering the rainfall component of climate only. Other components of climate and their impacts on various directions are not considered in this study. The trend of occurrence of rainfall at Mathanguri and the Manas River catchment has been creating changes in the hydrology of Manas River and threat to the biodiversity of Manas National Park.

2 Materials and Methods

The study is carried out based on the data collected from various secondary sources and personal field observation in different years. Secondary data have been collected from various governmental and non-governmental sources and other relevant publications. The Forest Dept., Water Resources Dept., Tourism Dept. of Govt. of Assam, Assam Science Technology and Environment Council (ASTEC), etc. are consulted.

A brief overview of the Manas River catchment and the Manas National Park is given in this chapter. The rainfall data of different periods of the study area are collected from Master Plan of Beki–Manas–Aie Sub-Basin, Brahmaputra Board, 2010 to discuss the change of climate. The change of rainfall and its effects on the hydrology of Manas River has been highlighted by using rainfall trend line, hydrographs of water level and water discharge and sediment load curve. The biodiversity of the park and its various threats from both side of nature and man are also discussed. The changes of hydrology and biodiversity have been assessed with the help of relevant data and information collected for different periods of time. Finally, some measures have been put forwarded to maintain the balance of ecosystem of the study area to cope up with the changing climatic condition.

3 Results and Discussion

3.1 The Scenario of the Study Area

Manas River is one of the major north bank tributaries of the Brahmaputra River which passes through the heart of the Manas National Park. The Manas River catchment occupies between 26° 00′N to 27° 15′N latitudes and 90° 45′E to 91° 10′E longitudes covering 4% of the Brahmaputra River basin. The Manas River is originated from Bhutan hills and its outfall falls in the Brahmaputra River traversing for about 87.5 km in its Indian part. The catchment area of this river is approximately 3,241 km^2. The Manas River is splitting up into three major streams known as Manas, Hakuwa and Beki to join the Brahmaputra River. These three and other five small rivers are running through the park carrying enormous amounts of silt and rock from the foothills as a result of heavy rainfall, steep gradiant and friable bedrock upstream. The average annual yield of sediment in Manas River is found to be 0.2920 mham, whereas the percentage of yield is high in Monsoon period (June–September), i.e., 52% in NH Crossing site.

Manas National Park is located at the foothills of the Bhutan Himalayas in between 26° 30′N to 27° 00′N latitudes and 90° 50′E–92° 00′E longitudes (Fig. 1) continuation with the Royal Manas National Park in Bhutan (1,02,300 ha). It spans on both sides of the Manas River and is restricted to the north by the international border of Bhutan, to the south by thickly populated villages of Baksa and Chirang districts of Assam and to the east and west by reserve forests. Elevation ranges from 61 to 110 m above mean sea level. This park occupies an area of 950 km^2, which forms the core area of the Tiger Reserve (2,837 km^2). The natural gradient of the land is gentle sloping southward and area along the southern boundary which is flatter and get waterlogged during the rainy season. The Manas basin in the west of the park is frequently flooded during the monsoon but never for very long duration due to the sloping relief.

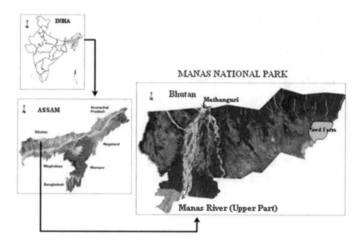

Fig. 1 Location map of Manas National Park and Manas River (Upper Part)

3.2 Climate Change

The climate of the study area is warm and humid with up to 76% relative humidity. It rains from mid-March to October with most rain falling during the monsoon months from June to September, flooding the western half of the park. From November to February is relatively dry when the smaller rivers dry up and the water level of Manas River dwindles. The mean maximum summer temperature is 37 °C and the mean minimum winter temperature is 5 °C. The climate of this area can be divided into four distinct seasons on the basis of variation in rainfall, temperature and wind. These are (i) winter (December–February), (ii) pre-monsoon (March–May), (iii) monsoon (June–September) and (iv) retreating monsoon (October–November).

Environmental safety and sustainability of the park area are greatly challenged by the impact of climate change and human interference. The rainfall is found to be decreasing over this region significantly during the last century at an approximate rate of 5 mm per decade (Talukdar 2012). Forest ecosystem, especially wetland and grassland has been changing on the banks of the Manas River in particular and the park in general which may have significant detrimental effects on the environment characterized by temperature, soil moisture and humidity. The impacts of these changes on the hydrology of Manas River and wild flora and fauna are visible. The sediment load on the river flow is also fluctuating as a result of variations in runoff of water and change of environmental situation in the upper part. The glacial recession is linked to increasing the sediment load in the Manas River. A number of major flash floods have occurred during the period 2001–2009 due to heavy rainstorms or cloudbursts in the upper catchment area of the Manas River. The flash floods occurred in this river have caused hundreds of deaths of wild animals, great damage to flora, huge economic loss and colossal damage to the infrastructure of the park. The recession of glaciers caused by climate change have created more glacial lakes in the Bhutan Himalaya and outburst floods of these glacial lakes have caused more flash floods in the Manas National Park. The geo-environmental situation has been changing in Bhutan in the last few decades. The construction of mega dams for the purpose of generation of hydroelectricity is one of the important factors of such happenings. Government of India is also promoting mega dam-based hydro projects in Bhutan like the Mangdechhu Hydroelectric Project in the Manas River catchment in Bhutan. Such mega dams will enhance the probability of disturbance of normal flow of Manas River which may increase the intensity of landslide, flash flood, bank erosion, sedimentation, etc. It also increases the threats to the biodiversity of the Manas National Park. Human interference is also responsible for the environmental change in the forest cover area of the park. The study reveals that shrinkage of forest cover area, encroachment into the forest land and reclamation of forest area for the agricultural purpose are much more prominent in the southern and eastern end of the study area. Tremendous changes have been noticed in the land use pattern in Baksa district of BTAD over the years. The forest cover of the southern peripheral region of the park is now nearly the step of extinction. The areas which were previously covered by evergreen and semi-evergreen forest now have been wiped out and modified into croplands. In view of the above, certain other changes have also been taken place in the social environment of the study area.

Mathanguri is the place located on the border between India and Bhutan from where the Manas River enters into Indian landmass originating from the highlands of Bhutan Himalayas. The downstream behavior of Manas River is highly influenced by the environment existed in Mathanguri. The climatic conditions prevailed in Mathanguri has been changing due to change in the environment of its surrounding. The trend of occurrence of annual and monsoonal rainfall has been gradually decreasing during 1982–2006 (Table 1). The maximum and minimum

Table 1 Distribution of rainfall (mm) at Mathanguri (1982–2006)

Year	June–September (monsoon season)	Annual total	Year	June–September (monsoon season)	Annual total
1982	2,829.75	3,625.50	1998	2,037.00	3,807.60
1983	3,745.75	4,519.00	1999	1,851.50	2,235.90
1984	3,204.50	4,255.50	2000	1,601.20	2,100.90
1985	2,062.75	3,054.25	2001	1,486.50	2,642.20
1986	1,892.90	2,480.35	2002	1,224.69	1,662.03
1987	2,260.00	2,850.39	2003	1,712.10	2,061.30
1988	2,407.90	3,529.00	2004	1,550.20	2,484.60
1995	1,931.10	2,523.10	2005	1,702.10	2,428.10
1996	1,569.80	2,054.50	2006	1,509.40	1,842.50
1997	2,069.48	2,493.42			

Source Master Plan of Beki–Manas–Aie Sub-Basin, Brahmaputra Board (2010)

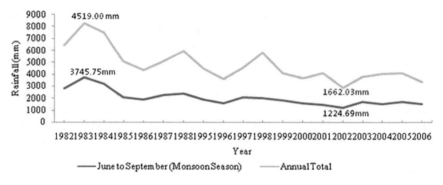

Fig. 2 Trend of rainfall at Mathanguri (1982–2006)

annual rainfalls are recorded 4,519 and 1,662.03 mm in 1983 and 2002 respectively (Fig. 2). In the other hand, the maximum and minimum monsoon rainfalls are recorded 3,745.75 and 1,224.69 mm in the same years respectively.

River hydrology is one of the important facets of hydrology. Water level, water discharge, velocity of water, quality of water, etc., are some of the aspects of river hydrology which are influenced by the occurrence and amount of rainfall along with some other elements. Changes in rainfall due to climate change resulted in variation in the water level of Manas River during the period 1972–2007 (Fig. 3). The maximum and minimum water levels are recorded 49.79 and 47.56 m in 1996 and 2002 respectively. The changes in water levels of the Manas River and its variation over the last few decades are due to the climate change. The recorded water discharge exhibits a strong decline after 1972–2007 due to the shortfalls of rainfall (Table 2). However, the exception in case of water discharge is noticed in the year 1988. The maximum and minimum water discharge are recorded during this period are 881.88 and 95.48 cumec in 1972 and 2003 respectively (Fig. 4). Changes in future climate may, therefore, have far-reaching effect on the hydrology of this river. The sediment load in the form of coarse, medium and fine of Manas River has been highly fluctuating during 1971–1978 and again in between 1988 and 1994 (Fig. 5). But it is more or less stable in the period 1979–1987. The main cause behind such happenings is the reduction of rainfall in the upstream of Manas River, i.e., nearby Mathanguri and Bhutan.

The mean monthly rainfall of the Manas River catchment recorded during 1983–2008 is given in Table 3. The maximum rainfall is normally recorded in the month of July. During the month of July, the maximum rainfall is recorded in 2004, i.e., 1,426.2 mm and minimum is recorded in 2000, i.e., 288.4 mm. The recorded maximum monsoonal rainfall is 3,858.4 mm in the 1990 and minimum is 1,572.6 mm in 1999. There is a fluctuation in the occurrence of monsoonal rainfall during 1983–2008 due to change of climate (Fig. 6), which has been prepared to show the trends of occurrence of rainfall separately during the months of June, July, August, September and monsoon season.

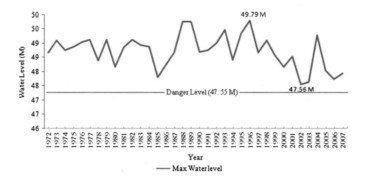

Fig. 3 Yearly maximum stage hydrograph of Manas River (1972–2007)

Table 2 Yearly maximum water discharge (cumec) of Manas River (1972–2007)

Year	Max water discharge	Year	Max water discharge	Year	Max water discharge	Year	Max water discharge
1972	881.88	1981	374.00	1990	304.55	1999	116.13
1973	527.29	1982	351.02	1991	289.49	2000	115.75
1974	386.45	1983	334.50	1992	165.15	2001	104.12
1975	384.46	1984	222.75	1993	171.40	2002	98.04
1976	424.02	1985	377.01	1994	126.71	2003	95.48
1977	568.35	1986	196.27	1995	151.75	2004	101.41
1978	285.00	1987	243.99	1996	148.77	2005	100.82
1979	410.00	1988	673.82	1997	122.33	2006	116.40
1980	497.00	1989	348.79	1998	117.26	2007	119.67

Source Water Resources Department, Govt. of Assam

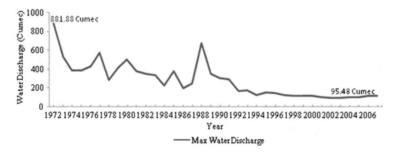

Fig. 4 Yearly maximum water discharge hydrograph of Manas River (1972–2007)

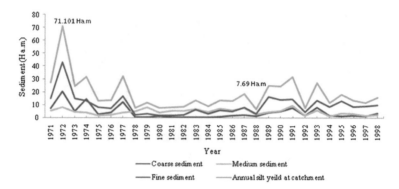

Fig. 5 Trend of sediment load of Manas River (1971–1998)

3.3 Status of Biodiversity in Manas National Park

The Manas National Park is very rich in floral as well as faunal diversity. There are two major biomes presents in this park, namely, (i) grassland biome and (ii) forest biome.

3.3.1 Floral Diversity

There are three main types of vegetation: (i) sub-Himalayan alluvial semi-evergreen forest, (ii) east Himalayan mixed moist and dry deciduous forests (the common type) and (iii) grassland. Alluvial grasslands cover almost 42.84% of the park. These are created and maintained by burning, and on a smaller scale, by elephants. The riparian grasslands are the best tiger habitat in India, and also well suited to the unique wild buffalo herds, gaur and barasingha, elephants, and waterbirds. There are 43 different grass species and also a variety of tree and shrub species. A wide variety of aquatic flora along the Manas River are predominant. Some 374 species of dicotyledons, including 89 trees, 139 species of 6 monocotyledons and 15 species of orchid have been identified (Project Tiger 2001). A total of 543 plants species have been recorded from the core zone. Of these, 374 species are dicotyledons (including 89 trees), 139 species monocotyledons and 30 are pteridophytes and gymnosperms.

3.3.2 Faunal Diversity

The park supports an impressive diversity and biomass of large wildlife species. A total of 55 mammals, 50 reptiles and 3 amphibians have been recorded, several species being endemic (Project Tiger 2001). It contains 22 mammals of the 41 Indian species of that are classified as highly endangered, i.e., under Schedule I

Table 3 Distribution of mean monthly rainfall (mm) of Manas River catchment (1983–2008)

Year	June	July	August	September	Monsoon season
1983	534.4	576.2	549.6	605.4	2,265.6
1984	552.7	993.6	207.3	134.9	1,888.5
1985	426.4	793.1	271.7	186.9	1,678.1
1986	629.8	329.3	356.7	414.2	1,730.0
1987	791.7	1,253.7	293.1	468.1	2,806.6
1988	311.1	627.3	1,076.2	381.5	2,396.1
1989	503.9	785.9	178.0	580.4	2,048.2
1990	1,762.2	1,379.4	428.4	288.4	3,858.4
1991	656.2	385.5	432.5	493.0	1,967.2
1992	456.9	507.6	528.9	294.8	1,788.2
1993	1,094.7	1,053.7	534.0	269.4	2,951.8
1994	570.8	638.9	444.0	481.6	2,135.3
1995	594.5	667.7	630.1	627.7	2,520.0
1996	391.8	1,220.3	238.9	265.6	2,116.6
1997	682.7	667.7	630.0	621.0	2,601.4
1998	607.0	689.0	522.2	224.2	2,042.4
1999	528.0	661.7	292.0	90.9	1,572.6
2000	542.2	288.4	659.8	450.8	1,941.2
2001	852.1	704.8	271.0	411.0	2,238.9
2002	1,061.7	7,91.2	197.1	388.6	2,438.6
2003	1,338.5	679.9	271.7	345.4	2,635.5
2004	825.3	1,426.2	313.6	304.8	2,869.9
2005	475.6	832.0	495.6	553.1	2,356.2
2006	529.5	725.5	266.7	697.7	2,219.4
2007	588.0	488.0	257.6	260.0	1,593.6
2008	517.0	387.0	626.2	242.0	1,772.2

Source Master Plan of Beki–Manas–Aie Sub-Basin, Brahmaputra Board (2010)

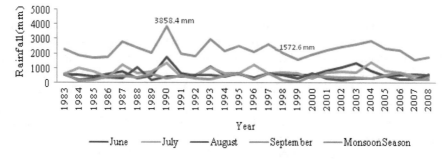

Fig. 6 Trend of mean monthly rainfall of Manas River catchment (1983–2008)

(Wildlife Protection Act 1972) mammals and at least 33 of its animals listed as threatened, by far the greatest number of any protected area in the country. Many are typical of the Southeast Asian rainforest and have their westernmost distribution there, while other species are at the easternmost point of their range. Over 450 species of birds including migrants have been recorded and about 350 breeds in the area, 16 being endemic (Deb Roy 1992). The diverse habitat of the park is an ideal home for a variety of specialized birds. An astounding variety of insects, butterflies and birds are also found in the park.

3.4 Threats to Biodiversity

Like any other national park, wildlife sanctuaries, reserve forest of Assam, the Manas National Park also has been experiencing a great loss of both floral and faunal species resulting from various natural and anthropogenic causes. The following are the basic causes of the loss of biodiversity in the park and these causes may be identified as threats to biodiversity in the recent years.

(i)   Change in climate: The decreasing trend of rainfall and increasing temperature in the study area have been reducing the density of forest and grass coverage which otherwise effect on biodiversity of flora and fauna. Lots of species have been died due to the changing as well as drying up the environmental situation in the area.

(ii) Habitat destruction: Destruction of natural habitat for the commercial feeling of trees, encroachment of forest area for cultivation and habitation and various developmental activities has become the most serious threat to biodiversity of the park. Natural habitat provided by the forest and riverine ecosystem spreading over the park area has witnessed some kind of destruction from the beginning of Bodo Movement in Assam. Construction of road, building, and dam along with some of the other activities such as forest fire, the rapid growth of population in the fringe villages, etc., of the park area are to be considered as destructive to habitats of its local flora and fauna.

(iii) Poaching: Poaching of the biological resource is another major threat to biological diversity in this park. Rhino, elephant and tiger are poached for international trade, whereas deer, pigmy hog, bird, various kinds of small animal and reptile are poached for domestic consumption. It is reported that most of the rhino of the park were killed by poachers in the past decade of the twentieth century. Poaching has been playing a key role in reducing the rhino population not only in this park but also other national parks of Assam.

(iv) Overgrazing: Overgrazing, especially in the nearby grassland areas of the park, is another major problem related to the loss of biodiversity. The cattle grazing not only destroys the grassland but also acts as an agent to transmit various diseases to the wild animals. So, the excessive grazing of domestic animals has been a threat to the grassland belt of the park.

(v) Insurgency and political problems: Insurgency and political problems of this area of Assam greatly affects the biodiversity mostly in the Manas National Park within 1989–2003. Besides creating a lawlessness atmosphere in and around the park area, the insurgency had made available modern sophisticated weapons and it helped the poaching activity.

It is noticed that the change in climate and human interference during the recent period largely effect to the loss of biodiversity in the park.

4 Conclusion

Changes in streamflow, flood, drought, water temperature and water quality due to climate change will make freshwater ecosystem highly vulnerable to loss of biodiversity and ecosystem services. Reduction of the amount of rainfall, change in peak flow intensity and fluctuation in sediment transport with consequences in downstream river morphology and ecology are seen. Climate change increases the stress on existing water resources. Changing rainfall pattern affected the replenishment of Manas River reducing the water flow during the summer season when the demand is the highest. The reduced flow of water has been affecting the dilution capacity of pollutants thus affecting water quality.

A number of rare and endemic species of flora and fauna are found in the Manas National Park may soon be extinct which have already been happening due to climate change and human interference unless conservation efforts are accelerated. The threats to biodiversity arising from climate change are more concern in this park on account of ecological fragility, economic marginality and richness of threatened and endemic species with restricted distributions. As a result of the ongoing pattern of climate change, wetlands, grasslands and forests are highly vulnerable to face shrinkage of habitat, loss of endemic species and proliferation of invasive species.

Through the combined efforts of various NGOs, institution, persons and government departments, viz., forest and environment, agriculture, soil conservation, tourism, etc., the biodiversity loss can be minimized. The success of any conservation efforts largely depends on a concerted effort made by the whole society which is born from the realization. The only need is to teach the people about the real situation of loss of biodiversity, their conservation and sustainable utilization for the present and future generations. The problem of biodiversity conservation is not likely to be solved very easily in the near future until and unless a large proportion of human population comes to share this view. Some appropriate measures should be taken by the government for saving the whole ecosystem. The Act on Legal Protection, Habitat Protection, etc., should be implemented properly. Policies and strategies on climate change should be implemented and the interaction and coordination among different sectors are essential. In order to minimize emerging problems and make the park livable for biological communities, all concern people must join hands and work together. Adapting to the impacts of climate change and meeting sustainable ecosystem, both technical and financial support is required at various levels.

The global community must be aware of the seriousness of the impacts of climate change and act now collectively to reduce them. The climate will probably change, which has constantly done during the 4.5 billion year history of the earth. More scientific studies are very much essential to provide guidelines for the people to maintain the balance of ecosystem of any area as far as possible for the survival of all living organisms in our living earth.

References

Arnell NW (1999) Climate change and global water resources. Glob Environ Change 9:31–49

Deb Roy S (1992) Manas—a world heritage site. A report to the 4th World Congress on Protected Areas, Caracas, Venezuela

Kunstmann H et al (2004) Impact analysis of climate change for an Alpine catchment using high resolution dynamic downscaling of ECHAM4 time slices. Hydrol Earth Syst Sci 8(6): 1030–1044

National Academy of Sciences (2012) Climate change: evidence, impacts and choices (Adobe digital editions). Retrieved from www.national-academies.org

Talukdar B (2012) Environmental changes and threats to biodiversity of the Manas National Park of Assam, an unpublished seminar paper

Increasing Extreme Temperature Events in the Guwahati City During 1971–2010

R. L. Deka, L. Saikia, C. Mahanta and M. K. Dutta

Abstract Extreme temperatures are changing worldwide together with changes in mean temperatures. This study investigates the daily maximum and minimum temperatures of Guwahati city of Northeast India for different intensity and frequency indices for the period 1971–2010. The trends were estimated by linear regression technique and statistical significance of the trends was determined by Kendall's tau statistic. Annual mean indices of extreme temperature events, viz, hottest day, hottest night, and coldest night showed increasing trends while coldest day showed decreasing trend. All frequency indices of hot events showed increasing trends while that of the cold events showed decreasing trends. Number of days above 35 °C is increasing significantly. On the other hand, number of nights above 25 °C is increasing while the number of nights below 10 °C is decreasing. Monthly intensity and frequency indices of hot and cold events during two sub-periods of 1971–1990 and 1991–2010 relative to the mean of the entire period (1971–2010) showed that the hottest day temperature was lower in the months of February, March, and April during 1991–2010 while in other months it was higher than the 1971–1990 sub-period. On the other hand, hottest and coldest night temperatures are more during the recent period. Similarly, both the number of

R. L. Deka (✉)
Department of Agrometeorology, Assam Agricultural University,
Jorhat 785013, Assam, India
e-mail: rldeka2011@gmail.com

L. Saikia · C. Mahanta
Department of Civil Engineering, Indian Institute of Technology Guwahati,
Guwahati, Assam, India
e-mail: lalit.s@iitg.ernet.in

C. Mahanta
e-mail: chandan@iitg.ernet.in

M. K. Dutta
Department of Humanities and Social Sciences, Indian Institute of Technology Guwahati,
Guwahati, India
e-mail: mkdutta@iitg.ernet.in

© Springer International Publishing AG 2018
A. K. Sarma et al. (eds.), *Urban Ecology, Water Quality and Climate Change*,
Water Science and Technology Library 84,
https://doi.org/10.1007/978-3-319-74494-0_13

hot days and hot nights showed increase in frequencies while number of cold days and cold nights showed decrease in frequencies in almost all months.

Keywords Extreme temperature · Trend · Kendall's tau statistics Guwahati

1   Introduction

Since the industrial revolution, there has been a marked increase in the emission of greenhouse gases, mainly CO_2 in the atmosphere. The Fourth Assessment Report of the Intergovernmental Panel on Climate Change (IPCC AR4) has concluded that the global mean surface temperatures rose by 0.74 ± 0.18 °C over the last 100 years (1906–2005). The rate of warming over the recent 50 years (0.13 ± 0.03 °C/ 10 year) is nearly twice that over the past 100 years (0.07 ± 0.02 °C/10 year). The warming is very likely the response of the main anthropogenic drivers, such as the population growth, deforestation, industrialization, changes in land use, and increasing atmospheric concentration of greenhouse gases (IPCC 2007). Continuous addition of gaseous and particulate pollutants and waste heat in the urban and industrial atmospheric environment has a potential of modifying the climate through changes in the components of radiation balance like the greenhouse effect and albedo (Rupa Kumar and Hingane 1988). Isolation and quantification of the impacts of urbanization, land use change, and industrialization on local climate is a challenging task. Investigations have been carried out by climatologists to find a possible link of climate change with anthropogenic activities by studying trends in different climatic parameters, particularly surface air temperature of densely populated cities (Colacino and Lavagini 1982; Rupa Kumar and Hingane 1988; Karl et al. 1988; Balling and Idso 1989; De and Rao 2004; Ha and Yun 2011). Colacino and Rovelli (2016), from an analysis of mean annual surface air temperatures in Rome during 1782–1975, found that the increasing trend in urban temperatures was more conspicuous in the minimum temperatures rather than in the maximum temperatures. A study by Jones et al. (1990) on urbanization and related temperature variation indicates that the impact of urbanization on the mean surface temperature would be no more than 0.05 °C per 100 years. Fujibe (2011) observed widespread urban warming around Tokyo and other megacities of Japan during afternoons of the warm season as a result of extensive urbanization. Hingane (1996) estimated rising trends of 0.84 and 1.39 °C per 100 years in the mean surface temperature of Mumbai and Kolkata, respectively. A similar study by Thapliyal and Kulshreshtha (1991) on temperature trends over Indian cities indicates a slight warming within the limits of 1 SD between 1901 and 1990. Until recently, most of the studies on long-term global climate change using observational temperature have focused on changes in mean values. During the past two decades, more attention has been given to study the extremes in daily temperatures and their variability due to their adverse socioeconomic impacts.

Urbanization, industrialization, and changing land use influence air temperature, particularly nighttime temperature. Local temperature is one of the major climatic elements to record environmental changes brought about by industrialization and urbanization. Within this context, the present study aims to investigate the trends of extreme temperature events during the 40-year period from 1971 to 2010 at Guwahati city, which, during the past three decades saw a phenomenal rise in urbanization and development.

2 Materials and Methods

2.1 Data

The daily maximum and minimum temperature data of Guwahati (26° 11′N, 91° 45′E, 55 m above mean sea level) covering the period from 1970 to 2010 were collected from National Data Centre of the India Meteorological Department (IMD).

2.2 Calculation of Extreme Indices

Different research groups may define different extreme indices for their purpose (Jones 1999; Plummer et al. 1999; Alexander et al. 2006). The Expert Team on Climate Change Detection, Monitoring and Indices (ETCCDMI) has been jointly established by the WMO Commission for Climatology and the Research Programme on Climate Variability and Predictability (CLIVAR). Sixteen of the 27 indices recommended by the ETCCDMI are temperature related (Alexander et al. 2006). The indices are derived from daily maximum and minimum temperature. Exact definitions of the extreme temperature indices are available in http://cccma. soes.uvic.ca/ETCCDMI. In this study, 13 extreme temperature indices are chosen and are listed in Table 1. The period 1971–2000 was chosen as the base period for the indices that represent counts of days crossing the climatological percentile thresholds. The trends of different extreme temperature indices were estimated following linear regression technique. Non-parametric Kendall's tau test (Press et al. 1986) was applied to determine the significance of the slopes.

2.3 Annual Cycles of Extreme Indices

It is of great interest to examine possible changes in annual cycles, to see whether there is any perceptible change in climatologically preferred annual patterns of the

Table 1 Definitions of indices

Indices	Description
TXx	Hottest day, highest maximum temperature (Tmax) in a year
TXn	Coldest day, lowest Tmax in a year
TNx	Hottest night, highest minimum temperature (Tmin) in a year
TNn	Coldest night, lowest Tmin in a year
TX90p	Hot days, % of days in a year with Tmax above 90th percentile
TN90p	Hot nights, % of days in a year with Tmin above 90th percentile
TX10p	Cold days, % of days in a year with Tmax below 10th percentile
TN10p	Cold nights, % of days in a year with Tmin below 10th percentile
TX-35	Number of days in year with Tmax > 35 °C
TN-25	Number of days in year with Tmin > 25 °C
TX-20	Number of days in year with Tmax < 20 °C
TN-10	Number of days in year with Tmin < 10 °C
DTR	Diurnal temperature range, annual mean of difference between Tmax and Tmin

occurrence of extreme temperature events. A simple way of determining whether marked changes have occurred in the annual cycles is to examine the mean annual cycles over two equal halves of the data period, following the semi-average method of estimating change (Revadekar et al. 2011). For this purpose, the annual cycles in terms of mean indices for two halves of the period consisting of 20 years each have been analyzed. Changes in mean indices of Guwahati city for two sub-periods, viz., 1971–1990 and 1991–2010 have been computed for each month with respect to data for the entire period 1971–2010.

3 Results and Discussion

3.1 Annual Cycle of Temperature

In order to understand the nature of extreme temperature events, it is important to consider the features of the mean annual cycle of temperature on a daily basis. The annual cycle of daily maximum and minimum temperatures of Guwahati, based on the average over the period 1971–2000 is depicted in Fig. 1. The temperature increases from January and attains a peak in the month of August and starts decreasing up to December. The increasing pace of maximum temperature is moderated from later part of April onwards due to pre-monsoon and monsoon rainfall activity.

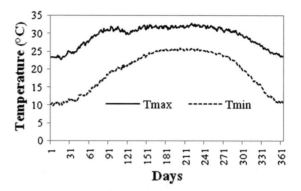

Fig. 1 Annual mean maximum and minimum temperature cycle at Guwahati

3.2 Trends in Annual Mean Temperature

The annual maximum and minimum temperature at Guwahati increased significantly during the past 40-year period (Fig. 2). The rate of increase of minimum temperature (0.38 °C/decade) is larger than that of maximum temperature (0.25 °C/decade). Annual mean maximum and minimum temperature anomalies (deviations from 1961 to 1980 mean) during the past three decades are shown in Fig. 3. It showed the relative dominance of minimum temperature over maximum temperature in the observed warming at Guwahati. The past decade was the warmest decade.

3.3 Trends in Annual Extreme Temperature

The trends of different extreme temperature indices, viz, intensity indices (TXx, TXn, TNx, and TNn), percentile-based indices (TX90p, TX10p, TN90p, and TN10p), frequency indices with fixed thresholds (TX-35, TX-20, TN-25, and TN-10), and range index (DTR) of Guwahati city for the period 1971–2010 are

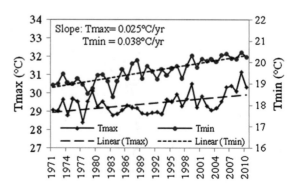

Fig. 2 Annual maximum and minimum temperature trend at Guwahati

Fig. 3 Annual mean
temperature anomalies
(deviations from 1961 to 1980
mean)

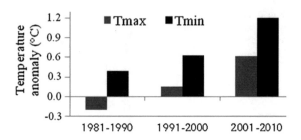

presented in Table 2. It is seen that hottest day, hottest night, and coldest night
showed increasing trends while the coldest day showed decreasing trend. The
increases in hottest (0.34 °C/decade) and coldest night (0.52 °C/decade) at
Guwahati are found to be statistically significant.

All frequency indices of hot events (above 90th percentile of 1971–2000 average) showed increasing trends while that of the cold events (below 10th percentile
of 1971–2000 average) showed decreasing trends (Table 2). The increase in the
frequency of hot days (3.14% per decade) and hot nights (4.75% per decade) and a
decrease in frequency of cold days (1.51% per decade) and cold nights (2.15% per
decade) are statistically significant.

Number of days above 35 °C (hot days) is increasing significantly by a rate of
4.6 days/decade. On the other hand, number of nights above 25 °C (hot nights) is
increasing while the number of nights below 10 °C (cold nights) is decreasing. The
trend magnitudes for the nighttime indices are found to be larger than their daytime
counterparts resulting lowering of annual diurnal temperature range by a rate of
0.103 °C per decade (Table 2). The diurnal temperature range (DTR) was found to
decrease at a rate of 0.103 °C per decade during the period of study.

Table 2 Trends of extreme
temperature indices at
Guwahati during 1971–2010

Index	Slope/decade	Kendall's tau
TXx	0.199	0.177
TXn	−0.276	−0.115
TNx	0.342	0.380*
TNn	0.524	0.374*
TX90p	3.142	0.360*
TX10p	−1.513	−0.379*
TN90p	4.750	0.669*
TN10p	−2.146	−0.420*
TX-35	4.563	0.373*
TX-20	0.402	0.174
TN-25	16.824	0.648*
TN-10	−5.886	−0.366*
DTR	−0.103	−0.131

*Significant at 5% level

3.4   Changes in Annual Cycle of Indices

Monthly mean indices of extreme temperature events (TXx, TNx, TXn, TNn) in the two sub-periods are presented in Fig. 4a–d. The hottest day temperature was lower in the months of February, March, and April during 1991–2010 while in other

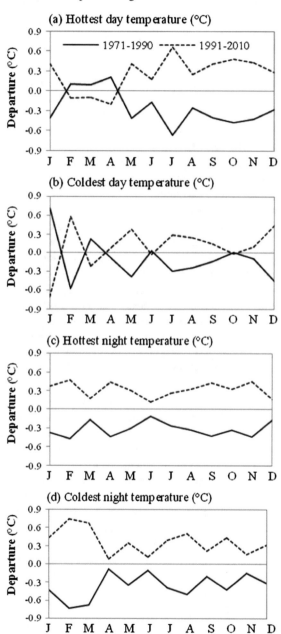

Fig. 4 Monthly departure of hottest and coldest day, hottest and coldest night

months it was higher than the 1971–1990 period. The magnitude of changes in hottest day temperature (TXx) are more in October, November, December, and January (0.2–0.4 °C) compared to other months.

The magnitude of changes in coldest day temperature was lower during monsoon and post-monsoon months. Hottest night and coldest night temperature are more during the sub-period 1991–2010.

The differences in annual cycles of the frequency indices for hot and cold events based on 10th and 90th Both number of hot days and hot nights (TX90p and TN90p) showed an increase in the frequencies in almost all months (except TX90p in March and June). On the other hand, both the number of cold days and cold nights (TX10p and TN10p) showed a decrease in frequencies in almost all months (except TX10p in March and June) (Fig. 5a–d).

The results clearly showed the significant warming as well as an increase in intensity and frequency of extreme temperature events in Guwahati city during the past four decades. The causes for such changes are thought to be mostly anthropogenic. Guwahati is a fast growing metropolis, which witnessed haphazard growth and population explosion after the establishment of the State capital in 1972. As per census report, the population of Guwahati in 1971 was 2.93 lakhs, increased to 8.91 lakhs in 2001 (Fig. 6). As the population became more than triple within a period of 30 years in a finite area of 216 km^2, hills and wetlands including other open spaces were encroached upon for different purposes, resulting in cutting down of hills, filling up of low-lying areas and deforestation. This might have many interactions and interventions in the local environment.

Moreover, the increasing population leads to a corresponding increase in traffic population. According to State Transport Department, Guwahati accounts for nearly 65% of the total automobile vehicle population of the State. The number of the city's automobile vehicles, including the motorcycles and scooters, stands at around 10 lakh in 2011. Though Guwahati is the major industrial center of Assam, industrial area shares only 5.2% of total area (City Development Plan Guwahati 2006).

The surface maximum and minimum temperature have been increasing during past four decades at Guwahati. A lower change in the increase in maximum temperature than minimum temperature may be a sign of the presence of solar dimming as stated by Wild et al. (2007). Moreover, due to high levels of fossil fuel consumption and friction, a significant portion of waste heat is stored in walls and buildings, streets, etc., which gets released during the night, thereby making the nights warmer. This has resulted in increased frequencies of hot nights and decreased frequencies of cold nights during different months over the period 1991–2010 compared to the previous 20-year period. From an analysis of surface reaching solar radiation data of 12 stations located at different cities of India, Kumari et al. (2007) reported an overall decreasing trend at a rate of 0.86 W/m^2 per year for the period 1981–2004. Significant decreasing trend of solar radiation at different locations of Northeast India have also been observed in a few recent studies (Jhajharia and Singh 2010; Deka et al. 2012). This points out to the fact that the amount of aerosols which reflect the incident solar radiation might be increasing

Fig. 5 Monthly departure of frequency of hot and cold days, hot and cold nights

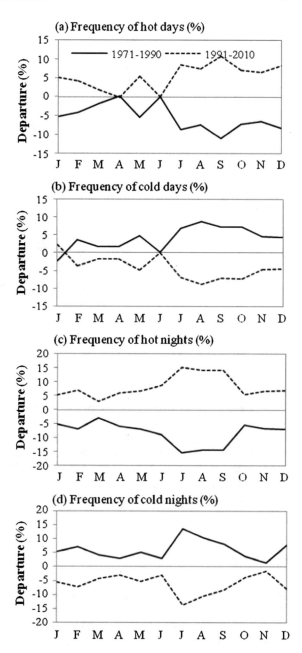

during the recent past, resulting in decreased solar radiation and increased minimum temperature. The recent spurt in construction works, local deforestation and forest fires in the neighboring countries might have contributed to the aerosol pool and has increased the reflected portion of the solar radiation. The monthly load of

Fig. 6 Trend of population
of Guwahati city during
1961–2011

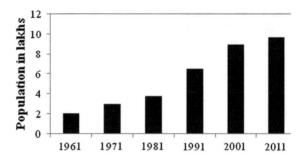

Fig. 7 Monthly SPM load at
Bamunimaidan, Guwahati
(mean data of 2004–2007)

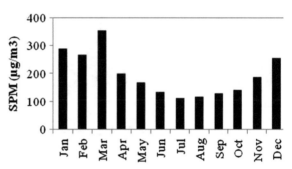

suspended particulate matter (SPM) measured at Bamunimaidan by State Pollution
Control Board, Guwahati is shown in Fig. 7.

The increased load of SPM during winter and pre-monsoon months might be one
of the reasons for the observed decreasing frequencies of cold days (TX10p) and
cold nights (TN10p) at Guwahati during the past 20-year period. As most of these
causal mechanisms are global or regional in nature undergoing complex interactions
and feedback processes with the local factors, the observed trend in extreme temperature events at Guwahati city need to be examined w.r.t. other variables over a
longer time series data.

4   Conclusions

The results of this study indicate significant warming trends in the Guwahati city.
The frequency of hot days and hot nights (cold days and cold nights) has increased
(decreased) significantly in the past 40 years. The highest and lowest minimum
temperature has also been increased significantly. The trend magnitudes for the
nighttime indices were larger than their daytime counterparts, resulting in
decreasing trend of DTR. Rapid urbanization after 1970s, land-use changes,
increased load of aerosol pool are thought to be the major anthropogenic factors
causing the observed variations in extreme temperature events at Guwahati city.

References

Alexander LV et al (2006) Global observed changes in daily climate extremes of temperature and precipitation. J Geophys Res 111:D05109. https://doi.org/10.1029/2005JD006290

Balling RC Jr, Idso S (1989) Historical temperature trends in the United States and the effect of urban population. J Geophys Res 94:3359–3363

Colacino M, Lavagini A (1982) Evidence of the urban heat island in Rome by climatological analyses. Arch Met Geophys BiocL Set B 31:871–897

Colacino M, Rovelli A (2016) The yearly averaged air temperature in Rome from 1782 to 1975. Tellus A: Dyn Meteorol Oceanogr 35(5):389–397

De US, Prakasa R (2004) Urban climate trend—the Indian scenario. J Ind Geophys Union 8(3): 199–203

Deka RL, Nath KK, Mahanta C, Sharma K (2012) Recent variability of solar radiation in the upper Brahmaputra valley of Assam. J Agrometeorol 14:145–150

Fujibe F (2011) Urban warming in Japanese cities and its relation to climate change. Int J Climatol 31(3):162–173

Ha KJ, Yun KS (2011) Climate change effects on tropical nigh days in Seoul, Korea. Theor Appl Climatol. https://doi.org/10.1007/s00704-011-0573-y

Hingane LS (1996) Is a signature of socio-economic impact written on the climate? Clim Change 32:91–102

IPCC (2007) Climate change 2007: the physical science basis. Contribution of working group I to the fourth assessment report of the IPCC. In: Solomon S, Qin D, Manning M, Chen Z, Marquis M, Averyt KB, Tignor M, Miller HL (eds) Cambridge. Cambridge University Press, UK, p 996

Jhajharia D, Singh VP (2010) Trends of temperature, diurnal temperature range and sunshine duration in Northeast India. Int J Climatol. https://doi.org/10.1002/joc.2164

Jones PD et al (1990) Assessment of urbanization effects in time series of surface air temperature over land. Nature 347:169–172

Jones PD (1999) The use of indices to identify change in climatic extreme. Clim Change 42: 131–149

Karl TR et al (1988) Urbanization: its detection and effect in the United States climate record. J Clim 1:1099–1123

Kumari BP et al (2007) Observational evidence of solar dimming: offsetting surface warming over India. Geophys Res Lett 34:L21810. https://doi.org/10.1029/2007GL031133

Plummer N et al (1999) Changes in climate extremes over the Australian region and New Zealand during the twentieth century. Clim Change 42(1):183–202

Press WH et al (1986) Numerical recipes: the art of scientific computing. Cambridge University Press, Cambridge, pp 488–493

Revadekar JV et al (2011) About the observed and future changes in temperature extremes over India. Nat Hazards. https://doi.org/10.1007/s11069-011-9895-4

Rupa Kumar K, Hingane LS (1988) Long term variations of surface air temperature at major industrial cities of India. Clim Change 13:287–307

Thapliyal V, Kulshreshtha SM (1991) Climate changes and trends over India. Mausam 42: 333–338

Wild M, Ohmura A, Makowski K (2007) Impact of global dimming and brightening on global warming. Geophys Res Lett 34:L04702. https://doi.org/10.1029/2006GL028031

Trends in Sunshine Duration in Humid Climate of Northeast India: A Case Study

D. Jhajharia, P. K. Pandey, T. Tapang, S. Laji, K. Dahal, R. R. Choudhary and Rohitashw Kumar

Abstract In the present study, trends in sunshine duration were investigated using the Mann–Kendall test in different timescales under the humid climatic conditions of Dibrugarh (Assam), Northeast India. Statistically significant decreasing trends were observed in the sunshine duration in the range of $(-)0.61$ to $(-)1.28$ h/decade over Dibrugarh in different timescales: annual; seasonal: winter; and monthly: January, June, and November during the past 28 years from 1981 to 2008. On the other hand, the significant increasing trend in sunshine duration was observed at the rate of $(+)0.69$ h/decade in the month of September over Dibrugarh. The observed decreasing trends in sunshine duration support the existence of the phenomenon of dimming over Northeast India, which may affect the water requirements in the region.

Keywords Trend · Mann–Kendall · Sunshine duration · Northeast India
Global dimming

D. Jhajharia (✉)
Department of Soil and Water Conservation Engineering, College of Agricultural
Engineering and Post-Harvest Technology, Ranipool, Gangtok 737135, Sikkim, India
e-mail: jhajharia75@rediffmail.com

P. K. Pandey · T. Tapang · S. Laji · K. Dahal
Department of Agricultural Engineering, North Eastern Regional Institute of Science
and Technology, Nirjuli, Itanagar 791109, Arunachal Pradesh, India

R. R. Choudhary
Department of Electronics Instrumentation and Control Engineering, Engineering College
Bikaner, Karni Industrial Area, Bikaner, Rajasthan, India

R. Kumar
Division of Agricultural Engineering, Sher-e-Kashmir University of Agricultural Sciences
and Technology of Kashmir, Shalimar Campus, Srinagar 191121, Jammu and Kashmir, India

© Springer International Publishing AG 2018
A. K. Sarma et al. (eds.), *Urban Ecology, Water Quality and Climate Change*,
Water Science and Technology Library 84,
https://doi.org/10.1007/978-3-319-74494-0_14

1   Introduction

The research on climate change due to the rising greenhouse gases has increased tremendously in the past couple of years. Any change in solar radiation and sunshine duration is of major importance for agricultural production in India as the majority of the population is dependent on agriculture or agriculture-related sectors for their livelihood. Also, the impact of any change in solar radiation and rising temperature will be severe for the country as the region has the second largest population in the world. Solar energy is the prime source of energy and has to be monitored to maintain the ecological balance of the earth. It is responsible for photosynthesis in plants which is the main essence of agriculture, and therefore any change in radiation pattern must be analyzed.

The available literature on trends in solar radiation has proved that the solar radiation and sunshine duration have decreased in the past few decades over different parts of the world. Global dimming refers to the occurrence of a significant reduction in global irradiance which is the flux of solar radiation reaching the earth's surface both in the direct solar beam and in the diffuse radiation scattered by the sky and clouds (Stanhill and Cohen 2001). The global dimming has interfered with the hydrological cycle by reducing evaporation and may have reduced rainfall in some areas. Global dimming has also created a cooling effect that may have partially masked the effect of greenhouse gases on global warming.

The main purpose of this research work is to check whether there is any sign of decrease in the solar radiation over the Northeast India. In the present study, the trends in sunshine duration, a proxy of solar radiation, were investigated on monthly (January–December), seasonal (winter, pre-monsoon, monsoon, and post-monsoon) and annual timescales through the Mann–Kendall nonparametric test over Dibrugarh located in the northeastern region of India. The study will help to tell us whether solar dimming or brightening is taking place over the Northeast India.

2   Materials and Methods

2.1   Study Area and Meteorological Data

Tea beside paddy, forest products like bamboo, different types of fruits and orchids grow extensively in Northeast (NE) India. The main ecosystem in NE India is a tropical wetland where annual rainfall is the heaviest in the world (Jhajharia et al. 2012a, b). The total annual rainfall in the region varies from place to place, i.e., annual rainfall of about 12,000 mm in Meghalaya to about 2000 mm over a few places in Assam (Dev and Dash 2007). The monsoon rainfall increases from south to north and also from west to east over subtropical Assam. Rainfall is quite low in winter and post-monsoon seasons (Barthakur 2004).

The solar radiation provides the energy for the carbon assimilation of plant canopies and their loss to the atmosphere (Stanhill and Cohen 2001). The data of solar radiation are recorded with the help of pyranometers, which usually are not readily available over most of the sites in India. However, the data of bright sunshine duration, considered as a proxy to the radiation, are available for more number of sites because of ease in measurements of sunshine duration with the Campbell–Stokes sunshine recorder. This is one of the major sources of information on changes in solar irradiance as this instrument automatically records the duration of direct solar beam irradiance above a threshold of 120 W m^{-2} (WMO 1997). Sunshine duration measurements are the most highly and linearly correlated with the global radiation of the widely available and used proxy measurements (Stanhill 1965). The bright sunshine duration data of Dibrugarh required for this study were obtained from India Meteorological Department (Pune) for a period of 28 years from 1981 to 2008. There were some missing values, which were filled with the respective normal value. Figure 1 shows the location of the Dibrugarh site in the Assam state in the northeastern region of India. The average monthly, seasonal, and

Fig. 1 Location of Dibrugarh site in Assam (Northeast India)

Table 1 Statistical properties of bright sunshine duration over Dibrugarh

Timescale	M (h)	S (h)	C_v	C_s
January	6.29	1.21	19.2	−0.1
February	5.67	1.15	20.3	−0.6
March	4.85	1.04	21.5	−0.1
April	4.52	0.93	20.6	−0.1
May	5.13	1.15	22.5	−0.1
June	4.08	1.21	29.7	−0.1
July	3.64	0.98	26.9	1.1
August	4.69	1.11	23.7	−0.4
September	4.21	0.83	19.6	0.1
October	6.36	0.84	13.1	0.0
November	7.80	0.94	12.0	−0.3
December	7.33	1.05	14.4	−0.6
Yearly	5.38	0.34	6.4	0.3
Winter	5.98	0.91	15.2	0.5
Pre-monsoon	4.83	0.66	13.6	0.2
Monsoon	4.59	0.45	9.8	0.3
Post-monsoon	7.57	0.62	8.2	−0.8

Note M, S, C_v, and C_s denote mean, standard deviation, coefficient of variation, and coefficient of skewness, respectively

annual actual bright sunshine duration over Dibrugarh located in Assam state of NE India are given in Table 1.

2.2 Methods of Trend Analysis

Trends in the data can be identified by using parametric or nonparametric methods. The nonparametric methods do not require normality of time series and are less sensitive to outliers. The nonparametric Mann–Kendall (MK) method (Mann 1945; Kendall 1975) is used for identifying trends as it is distribution free and has a higher power than other commonly used tests. The MK test is based on the test statistic, S, defined as follows:

$$S = \sum_{k=1}^{n-1} \sum_{j=k+1}^{n} \text{sgn}(x_j - x_k), \tag{1}$$

where n is the number of observations and x_j is the jth observation and $\text{sgn}(\cdot)$ is the sign function which can be defined as

$$\mathrm{sgn}(\theta) = \begin{cases} 1 & \text{if} & \theta > 0 \\ 0 & \text{if} & \theta = 0 \\ -1 & \text{if} & \theta < 0. \end{cases} \tag{2}$$

The mean and variance of the S statistic, under the assumption that the data are independent and identically distributed, are given by the following expressions:

$$E(S) = 0, \tag{3}$$

$$V(S) = \frac{n(n-1)(2n+5) - \sum_{i=1}^{m} t_i(t_i - 1)(2t_i + 5)}{18}, \tag{4}$$

where m is the number of groups of tied ranks, each with t_i tied observations. The MK statistic, designated by Z, can be computed as

$$Z = \begin{cases} \frac{S-1}{\sqrt{\mathrm{Var}(S)}} & S > 0 \\ 0 & S = 0 \\ \frac{S+1}{\sqrt{\mathrm{Var}(S)}} & S < 0. \end{cases} \tag{5}$$

The Z value is computed and if the value lies within the limits $(-)1.96$ and $(+)$ 1.96, i.e., $-Z_{1-\alpha/2} \leq Z \leq Z_{1-\alpha/2}$, then the null hypothesis of no trend can be accepted at the 5% (α) level of significance using a two-tailed test. Otherwise, the null hypothesis can be rejected and the alternative hypothesis can be accepted at the 5% significance level. If $Z > 1.96$, there is increasing trend; and if $Z < -1.96$, there is decreasing trend.

2.3 Modified Mann–Kendall Method

The application of the original MK procedure is not recommended for the data set in presence of serial correlation as the effect of lag-1 serial correlation poses a major source of uncertainty. So, the lag-1 serial correlation component was removed from the time series to eliminate the influence of serial correlation on trend prior to applying the MK test. This treatment is called "pre-whitening". The new time series as proposed by Kumar et al. (2009) can be obtained as

$$x_i' = x_i - (\beta \times i), \tag{6}$$

where β is Theil–Sen's estimator (Theil 1950; Sen 1968). The value of r_1 of the new time series is first computed and later used to determine the residual series as

$$y_i' = x_i' - r_1 \times x_{i-1}'. \tag{7}$$

The value of $\beta \times i$ was added again to the residual data set as

$$y_i = y_i' + (\beta \times i). \tag{8}$$

The y_i series was subjected to trend analysis.

2.4   Theil–Sen's Estimator

The slope of n pairs of data points was estimated using the Theil–Sen's estimator which is given as

$$\beta = \text{Median}\left(\frac{x_j - x_l}{j - l}\right) \quad \forall\, 1 < l < j. \tag{9}$$

The slope computed is a robust estimate of the magnitude of a trend and has been widely used in identifying the slope of a trend line in a hydrological time series (Yue et al. 2002).

3   Results and Discussion

The daily data of bright sunshine duration were used to compute monthly, seasonal (winter: January–February; pre-monsoon: March–May; monsoon: June–September and post-monsoon: October–December) and annual time series of sunshine duration over Dibrugarh.

3.1   Statistical Properties of Sunshine Duration Over Dibrugarh (Assam)

The statistical parameters; mean (M), standard deviation (S), coefficient of variation (C_V), and coefficient of skewness (C_S) of bright sunshine duration were calculated to describe the sunshine variability over Dibrugarh (see Table 1). The average annual sunshine duration is found to be 5.38 h for the considered period over Dibrugarh. The mean sunshine duration in pre-monsoon and monsoon seasons are found to be 4.83 and 4.59 h, respectively. However, the sunshine duration in winter and post-monsoon seasons are comparatively higher than the sunshine hours witnessed in the pre-monsoon and monsoon seasons over Dibrugarh. The mean bright sunshine hours in the winter and post-monsoon seasons over Dibrugarh are found to be 5.98 and 7.57, respectively. The C_V of sunshine duration in monsoon (9.8%) and post-monsoon (8.2%) seasons are quite low as compared to the C_V of winter and pre-monsoon seasons. The coefficient of variation of monthly sunshine duration is found to be the lowest, i.e., about 12% during the month of November. Similarly on the monthly timescale, mean bright sunshine duration is the highest (7.8 h) during the month of November and the lowest (3.64 h) during the month of July over Dibrugarh.

3.2 Trends in Sunshine Duration Over Dibrugarh

Table 2 shows the trend results in terms of Z statistics for different durations: all 12 months; all the four seasons; and annual timescale. It is important to emphasize that any insignificant Z presents a value due to random fluctuations, meaning not much in inferring the existence of a trend from the statistical standpoint (Dinpashoh et al. 2011). The Mann–Kendall test was carried out to identify the existence of trends in bright sunshine hours in all the timescales over Dibrugarh site located in the Upper Assam.

Table 2 shows that statistically significant trends are witnessed in bright sunshine hours at Dibrugarh in only five timescales (monthly: June, September, and

Table 2 Test statistics (Z) values obtained through Mann–Kendall test and magnitude of trends in sunshine duration over Dibrugarh

Timescale	Test statistics (Z) value	Magnitude of trends (hours/decade)
January	**−1.748**[*]	**−0.71**[*]
February	−0.846	−0.40
March	−1.241	−0.46
April	0.508	0.18
May	−0.846	−0.58
June	**−2.594**	**−1.10**
July	0.113	0.09
August	−0.959	−0.33
September	**2.143**	**+0.69**
October	−0.395	−0.13
November	**−3.102**	**−1.28**
December	0.395	0.06
Yearly	**−2.199**	**−0.30**
Winter	**−2.481**	**−0.61**
Pre-monsoon	−1.184	−0.24
Monsoon	−0.790	−0.15
Post-monsoon	−1.640	−0.18

Bold number and Bold* Denote statistically significant trends at 5% and 10% levels of significance, respectively

November; annual; and seasonal: winter) at 5% level of significance. Out of these five statistically significant trends at 5% level of significance, all but one (September) were cases of significant decreases in sunshine duration. It is pertinent to mention here that significant decreasing trend in sunshine duration is also observed in one timescale, i.e., January at 10% level of significance (test statistic value -1.75). On the other hand, "no trend" is witnessed in bright sunshine hours over Dibrugarh in all the remaining timescales, i.e., monthly: February to May, July to August, October and December; and seasonal; Pre-monsoon, monsoon, and post-monsoon.

Figure 2 shows a sample time series of the bright sunshine duration in annual timescale over Dibrugarh station in the period of 1981–2008. The bold curve and the broken line denote the annual time series and the trend line for the bright sunshine duration, respectively.

Trends were obtained through the nonparametric Mann–Kendall (MK) test after carrying out the pre-whitening in the bright sunshine duration data. The bold number and the bold number* denote the statistically significant trends at 5 and 10% levels of significance, respectively.

The slope of the trend lines was computed using the Thiel–Sen's slope estimator. The magnitude of the trends in sunshine duration (hours per decade) in all the timescales are given in Table 2. Statistically significant decreasing trends were observed in the sunshine duration in the range of $(-)0.61$ to $(-)1.28$ h/decade over Dibrugarh in different timescales: annual; seasonal: winter; and monthly: January, June, and November during the past 28 years from 1981 to 2008. On the other hand, significant increases were observed in sunshine duration at the rate of $(+)$ 0.69 h/decade in September. The decreasing trends witnessed in sunshine duration under the humid climatic conditions of Dibrugarh support the existence of the phenomenon of dimming over Northeast India.

Fig. 2 Time series of bright sunshine hours in annual duration over Dibrugarh

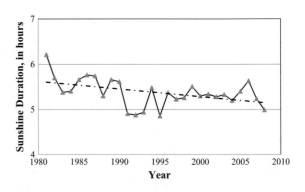

3.3 Climate Change Over the Northeast Region of India

Various researchers have carried out various climate change studies for the northeastern region of India. Jhajharia et al. (2007) reported decreasing trends in annual rainfall at the rate of 2.4 mm/year at Agartala (Tripura). The trends in annual rainfall at Agartala were in agreement with the decreasing trends observed in the total cloud amount over Agartala. Jhajharia et al. (2009) have reported no significant trends in yearly rainfall at Chuapara (Nagrakata) and mixed trends in seasonal rainfall at Agartala.

Jhajharia et al. (2012b) studied the trends in rainfall and rainy days on monthly and seasonal scales over 24 sites of Assam. Jhajharia et al. (2012b) reported that most of the stations selected from the four different regions of Assam witnessed no trend in rainfall and rainy days at 5% significance level on annual, seasonal and monthly timescales. However, a few sites of Assam witnessed decreasing trends in total rainfall in different durations: annual; monthly-May (five); and seasonal: pre-monsoon, monsoon, and post-monsoon. On the other hand, about half of the total 24 sites witnessed decreasing trends on rainy days in the month of December and January. Similarly, all but two (Rangia and Haflong) sites witnessed no trend (decreasing trends) in 24 h maximum rainfall.

Temperature increases were observed mainly in the monsoon and post-monsoon seasons in NE region (Jhajharia and Singh 2011). Recently McVicar et al. (2012) reported decreasing trends in wind speed on annual timescale over 11 sites of NE India. Also, Jhajharia et al. (2009) reported decreasing trends in wind speed on annual and seasonal timescales over different sites of NE India. The climate change studies relating to rising temperature (Jhajharia et al. 2007; Chattopadhyay et al. 2011) and decreasing and rising diurnal temperature range (Jhajharia and Singh 2011) over the NE region India are also available in the existing literature.

Decreasing trends in pan evaporation and reference evapotranspiration over different sites of NE India are reported by Jhajharia et al. (2009) and Jhajharia et al. (2012a, b), respectively. The decreases in reference evapotranspiration have been more significant in the pre-monsoon, which indicate the presence of an element of a seasonal cycle over Northeast India (Jhajharia et al. 2012a, b). The decreasing trends in pan evaporation and evapotranspiration in NE India in spite of the rising temperature indicate that other climatic parameters, such as wind speed, sunshine duration, etc., may have played crucial roles in the changes in observed decreases in evaporation in the northeastern region of India.

4 Conclusions

The daily data of bright sunshine duration over Dibrugarh (Assam), Northeast India were used to investigate trends in different timescales through the Mann–Kendall test. Statistically significant decreasing trends were observed in the bright sunshine duration under the humid climatic conditions of Dibrugarh in the range of $(-)0.61$ to $(-)1.28$ h/decade in following timescales: annual; winter season and in the months of January, June, and November. On the other hand, increasing trend in sunshine duration was witnessed at the rate of 0.69 h/decade in the month of September that suggests brightening under the humid climatic conditions of Dibrugarh. The observed decreasing trends in sunshine duration under the humid climatic conditions of Dibrugarh support the existence of the phenomenon of dimming over Northeast India, which ultimately may influence the irrigation water requirements in the region as the decreases in solar radiation are most likely to affect the rate of evapotranspiration. Jhajharia et al. (2009, 2012a, b) have also reported significant decreases in pan evaporation and reference evapotranspiration over different parts of Northeast India.

Acknowledgements The authors thank the India Meteorological Department (Pune) for providing the bright sunshine duration data. The authors are grateful to the reviewer for critical comments that have helped in the improvement of the paper. Also, the authors acknowledge the financial support from the Director, NERIST, Nirjuli to attend the international conference.

References

Barthakur M (2004) Weather and climate. In: Singh VP, Sharma N, Ojha CSP (eds) The Brahmaputra basin water resources. Kluwer Academic Publishers, pp 17–23

Chattopadhyay S, Jhajharia D, Chattopadhyay G (2011) Univariate modeling of monthly maximum temperature time series over northeast India: neural network versus Yule-walker equation based approach. Meteorol Appl 18:70–82

Dev V, Dash AP (2007) Rainfall and malaria transmission in north-eastern India. Ann Trop Med Parasitol 101:457–459

Dinpashoh Y, Jhajharia D, Fakheri-Fard A, Singh VP, Kahya E (2011) Trends in reference evapotranspiration over Iran. J Hydrol 399:422–433

Jhajharia D, Roy S, Ete G (2007) Climate and its variation: a case study of Agartala. J Soil Water Conserv 6(1):29–37

Jhajharia D, Shrivastava SK, Sarkar D, Sarkar S (2009) Temporal characteristics of pan evaporation trends under the humid conditions of northeast India. Agric Forest Meteorol 149:763–770

Jhajharia D, Singh VP (2011) Trends in temperature, diurnal temperature range and sunshine duration in northeast India. Int J Climatol 31:1353–1367

Jhajharia D, Dinpashoh Y, Kahya E, Singh VP, Fakheri-Fard A (2012a) Trends in reference evapotranspiration in the humid region of northeast India. Hydrol Process. 26:421–435. https://doi.org/10.1002/hyp.8140

Jhajharia D, Yadav BK, Maske S, Chattopadhyay S, Kar AK (2012b) Identification of trends in rainfall, rainy days and 24 hours maximum rainfall over sub-tropical Assam in northeast India. CR Geosci. 344:1–13. https://doi.org/10.1016/j.crte.2011.11.002

Kendall MG (1975) Rank correlation methods, 4th edn. Charles Griffin, London

Kumar S, Merwade V, Kam J, Thurner K (2009) Streamflow trends in Indiana: effects of long term persistence, precipitation and subsurface drains. J Hydrol 374:171–183

Mann HB (1945) Non-parametric tests against trend. Econometrica 33:245–259

McVicar TR, Roderick ML, Donohue RJ, Li LT, Van Niel TG, Thomas A, Grieser J, Jhajharia D, Himri Y, Mahowald NM, Mescherskaya AV, Kruger AC, Rehman S, Dinpashoh Y (2012) Global review and synthesis of trends in observed terrestrial near-surface wind speeds: implications for evaporation. J Hydrol 416–417:182–205

Sen PK (1968) Estimates of the regression coefficients based on Kendall's tau. J Am Stat Assoc 63:1379–1389

Stanhill G (1965) A comparison of four methods of estimating solar radiation. In: Eckardt FE (ed) Methodology of plant eco-physiology. Proceedings of the Montpellier symposium, Arid Zone Research, Vol XXV UNESCO, Paris, pp 55–61

Stanhill G, Cohen S (2001) Global dimming: a review of the evidence for a widespread and significant reduction in global radiation with discussion of its probable causes and possible agricultural consequences. Agric Forest Meteorol 107:255–278

Theil H (1950) A rank invariant method of linear and polynomial regression analysis, Part 3. Netherlands Akademie van Wettenschappen, proceedings 53:1397–1412

WMO (1997) Measurement of radiation guide to meteorological instruments and methods of observation, 6th edn. World Meteorological Organization, Geneva (Chapter 7)

Yue S, Pilon P, Phinney B, Cavadias G (2002) The influence of autocorrelation on the ability to detect trend in hydrological series. Hydrol Process 16:1807–1829

Part III
Water Quality Concerns in Urban Environment

Factors Affecting the Quality of Roof-Harvested Rainwater

V. Meera and M. Mansoor Ahammed

Abstract Rooftop rainwater harvesting (RWH) is receiving increased attention as an alternative source of drinking water. In urban areas, RWH can bring many benefits and may serve to cope with current water shortages, urban stream degradation, and flooding. The present study examined the effect of some of the factors that affect the quality of roof-harvested rainwater, namely, type of the roofing material, age of the roof and characteristics of rainfall event, on the quality of roof-harvested rainwater. More than 100 roof-harvested rainwater samples collected from newly constructed pilot-scale roofs and existing full-scale roofs of different materials were analysed for different physico-chemical and microbiological water quality parameters. Enrichment factor, correlation analysis and other statistical techniques were used to arrive at conclusions. Results showed that, in general, the physico-chemical characteristics except pH, turbidity and some heavy metals conformed to the drinking-water quality standards. Bacteriological contamination was found for most of the runoff samples from roofs. The results showed high variability in the concentration of different water quality parameters between different roofing materials. Enrichment factor (that is the ratio of concentration in roof runoff to that in free fall) was found to be different for different roofing materials and its value was >1 for most of the parameters tested indicating the significant effect of roofing material on the concentration of various parameters by weathering and/or by deposition. The enrichment factor was high for old roofs compared to new roofs. Correlation analysis revealed a positive correlation between the concentration of different water quality parameters and the dry period, and a negative correlation between some water quality parameters and the amount of rainfall. Thus, roofing material and characteristics of rainfall event were found to have a significant influence on the quality of roof-harvested rainwater.

V. Meera (✉)
Department of Civil Engineering, Government Engineering College, Thrissur 680009, India
e-mail: meerav17@hotmail.com

M. Mansoor Ahammed
Civil Engineering Department, SV National Institute of Technology, Surat 395007, India
e-mail: mansoorahammed@gmail.com

© Springer International Publishing AG 2018
A. K. Sarma et al. (eds.), *Urban Ecology, Water Quality and Climate Change*,
Water Science and Technology Library 84,
https://doi.org/10.1007/978-3-319-74494-0_15

Keywords Rainwater harvesting · Roofing materials · Roof runoff
Water quality

1 Introduction

Many rural communities in developing countries use roof-harvested rainwater as
their primary water source for drinking. Traditionally, this was the major option to
people in water-scarce regions, where people had to manage to fulfil drinking water
and household water needs by rainwater harvesting. Due to ubiquitous contamination of surface and groundwater resources by microbial and chemical contaminants, rainwater harvesting has become more relevant now even in areas which
enjoy high rainfall.

Rainwater is generally considered as nonpolluted, or at least not significantly
polluted, but may be acidic, contain traces of lead, pesticides, etc., depending on the
locality and prevailing winds. Contamination occurs when it falls on the roof, col-
lects dirt, dissolves some heavy metals in the case of metal surfaces, and then flows
into storage. Chang et al. (2004) considered roof runoff as a source of nonpoint
pollution. The acidic nature of rainwater will make it react with compounds retained
in or by the roof and cause many elements in roof runoff to leach out. Roof inter-
ception causes enrichment (compared to free-fall rainwater) in virtually all the water
quality parameters (Adeniyi and Olabanji 2005). A number of studies reported from
different parts of the world reveal the prevalence of microbiological and chemical
contaminants in roof-collected rainwater (Yaziz et al. 1989; Forster 1996, 1998;
Ghanayem 2001; Simmons et al. 2001; Lye 2002, 2009; Meera and Ahammed 2006;
Abbott et al. 2006; Sazakli et al. 2007). Several factors influence the quality of roof
runoff which includes characteristics of roof materials, the rainfall event, meteorological factors, chemical properties of the substances and location of the roof (Forster
1996). The objective of the present study was to assess the influence of some of these
factors on the quality of roof-harvested rainwater. The study focussed on the
influence of roofing materials, age of roofs, and characteristics of rainfall events on
the quality of harvested rainwater. For this, harvested rainwater from different roof
materials was analysed during a number of rainfall events. More than 100 water
samples were analysed for different physico-chemical and microbiological water
quality parameters to arrive at conclusions.

2 Materials and Methods

2.1 Site Selection and Roof Catchment Details

Rainwater was harvested both from newly constructed pilot-scale roofs and existing
full-scale roofs of different materials. The pilot-scale roofs were constructed over

the terrace of Environmental Engineering Laboratory block of the Government Engineering College, Thrissur, India. The college campus is situated approximately 6 km from the city and 1 km from a state highway. The roofs were approximately 8.5 m above the ground level. There were no tree branches overhanging the roofs. The roofs were constructed with the following materials: clay tile, asbestos, fibre, tarsheet, galvanised iron (GI) and aluminium. The area of each roof was 2.06 m^2. PVC gutter and downspouts were used in all the roofs. Samples were collected and analysed during three rainfall events during southwest monsoon season (June–August) in the year 2004.

The full-scale existing roofs were located in Nedumbassery, Kochi, India. Four roofs of different materials—concrete, tile, asbestos and GI—were at heights of 3.6, 2.5, 2.5 and 3.0 m, respectively, above the ground, while the areas of the roofs were 12.0, 4.8, 2.6 and 2.6 m^2 respectively. The approximate ages of the roofs were: concrete—5 years, tile—25 years, GI and asbestos—15 years. Samples were collected during nine rainfall events during summer and southwest monsoon in the year 2004.

2.2 Sample Collection and Storage

For analysing the quality of roof runoff from pilot-scale and existing roofs, samples were taken from plastic storage tanks during rainfall events. The storage tanks were fitted with the first-flush device to discard the first 2 mm of rainfall. Samples were taken in pre-cleaned polyethylene containers. Great care was taken to ensure the integrity of samples. For heavy metal analysis, the samples were acidified using nitric acid. Samples were stored at 4 °C if the analysis was not done immediately after sampling. Rainwater was also collected from the open (free fall) by placing plastic containers on the terrace of the Environmental Engineering Laboratory block.

2.3 Water Analysis

The analytical determination of the selected physico-chemical parameters was carried out within the holding time of each parameter according to Standard Methods (APHA 1998). pH was measured by electrometric method using pH meter, turbidity by nephelometer and conductivity by water quality analyser. Heavy metals, lead and zinc, were determined by atomic absorption spectrophotometer (ECIL 4139, India). Samples were analysed in the Environmental Engineering Laboratory of Government Engineering College, Thrissur, India. All samples were tested for total coliforms by a multiple-tube fermentation method (most probable number method) by employing Lauryl tryptose broth (HiMedia Laboratories, Mumbai). The inoculated samples were incubated at 37 °C for 48 h. The result is expressed as most probable number per 100 mL (MPN/100 mL). The test was conducted in accordance with the techniques described by American Public Health Association (APHA 1998).

2.4 Statistical Analysis

The data generated were subjected to statistical analyses to find the mean, significant level, and Pearson's correlation coefficient. Pearson's correlation coefficient was estimated to assess the degree of association between microbiological quality and other chemical parameters in free-fall and in various runoffs from different roofs. The Pearson's correlation coefficients of different parameters in various roof runoffs with a dry period, and amount of rainfall were also determined. The statistical significant level was set at $\alpha = 0.05$.

3 Results and Discussion

Mean values of different water quality parameters in runoff collected from three pilot-scale roofs and three full-scale existing roofs are given in Table 1. Comparison with Indian drinking-water quality guideline values indicates a few departures for parameters like pH and bacteriological quality for harvested water from newly constructed pilot-scale roofs, and pH, turbidity, iron and bacteriological quality for existing roofs. This table also shows that water quality is influenced by the roof materials since great variation was observed between the different roof

Table 1 Mean concentration for different water quality parameters for full-scale and pilot-scale (new) roofs

Parameters	Full-scale existing roofs			Pilot-scale roofs		
	GI	Asbestos	Tile	GI	Asbestos	Tile
pH	6.59	6.73	6.23	6.11	6.44	5.79
Turbidity (NTU)	15.50	10.90	12.40	3.20	4.80	3.20
Conductivity at 25 °C (μS/cm)	187.90	119.16	54.88	94.48	45.94	15.54
DO (mg/L)	4.50	5.50	4.60	7.50	7.30	7.30
Iron (mg/L)	0.91	0.32	0.27	0.14	<0.05	<0.05
Calcium (mg/L)	65.40	45.50	23.50	<0.10	9.70	<0.10
Zinc (mg/L)	1.33	0.08	0.10	0.92	0.03	0.03
Total coliforms (MPN/100 mL)	7300	1258	2536	120	410	3070

The values represent mean of nine and three rainfall events in the case of existing roofs and pilot-scale roofs, respectively. In both cases, 2 mm of first flush was discarded before sampling

materials. Bacteriological contamination was found for most of the runoff samples from new and old roofs. In general, most parameters showed a significantly lower value in the case of new (pilot-scale) roofs compared to old roofs, indicating the influence of age of the roof on the harvested water quality.

Table 2 presents the mean values of enrichment factor (that is the concentration in roof runoff to that in free fall) for different water quality parameters for six pilot-scale roofs. Enrichment factors were calculated for four existing roofs also (Fig. 1). The enrichment factor ranged between ∼0 (for chloride in new tile roof) and 46 (for Zn in new GI roofs). Enrichment factor indicates the role of roof as a

Table 2 Enrichment factors of different quality parameters for pilot-scale roofs

Parameters	GI	Asbestos	Tarsheet	Aluminium	Tile	Fibre
Alkalinity	0.70	1.60	0.80	0.70	0.70	1.00
Turbidity	1.78	2.70	2.30	0.28	1.80	1.50
Hardness	0.90	1.67	0.81	0.95	1.14	1.10
Chloride	1.00	2.01	3.51	1.00	∼0.00	0.50
Conductivity	2.44	1.18	0.55	0.44	0.40	0.49
DO	1.01	0.97	1.05	1.03	0.97	0.98
Nitrate	5.28	7.56	5.56	6.22	3.11	6.56
Sulphate	0.42	0.90	0.37	0.47	0.58	0.32
Sodium	0.14	0.24	0.16	0.14	0.08	0.16
Calcium	<0.03	2.72	1.00	<0.03	<0.03	<0.03
Zinc	46.00	1.50	3.00	1.50	1.50	1.50

The values are mean of three rainfall events

Fig. 1 Enrichment factors of different parameters for full-scale roofs

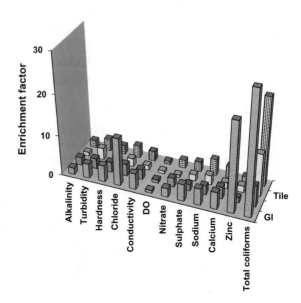

source or sink of pollutants. A value greater than 1 indicates that the roof is acting as a source of pollutants either due to weathering or deposition during the antecedent dry time, and a value less than 1 shows that it is acting as a sink by retaining the pollutants. For pilot-scale roofs, a value around 1 was observed for parameters like alkalinity, hardness and DO indicating very little influence of roofs on these parameters whereas for parameters like sulphates and sodium the value was less than 1. The enrichment factor for most of the parameters in runoff from old roofs was greater than 1 and was higher than that for new roofs. This indicates the role of roofs in contributing various contaminants to the roof runoff by weathering and/or by deposition. A high enrichment factor of 46 was observed for zinc in new GI roof, though the concentration did not exceed the drinking-water guideline value. Variation reflects differences in roofing materials and its treatment, air quality of region, characteristics of precipitation, etc. (Forster 1996; Wu et al. 2001; Chang et al. 2004). The enrichment factor for all physico-chemical parameters except DO, zinc and nitrate was greater for old roofs when compared to new (pilot-scale) roofs. In general, the mean concentration and enrichment factors were higher for most of the parameters for old roofs. The higher values of enrichment factor for old roofs can be attributed to increased dry deposition and/or weathering of roof material. Adeneyi and Olabanji (2005) also found the enrichment factor for most of the physico-chemical parameters in old roofs higher when compared to relatively new roofs.

The Pearson correlation coefficients for a dry period or amount of rainfall with various water quality parameters in free-fall rain and runoff were calculated (data not shown). In general, a positive relation was observed between concentrations of various substances in roof runoff and dry period, and a negative relation with the amount of rainfall. A high positive correlation ($r > 0.8$) was found between total coliforms and dry period in all runoff except asbestos. The positive correlation between the water quality parameters and dry period is presumably due to increased dry deposition and dissolution of roofing material during an antecedent dry time when the higher roof temperature accelerates chemical reaction and organic decomposition of materials that could have accumulated on rooftops. The negative correlation between rainfall amount and quality parameter may be due to the diluting effect of rainwater.

A comparison of the concentration of different parameters found in the present study with the values reported in the literature from different parts of the world showed that the values in free-fall samples found in the present study are comparable with the values reported, with the exception of heavy metals. Heavy metals were not detected in free-fall samples in the present study. In reality, rainfall ion concentrations obtained from different studies are not practically comparable due to sample size, sample frequency, method of analysis, and detection limits. Concentration in roof-intercepted rainwater from different roof surfaces showed wide variation with location. There are several factors which influence the quality of roof runoff, viz., roof material, physical boundary condition of roofs, characteristics of rainfall, other meteorological factors, and proximity to pollution sources. Due to the influence of these numerous factors, high variability is observed among different studies. Another significant factor which affects the concentration of various parameters is the first-flush device. A first-flush device was used in the present study.

4 Conclusions

Analysis of more than 100 water samples for different water quality parameters revealed that the quality of roof-harvested rainwater often did not meet drinking-water quality guidelines especially with regard to microbiological quality and pH. The results showed high variability in characteristics between different roofing materials with significant differences in the mean values of concentration for different water quality parameters. Enrichment factor (that is the ratio of concentration in roof runoff to that in free fall) was also found different for different roofing materials and its value was >1 for most of the parameters indicating the significant effect of roofing material on the concentration of various parameters by weathering and/or by deposition. The enrichment factor was also found to increase with age of the roof. Correlation analysis revealed a positive correlation between the concentration of different water quality parameters and the dry period, and a negative correlation between some water quality parameters and the amount of rainfall. Thus, roofing material and characteristics of rainfall event were found to have a significant influence on the quality of roof-harvested rainwater.

References

Abbott SE, Douwes J, Caughley BR (2006) A survey of the microbiological quality of roof-collected rainwater of private dwellings in New Zealand. In: Proceedings of water 2006, Auckland, New Zealand, pp 1–24

Adeniyi IF, Olabanji IO (2005) The physico-chemical and bacteriological quality of rainwater collected over different roofing materials in Ile-Ife, southwestern Nigeria. Chem Ecol 3: 149–166

American Public Health Association (APHA) (1998) Standards methods for examination of water and waste water. American Water Works Association and Water Pollution Control Federation, Washington DC., USA

Chang M, McBroom MW, Beasley RS (2004) Roofing as a source of nonpoint water pollution. J Environ Manage 73:307–315

Förster J (1996) Patterns of roof runoff contamination and their potential implications on practice and regulations of treatment and local infiltration. Water Sci Technol 33:39–48

Förster J (1998) The influence of location and season on the concentration of microions and organic trace pollutants in roof runoff. Water Sci Technol 38:83–90

Ghanayem M (2001) Environmental considerations with respect to rainwater harvesting. In: Proceedings of 10th international rainwater catchment systems conference. Weikersheim, Germany. International Rainwater Catchment Systems Association, pp 167–171

Lye DJ (2002) Health risks associated with consumption of untreated water from household roof catchment systems. J Am Water Res Assoc 38(5):1301–1306

Lye DJ (2009) Rooftop runoff as a source of contamination: a review. Sci Total Environ 407:5429–5434

Meera V, Ahammed MM (2006) Water quality of rooftop harvesting systems: a review. J Water Supply Res Technol Aqua 55:257–268

Sazakli E, Alexopoulos A, Leotsinidis M (2007) Rainwater harvesting, quality assessment and utilization in Kefalonia Island, Greece. Water Res 41:2039–2047

Simmons, Hope V, Lewis G, Whitmore J, Gao W (2001) Contamination of roof-collected rainwater in Auckland, New Zealand. Water Res 35:1518–1524

Wu C, Huizhen W, Junqi L, Hong L, Guanghu, M (2001) The quality and major influencing factors of runoff in Beijing's urban area. In: Proceedings of 10th conference of International Rainwater Catchments Systems Association. Weikersheim, Germany, International Rainwater Catchment Systems Association, pp 13–16

Yaziz MI, Gunting H, Sapari N, Ghazali AW (1989) Variations in rainwater quality from roof catchments. Water Res 23:761–765

Hydro-physico-chemical Grouping of Cachar Paper Mill Effluents in Assam Using Multivariate Statistical Model

Sangeeta Dey, Manabendra Dutta Choudhury and Suchismita Das

Abstract Effluents from the paper mill are highly toxic and are one of the major sources of aquatic pollution. Their toxic nature is derived from the presence of several naturally occurring and xenobiotic compounds, which are formed and released during various stages of papermaking. In the present study, mill effluents are collected from three different stations around the Cachar paper mill (CPM) in Assam and samples are analyzed for 17 parameters such as appearance, color, odour, conductivity, total dissolved solids, pH, total alkalinity, hardness (total and Ca), chloride, nitrite, BOD and COD along with the heavy metals As, Cu, Cd, Cr, Fe, Pb and Zn. The results are compared with water quality standards prescribed by World Health Organization. Multivariate statistical analyses are performed on data sets obtained to gain a better understanding of the nature of each parameter in the sampling stations. The results show variation amongst stations that may be due to spatial variations amongst them. The results from this study can be used for water quality monitoring purposes.

Keywords Paper mill effluent · Physico-chemical parameters · Multivariate statistics

1 Introduction

Pulp and paper mill effluents are considered as hazardous waste due to the presence of multiple contaminants of chemical and/or biological origin, which are predominantly derived from the raw materials required for its production (Lacorte et al. 2003). There are reports from different parts of India that discharge of pulp and paper mill effluents into water bodies, agricultural fields and irrigation channels, jeopardize the health of human and aquatic life (Singh et al. 2009; Rana et al. 2004).

S. Dey · M. D. Choudhury · S. Das (✉)
Aquatic Toxicology and Remediation Laboratory, Department of Life Science
and Bio Informatics, Assam University, Silchar 788011, India
e-mail: drsuchismita9@gmail.com

© Springer International Publishing AG 2018
A. K. Sarma et al. (eds.), *Urban Ecology, Water Quality and Climate Change*,
Water Science and Technology Library 84,
https://doi.org/10.1007/978-3-319-74494-0_16

203

Cachar Paper Mill (CPM) in South Assam, India, discharges its effluents through small channels into rivers, *Dolasor* and *Barak*. However, there is limited data available on the quality of these effluents. In addition to evaluation of the physico-chemical quality of these effluents, there is a need for a systematic approach for hydro-physico-chemical grouping of these effluents in the context of effluent quality and their potential to perturb the receiving aquatic ecosystem. Multivariate statistical analyses (MVA), such as, factor analysis (FA), principal component analysis (PCA) and cluster analysis (CA), had been employed by various researchers as effective tools for monitoring water qualities that helped in the past in formulating regulatory decisions (Kim et al. 2014; Lin et al. 2012). Our main aim is to evaluate the efficiency of different MVA methods to characterize effluents from CPM and group them into different hydro-physico-chemical categories. In the present work, we have applied FA, PCA and CA to evaluate spatial variations in the water quality of effluent discharge channels of CPM.

2 Materials and Methods

2.1 Study Site

CPM is situated in the southern Assam by the side of the Barak River in the Hailakandi district of Assam on the National Highway No. 57 between Silchar and Guwahati (24° 51′ 10.3″N, 92° 35′ 33.8″E). The mill is located at a distance of 25 km from Silchar (Fig. 1). The main raw material used in the mill is mostly bamboo and the sulphate cooking process is used for pulping. The treatment of effluents comprised both primary and secondary (biological) treatments, prior to its discharge into the *Barak* and *Dolasor* Rivers. The mill produced bleached (elemental chlorine and H_2O_2) pulp for a production of about 103,155 tonnes/year of paper during the year 2006–2007 (http://www.hindpaper.in/mills/cachar.htm).

2.2 Sampling

Effluent samples has been collected at 3 different sites from 3 different outlets such that Station 1 (24° 51′ 47.9″N 92° 36′ 20.8″E) is at a distance of 1.6 km, Station 2 (24° 51′ 09.8″N 92° 35′ 46.7″E) at a distance of 0.085 m and Station 3 (24° 51′ 17.0″N 92° 35′ 55.0″E) is at a distance of 0.4 km from the paper mill (Fig. 1). Station 1 is near the River *Barak*, while, Station 2 and 3 are near the River *Dolasor*. The effluent samples are collected in sterile plastic bottles. Collected samples are transported to the laboratory for physico-chemical analysis.

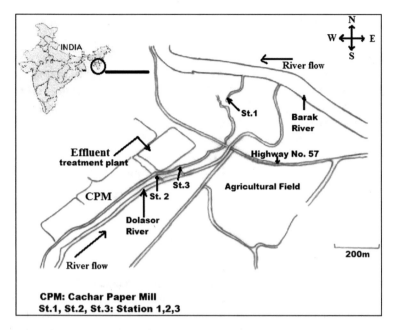

Fig. 1 Study site showing effluent collection Stations 1, 2 and 3

2.3 Physico-chemical Analysis of Effluent

The pH and conductivity are measured by a Metller Toledo pH metre and Systronic conductivity metre respectively. Biological oxygen demand (BOD), chemical oxygen demand (COD), total dissolved solids (TDS) and electrical conductivity (EC) are estimated as per APHA (2005). The total alkalinity, total hardness, free chlorine, chloride, sulphite and nitrite are estimated by respective water estimation kits (procured from Himedia). Water samples for heavy metal analyses are filtered and digested with 10 mL of concentrated analytical grade nitric acid to 250 mL of the water sample. The solutions are evaporated in a crucible to approximately 5 mL, then filtered into 20 mL standard flask and made up to the mark with distilled water. The water extract is analyzed for metals such as arsenic (As), cadmium (Cd), chromium (Cr), copper (Cu), iron (Fe), lead (Pb) and zinc (Zn). Analysis of the heavy metal content of water samples is carried out with a Graphite Furnace Atomic Absorption Spectrometry (GF-AAS), model Vario 6, Analytikjena, performed in the Bose Institute, Main Campus, Kolkata.

2.4 Data Analyses

The multivariate analyses of the data are performed by CA, PCA and FA and these tools are standardized through z-scale transformations due to wide differences in data dimensionality (Liu et al. 2003). CA is performed on normalized data set by Ward's method using squared Euclidean distances as a measure of similarity. PCA is designed to transform the original variables into new, uncorrected variables, called principal component, which are linear combinations of original variables. PCA is followed by FA. The purpose of FA is to extract a lower dimensional linear structure from the data sets. All the statistics are performed using SPSS 18 statistical software for windows.

2.5 QC/QA

For quality control and quality assurance, the metal standards (Merck, Germany) are used for the calibration and quality assurance for each analytical batch. Analytical data quality of metals are ensured with repeated analysis of quality control samples ($n = 3$) and the results has been found within the certified values. The detection limits for As, Cd, Cr, Cu, Fe, Pb and Zn, are 0.4, 0.007, 0.1, 0.19, 0.1, 0.08 and 0.002 $\mu g\ L^{-1}$ respectively.

3 Results and Discussion

Surface water quality is vulnerable, due to a multitude of contaminants released from industrial, domestic/sewage, and agricultural effluents, which finally compromise its pristine composition (Qadir et al. 2008). Wastewater/effluent comprises the main pollution sources affecting water quality and hence, analysis of the physico-chemical parameters of the wastewater is the best means of assessing its impact. Further, the long-term management of water quality requires a fundamental understanding of the analysis of the polluting sources, which can provide the basic data for monitoring purposes. Table 1 depicts the physico-chemical characteristics of the effluent collected from 3 stations near a paper mill in Assam. The effluent shows dark brown color and is against the international standard limit for discharge into natural water bodies (WHO 2006). The dark brown color of the effluent is probably due to the presence of lignin compound in the raw material used for paper production (Chandra et al. 2007). However, Livernoche et al. (1983) reported that the color of the effluent is pH sensitive and the effect of the pH on the color of the effluent is reversible. The maximum value of pH of the effluent sample is recorded as 7.6 at Station 2 and the minimum value of pH is recorded as 7.43 at Station 1. The total alkalinity, total hardness, Ca-hardness for Station 1, 2 and 3 are found in

Table 1 Physico-chemical parameters of paper mill effluent sample collected from 3 different stations of effluent discharge sites

Parameters	Station-1	Station-2	Station-3	WHO (2006)[a]
Color	Reddish brown	Reddish brown	Reddish brown	Colourless
Odour	Pungent	Pungent	Pungent	Odourless
Electrical conductivity (EC) (µS)	2860 ± 1.00	2860.33 ± 0.57	2829.66 ± 0.57	NA
TDS (mg/L)	1830.4 ± 0.91	1830.4 ± 0.40	1811.2 ± 0.86	NA
p^H	7.43 ± 0.44	7.61 ± 0.54	7.53 ± 0.55	NA
Total alkalinity (mg/L)	648 ± 1.00	696.33 ± 2.51	598.33 ± 0.45	NA
Total hardness (mg/L)	81.66 ± 0.39	66.7 ± 2.45	80 ± 2.64	NA
Ca-hardness (mg/L)	1466.64 ± 3.52	466.36 ± 9.30	500 ± 11	NA
Free chlorine (mg/L)	0	0	0	NA
Sulphite (mg/L)	0	0	0	NA
Chloride (mg/L)	600.33 ± 2.5	600.33 ± 2.08	683.33 ± 1.80	NA
Nitrite (mg/L)	5245.67 ± 5.96	3245 ± 1.00	4496 ± 3	3
BOD (mg/L)	150 ± 0.57	210 ± 0.28	200 ± 0.03	NA
COD (mg/L)	1760 ± 0.50	1440 ± 0.57	1760 ± 1	NA
Cu (mg/L)	5.4 ± 0.35	3.5 ± 1.23	2.4 ± 0.15	2
Cd (mg/L)	0.186 ± 0.01	0.812 ± 0.01	1.28 ± 0.02	0.003
Cr (mg/L)	37.17 ± 1.18	29.05 ± 0.18	47.42 ± 0.51	0.05
Fe (mg/L)	416.31 ± 1.14	446.97 ± 0.29	567.063 ± 0.49	NA
Zn (mg/L)	31.59 ± 0.062	35.28 ± 0.064	44.46 ± 0.092	NA
Pb (mg/L)	4.62 ± 0.002	2.1 ± 0.005	2.73 ± 0.004	0.01
As (mg/L)	3.42 ± 0.007	4.86 ± 0.05	0.69 ± 0.001	0.01

[a]Drinking water guidelines of World Health Organization (2006), NA—guideline value not available

the range of (598–696 mg/L), (66.7–82 mg/L) and (466–1466 mg/L) respectively. The concentration of BOD, COD, TDS and EC far exceeds the regulatory value (WHO 2006). The elevated level of some these parameters may be due to the utilization of various cellulose-based raw materials and chemicals used during various manufacturing process (Antony et al. 2012). Other parameters such as chloride and nitrite are 600 mg/L and 5245 mg/L for Station 1; 600 mg/L and 3245 mg/L for Station 2; 683 and 4496 mg/L for Station 3, respectively. The high BOD and COD value of the mill effluents suggest the presence of organic and inorganic pollutants in higher quantities (Devi et al. 2011). The average mean value of various heavy metals range from 0.186 to 1.28 mg Cd/L, 2.4–5.4 mg Cu/L,

416–567 mg Fe/L, 29–47.4 mg Cr/L, 2–4.6 mg Pb/L, 31.6–44.4 mg Zn/L and 0.69–4.86 mg As/L in the three stations. Iron content is relatively higher in concentration compared to others heavy metals. High levels of metals/metalloid can have a potential adverse impact on biota.

Understanding which parameters in effluents form the dominating attributes and the spatial variations, if any, in the effluent quality collected from three different sampling stations are the key goals of the multivariate statistical analysis that has been carried out in this study. CA is particularly helpful in assembling parameters based on their interdependence or similarity and hence each parameter in a cluster shall be a homogenous aggregation (McKenna 2003). Cluster hierarchy starts with the most similar pair of parameters and forms higher clusters step by step. The results of such hierarchical clusters give rise to dendrogram that provides a visual synopsis of the clustering process (Shrestha and Kazama 2007). Figure 2 depicts the dendrogram or CA performed on different water quality parameters on Station 1, Station 2 and Station 3. For Station 1, all the 17 water quality parameters are grouped into 3 statistically distinct groups (A, B, C) at a cutoff of (Dlink/Dmax) $\times$ 100 < 10 (Fig. 2a). Group A links total hardness, alkalinity, conductivity, Ca-hardness, pH, chloride to metal/metalloid Cr, Cd, Cu and As. Group B includes Zn, Pb and Fe, while group C includes BOD, COD, nitrate and TDS. For Station 2, two statistically distinct groups (A and B) at a cutoff of (Dlink/Dmax) $\times$ 100 < 10 (Fig. 2b) are formed. Group A links nitrate, BOD, total hardness, alkalinity, Ca-hardness, pH, chloride to Fe, Zn, Cd and Cu. Group B links Pb, As and Cr to TDS, COD and conductivity. For Station 3, three clusters viz., A, B and C are formed at a cutoff of (Dlink/Dmax) $\times$ 100 < 10 (Fig. 2c). Group A has TDS, nitrate, COD, alkalinity, Ca-hardness linked to As, Fe, Cu, Pb, Cr and Cd. In Group B, chloride, Zn and pH are linked; while, in group C, conductivity and BOD are linked. The results from the CA points to the fact that although the 17 water quality criteria are selected and several clusters are formed in each station, they have very little heterogeneity between them as all the possible clusters are within (Dlink/Dmax) $\times$ 100 < 25. This is quite pertinent as the effluents are from the same mill. However, on minimizing the cluster distance to Dlink/Dmax) $\times$ 100 < 10, we have found several clusters. For Station 1, BOD, COD, nitrate and TDS forms the closest related parameters. This implies high organic contents in the wastewater. For Station 2, the closest related parameters include Pb, As, Cr, TDS, COD and conductivity. For Station 3, conductivity and BOD are the closely related parameters. The clustering technique offers a superior understanding of the water pollutants and helps in better decision making. There are other reports where this approach has successfully been applied in water quality monitoring (Parashar et al. 2008; Simeonov et al. 2003; Wunderlin et al. 2001).

PCA/FA are performed on the correlation matrix (data not shown) of rearranged datasets separately for three sampling stations to assume the potential influences and evaluate the compositional patterns of analyzed water samples (Table 2). For Factor analysis, eigenvalues > 1 are chosen as they account for most of the total variance, which is sufficient to give a good idea of data structure (Masoud 2014). For Station 1, high factor loadings are observed for total alkalinity and trace metals

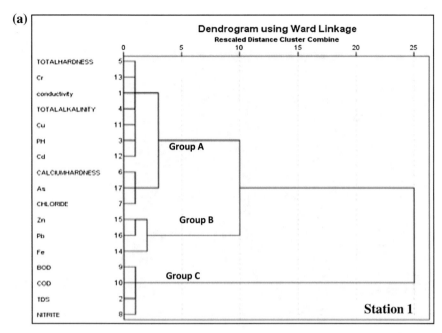

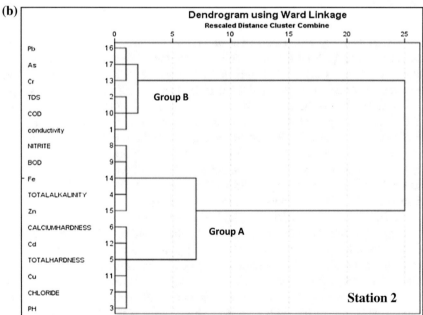

Fig. 2 Dendrogram showing clustering of various physico-chemical parameters of paper mill effluent of Station 1, 2 and 3 using Ward linkage

(c)

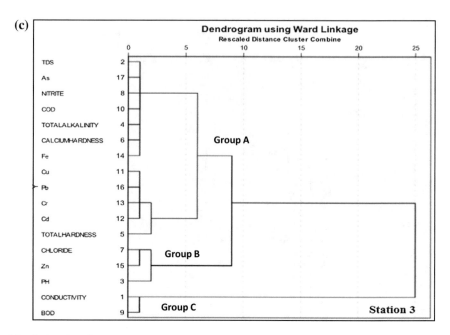

Fig. 2 (continued)

Cu, Fe Zn and Pb, which comprise of PC1 accounting for 92.7% of the total variance. Similarly, total dissolved solid, Ca-hardness, nitrite, BOD and COD are highly correlated in the same station, and comprised of PC2 with 7.2% of the total variance (Table 2a). For Station 2, high factor loadings are observed between pH, total hardness, Ca-hardness, chloride and trace metals Cd and Cu, which comprised of PC1 accounting for 90.7% of the total variance. Similarly, conductivity, total dissolved solids, COD, Cr, Pb and As are highly correlated in the same station, and comprised of PC2 with 9.3% of the total variance (Table 2b). Further, for Station 3, high factor loadings are observed between all but conductivity and BOD, which comprises of PC1, accounting for 85% of the total variance. In PC2 (15% of total variance), high factor loadings are observed only between conductivity and BOD (Table 2c). It has to be noted that Station 2 and 3 are nearer to the mill effluent discharge sites and also close to each other. Further, these two sites also have plenty of scope for dilution with river *Dolasor*. For Station 1, the effluents are accumulated over time in a shallow ditch and due to nearness to river *Barak* has the potential to impact this river water quality. The MVA study points to the fact that the effluents from the three stations are of the same origin but of different composition, probably arising due to the spatial difference in quality. This can variably affect the water quality of both the rivers. PCA/FA has been successfully used in monitoring the spatial difference in water qualities by prior workers (Jung et al. 2016; Wang et al. 2013).

Table 2 Total variance and component matrix (two principal components selected) of Station 1, 2 and 3 for effluent physico-chemical parameters ($n = 9$)

Initial eigenvalues			Rotated Varimax		
	Total	% of variance	Cumulative %	Parameters	PC1	PC2
Station 1 (a)						
1	15.770	**92.766**	**92.766**	Conductivity	0.755	0.655
2	1.230	**7.234**	**100.000**	TDS	0.428	**0.904**
3	8.436E−16	4.962E−15	100.000	pH	0.778	0.629
4	5.608E−16	3.299E−15	100.000	Total alkalinity	**0.812**	0.583
5	3.168E−16	1.863E−15	100.000	Total hardness	0.755	0.655
6	2.365E−16	1.391E−15	100.000	Ca-hardness	0.580	**0.815**
7	1.158E−16	6.811E−16	100.000	Chloride	0.675	0.738
8	8.531E−18	5.018E−17	100.000	Nitrite	0.499	**0.866**
9	6.409E−19	3.770E−18	100.000	BOD	0.327	**0.945**
10	−2.371E−32	−1.395E−31	100.000	COD	0.327	**0.945**
11	−9.736E−18	−5.727E−17	100.000	Cu	**0.826**	0.564
12	−3.795E−17	−2.232E−16	100.000	Cd	0.799	0.601
13	−9.351E−17	−5.501E−16	100.000	Cr	0.755	0.655
14	−1.726E−16	−1.015E−15	100.000	Fe	**0.980**	0.198
15	−2.660E−16	−1.565E−15	100.000	Zn	**0.916**	0.401
16	−4.567E−16	−2.687E−15	100.000	Pb	**0.878**	0.478
17	−5.753E−16	−3.384E−15	100.000	As	0.615	0.788
Station 2 (b)						
1	15.418	90.692	90.692	Conductivity	0.325	**0.946**
2	1.582	9.308	100.000	TDS	0.325	**0.946**
2	9.299E−16	5.470E−15	100.000	pH	**0.957**	0.289
4	4.442E−16	2.613E−15	100.000	Total alkalinity	0.674	0.739
5	3.470E−16	2.041E−15	100.000	Total hardness	**0.939**	0.345
6	2.069E−16	1.217E−15	100.000	Ca-hardness	**0.878**	0.480
7	1.165E−16	6.852E−16	100.000	Chloride	**0.907**	0.421
8	6.722E−17	3.954E−16	100.000	Nitrite	0.754	0.656
9	3.592E−51	2.113E−50	100.000	BOD	0.754	0.656
10	−1.956E−34	−1.150E−33	100.000	COD	0.325	**0.946**
11	−1.344E−18	−7.907E−18	100.000	Cu	**0.924**	0.383
12	−4.151E−17	−2.442E−16	100.000	Cd	**0.855**	0.518
13	−1.493E−16	−8.785E−16	100.000	Cr	0.471	**0.882**
14	−2.369E−16	−1.394E−15	100.000	Fe	0.762	0.648
15	−2.965E−16	−1.744E−15	100.000	Zn	0.614	0.789
16	−4.886E−16	−2.874E−15	100.000	Pb	0.543	**0.840**
17	−7.565E−16	−4.450E−15	100.000	As	0.543	**0.840**

(continued)

Table 2 (continued)

Initial eigenvalues			Rotated Varimax		
	Total	% of variance	Cumulative %	Parameters	PC1	PC2
Station 3 (c)						
1	14.469	85.110	85.110	Conductivity	0.028	**1.000**
2	2.531	14.890	100.000	TDS	**1.000**	0.007
3	7.517E−16	4.422E−15	100.000	PH	**0.896**	−0.444
4	6.286E−16	3.697E−15	100.000	Total alkalinity	**1.000**	−0.028
5	4.537E−16	2.669E−15	100.000	Total hardness	**0.954**	0.301
6	1.506E−16	8.858E−16	100.000	Ca-hardness	**1.000**	−0.028
7	7.877E−17	4.634E−16	100.000	Chloride	**0.956**	−0.294
8	4.699E−18	2.764E−17	100.000	Nitrite	**1.000**	−0.028
9	9.274E−20	5.455E−19	100.000	BOD	0.028	**1.000**
10	−9.348E−22	−5.499E−21	100.000	COD	**1.000**	−0.028
11	−6.266E−20	−3.686E−19	100.000	Cu	**0.988**	0.154
12	−1.734E−17	−1.020E−16	100.000	Cd	**0.996**	0.087
13	−5.628E−17	−3.310E−16	100.000	Cr	**0.984**	0.176
14	−1.547E−16	−9.101E−16	100.000	Fe	**0.999**	−0.051
15	−3.493E−16	−2.055E−15	100.000	Zn	**0.965**	−0.263

Boldface indicates significant factor loading; PC 1 is the principal components 1 and PC 2 is the principal components 2

The result of this study, thus, proves the fact that multivariate statistical methods such as CA and PCA/FA can be applied to infer and interpret water quality data in terms of probable spatial variations, and identify probable pollution sources/factors. Further, the results of this study will be useful in developing suitable long-term monitoring strategies for water qualities this region.

References

Antony A, Bassendeh M, Richardson D, Aquilina S, Hodgkinson A, Law I, Leslie G (2012) Diagnosis of dissolved organic matter removal by GAC treatment in biologically treated papermill effluents using advanced organic characterisation techniques. Chemosphere 86: 829–836

APHA (2005) Standard methods for the examination of water and wastewater, 21st edn. American Public Health Association, AWWA, WPCP, Washington, DC

Chandra R, Raj A, Purohit HJ, Kapley A (2007) Characterisation and optimisation of three potential aerobic bacterial strains for kraft lignin degradation from pulp paper waste. Chemosphere 67:839–846

Devi NL, Yadav IC, Shihua QI, Singh S, Belagal SL (2011) Physicochemical characteristics of paper industry effluents—a case study of South India Paper Mill (SIPM). Environ Monit Assess 177:23–33

Jung KY, Lee KL, Im TH, Lee IJ, Kim S, Han KY, Ahn JM (2016) Evaluation of water quality for the Nakdong River watershed using multivariate analysis. Environ Technol Innov 5:67–82

Kim KH, Yun ST, Park SS, Joo Y, Kim TS (2014) Model-based clustering of hydrochemical data to demarcate natural versus human impacts on bedrock groundwater quality in rural areas, South Korea. J Hydrol 519:626–636

Lacorte S, Latorre A, Barcelo′ D, Rigol A, Malmqvist A, Welander T (2003) Organic compounds in paper-mill process waters and effluents. Trends Anal Chem 22:725–737

Lin CY, Abdullah MH, Praveena SM, Yahaya AHB, Musta B (2012) Delineation of temporal variability and governing factors influencing the spatial variability of shallow groundwater chemistry in a tropical sedimentary island. J Hydrol 432–433:26–42

Liu CW, Lin KH, Kuo YM (2003) Application of factor analysis in the assessment of ground water quality in a Blackfoot disease area in Taiwan. Sci Tot Environ 313:77–89

Livernoche D, Jurasek L, Desrochers M, Dorica J, Veliky IA et al (1983) Removal of color from Kraft mill wastewaters with cultures of white-rot fungi and with immobilized mycelium of *Coriolus versicolor*. Biotechnol Bioeng 25:2055–2065

Masoud AA (2014) Groundwater quality assessment of the shallow aquifers west of the Nile Delta (Egypt) using multivariate statistical and geostatistical techniques. J Afr Earth Sci 95:123–137

McKenna JE (2003) An enhanced cluster analysis program with bootstrap significance testing for ecological community analysis. Environ Model Softw 18:205–220

Parashar C, Verma N, Dixit S, Shrivastava R (2008) Multivariate analysis of drinking water quality parameters in Bhopal. Environ Monit Assess 140:119–122

Qadir A, Malik RN, Husain SZ (2008) Spatio-temporal variations in water quality of Nullah Aik-tributary of the river Chenab, Pakistan. Environ Monit Assess 140:43–59

Rana T, Gupta S, Kumar D, Sharma S, Rana M, Rathore VS, Pereira BMJ (2004) Toxic effects of pulp and paper-mill effluents on male reproductive organs and some systemic parameters in rats. Environ Toxicol Pharmacol 18:1–7

Shrestha S, Kazama F (2007) Assessment of surface water quality using multivariate statistical techniques: a case study of the Fuji river basin, Japan. Environ Model Softw 22:464–475

Simeonov V, Stratis JA, Samara C, Zachariadis G, Voutsa D, Anthemidis A, Sofoniou M, Kouimtzis T (2003) Assessment of the surface water quality in Northern Greece. Water Res 37:4119–4124

Singh A, Sharma RK, Agrawal M, Marshall F (2009) Effects of wastewater irrigation on physicochemical properties of soil and availability of heavy metals in soil and vegetables. Commun Soil Sci Plant Anal 40:3469–3490

Wang Y, Wang P, Bai Y, Tian Z, Li J, Shao X, Mustavich LF, Li BL (2013) Assessment of surface water quality via multivariate statistical techniques: a case study of the Songhua River Harbin region, China. J Hydro-Environ Res 7:30–40

World Health Organization (WHO) (2006) Guidelines for drinking-water quality, vol. 1—Recommendations, 3rd edn. WHO, Geneva

Wunderlin DA, Diaz MP, Ame MV, Pesce SF, Hued AC, Bistoni MA (2001) Pattern recognition techniques for the evaluation of spatial and temporal variations in water quality. A case study: Suquia river basin (Cordoba, Argentina). Water Res 35:2881–2894

Heavy Metal Tolerance Exhibited by Bacterial Strains Sourced from Adyar River

**K. C. Ramya Devi, Krishnan Mary Elizabeth Gnanambal
and Yesupatham Babu**

Abstract A total of five water samples were collected from Adyar River during the months of November 2010–2011 and were analyzed for different physico-chemical parameters like, temperature, pH, DO, Pb, Hg, Ni, Cd, and Cr levels. The results of the study indicate that the temperature was in the range 29–31 °C and pH 7.56–8.18 along the entire sampling time. The temperature and pH range was well within the limits as specified by the Tamil Nadu Pollution Control Board (TNPCB) for Inland surface waters and coastal areas. The Dissolved Oxygen levels were found to be very low, which did not exceed 2 mg/L, indicating a very poor water quality. The concentration of Pb was found (greater than 10 mg/L) to have crossed the limit of tolerance as prescribed by TNPCB (Pb-0.1 mg/L). Around 30 bacterial isolates from the Adyar waters were adapted for metal tolerance test by amending with heavy metals at different concentration (5–50 mg/L). The results of bioremediation/biosorption process indicated that, out of 30 strains, 15 strains reduced/absorbed lead effectively. Pre and post-treated waters were analyzed by AAS and confirmed by ICP-OES. Biochemical characterization of the potent strain, "L" indicated that it may be Actinomycetes.

Keywords Heavy metals · River waters · Bacteria · Bioremediation

1 Introduction

River water studies have received wide attention due to the need to understand the weathering, hydrological, seasonal, and anthropogenic factors which influence the water quality. Urban rivers many have been associated with poor water quality and

K. C. Ramya Devi · K. Mary Elizabeth Gnanambal (✉)
Department of Biotechnology, Sri Ramachandra Medical College and Research Institute,
Deemed University, Porur, Chennai 600 116, India
e-mail: drelizabethrajesh@sriramachandra.edu.in

Y. Babu
Department of Zoology, Bishop Heber College, Trichy 620 017, India

© Springer International Publishing AG 2018
A. K. Sarma et al. (eds.), *Urban Ecology, Water Quality and Climate Change*,
Water Science and Technology Library 84,
https://doi.org/10.1007/978-3-319-74494-0_17

215

there has been a serious concern of discharging untreated domestic and industrial waste into the water course. High population density in Chennai increases the pressure on the drainage system and sewage treatment plant so that proper treatment facilities are to be implemented (Venugopal et al. 2009). Knowledge on the distribution of pollution, in particular, heavy metals in the river environment is important in studying the aquatic pollution, since such elements can be toxic even in traces. Heavy metals are biologically nondegradable and may finally pass on to man through food web. Also different studies have revealed that the presence of toxic heavy metals like Fe, Pb, Hg reduce soil fertility and agricultural output (Lokhande and Kelkar 1999). There has been a great importance in the study of heavy metal distribution relative to aqueous ecosystem (Thomas and Jaquet 1976).

Among the microorganisms, bacteria, yeast, and protozoa are generally the first categories to be exposed to heavy metals present in the environment. Hence, microorganisms have acquired a variety of mechanisms for adaptation to the presence of toxic heavy metals. Among the various adaptation mechanisms, metal sorption, mineralization, uptake and accumulation, extracellular precipitation and enzymatic oxidation or reduction to a less toxic form, and efflux of heavy metals from the cell has been reported (Mergeay 1991; Hughes and Poole 1991; Nies 1992; Urrutia and Beveridge 1993; Joshi-Tope and Francis 1995). Heavy metal resistant microorganisms have a significant role in wastewater treatment system. The detoxifying ability of these resistant microorganisms can be used for bioremediation of heavy metals in wastewater. The present study, therefore, is aimed at analyzing the presence of heavy metals in the river Adyar which meanders through Chennai and finding out effective bioremediation strategies toward the alleviation of heavy metal pollution.

2   Materials and Methods

2.1   Site of Collection and Sampling of Water

The study area lies between Kanchipuram district and Chennai city of Tamil Nadu, India. River Adyar starts from Malaipattu tank (80.00° latitude and 12.93° longitude) near Manimangalam village, Sriperumbathur taluk, at about 15 km west of Tambaram near Chennai and enters into the Bay of Bengal near Adyar. Water samples were collected once in every month from November 2010 to 2011 at five different locations namely Nandambakkam (Station I), Saidapet Parsan Nagar (Station II), Nandanam (Station III), Adyar Island (Station IV) and Foreshore Estate Beach (Station V) as shown in Fig. 1.

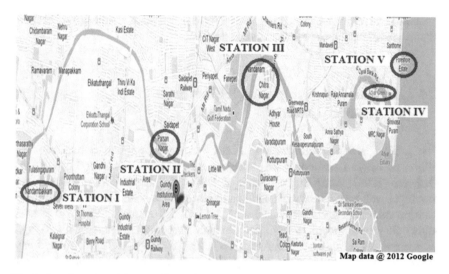

Fig. 1 Location map of the sampling points across Adyar river

2.2 Physicochemical Parameters and Heavy Metal Analysis

Physico-chemical parameters like temperature, pH and dissolved oxygen (DO) have been studied. Temperature and pH were measured using thermometer and digital pH meter respectively and DO was determined using Winkler's method (1888). The concentration of heavy metals like Pb, Hg, Ni, Cr, and Cd was estimated using an Atomic Absorption Spectroscopy (AAS) and Inductively Coupled Plasma Optical Emission Spectrometry (ICP-OES) based on the procedure described by APHA (1998).

2.3 Isolation of Bacteria from Water Sampling and Metal Tolerance Tests

Water sample was collected under aseptic condition in sterilized bottles from Station II (Saidapet Parsan Nagar), Adyar river and plated onto peptone yeast extract (PYE) medium by conventional spread plate method. Plates were incubated at 37 °C for 24 h and colonies were randomly picked, isolated, and purified and stored in labeled slants at 4 °C until usage. The bacterial cultures were adapted to metal tolerant test with metals like Pb, Hg, Ni, Cd, and Cr following the methodology of Uzel and Ozdemir (2009) with slight modification. The bacterial cultures were point-inoculated on PYE plates supplemented individually with five different metals (Pb, Ni, Hg, Cr, and Cd) in the concentration range of 1–50 mg/L. The inoculated plates were incubated at 37 °C for 2 days and after incubation, the colonies from metal-containing plates were observed and well grown bacterial strains were selected for bioremediation process.

2.4   Protocols for Heavy Metal Adaptation by Bacterial Cultures

The isolated bacterial colonies were adapted for metal sorption or removal capabilities on heavy metal containing water as per the modified method of Uzel and Ozdemir (2009). Accordingly 10 mL of PYE-based Broth was prepared and a loopful of well grown heavy metal degrading bacteria was inoculated and incubated on a microbiological shaker at 160 rpm at 37 °C for overnight. Five mL of overnight culture was inoculated into half strength PYE broth amended with 50 mg/L Pb acetate and incubated on shaker at 160 rpm for 5 days at 37 °C. After 5 days of incubation, the culture was centrifuged at 8000 rpm for 15 min and supernatant was collected and the metal concentration in the treated and control supernatants were analyzed using ICP-OES.

2.5   Identification and Characterization of Potent Strain

Strains possessing potent tolerance were further characterized based on biochemical features. Biochemical tests including Gram staining, IMViC, Triple Sugar Iron Agar, urease, catalase, and motility tests were performed.

3   Results and Discussion

The results of the analysis of various physical (Temperature) and chemical (pH, DO) parameters from water samples of Adyar river at five different locations mentioned above are shown in Figs. 2, 3 and 4, respectively. Maximum temperature was recorded at Station II (34 °C) and the minimum was observed to be 27 °C at Station V during the months of January and November 2011, respectively. Rajendran et al. (2004) recorded the temperature of Adyar river to be ranged between 26.0 and 32.6 °C, the maximum during May (32.6 °C) and minimum (26.0 °C) during December. It is opinioned that fluctuations in the river temperature usually depend on the season, sampling time as well as the temperature of the effluents pouring into the river (Koshy and Nayar 1999). The pH values of water samples during the period of investigation were in the alkaline range. The pH of water samples ranged from 7.11 to 8.27. In Fig. 3, Station V recorded higher pH values than the rest of the stations, during the entire study period, which may be associated with strong buffering capacity of the sea water. However, it should be

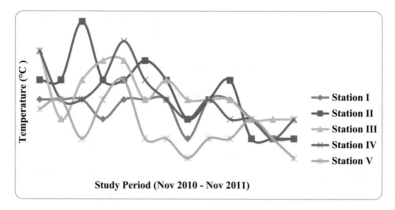

Fig. 2 Seasonal variation of temperature at various points across Adyar river

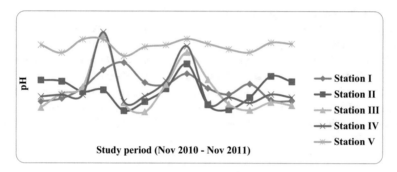

Fig. 3 Seasonal variation of pH at various points of Adyar river

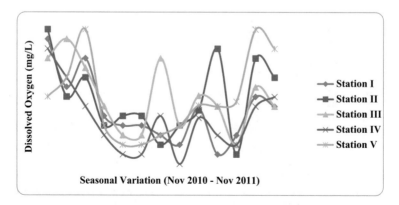

Fig. 4 Seasonal variation of dissolved oxygen level at various points across Adyar river

observed that parameters like temperature and pH were well within the specified limits indicated by the Tamil Nadu Pollution Control Board (TNPCB) for Inland surface waters. In Fig. 4, the DO concentration appears to be very low not even exceeding values of 2 mg/L. Shanmugam et al. (2006) had done extensive works on Adyar waters and concluded that the lower levels of DO are attributed mainly due to high nutrient loading and disposal of other oxygen-demanding substances.

Metals enter into river water from a variety of sources, such as chemical weathering of rocks and soils, dead and decomposing vegetation and animal matter and from anthropogenic sources, such as discharge of various domestic and industrial effluents (Nurnberg 1982). Adyar river is the starting point for a new study on quality of water being taken up by the Anna University in 2007 (Prabhahar et al. 2011). In addition, works on the flow of various effluents into the river Adyar were done at length by Venugopal et al. (2009) and commented that domestic waste discharges directly directed into the river water results in decline of the river velocity to almost stagnation in urban areas. The quality of water with regard to the concentration of heavy metals (Pb, Hg, Ni, Cd, and Cr) is assessed in the current study and the seasonal variation of the same is shown in Fig. 5. However, as a rule, Pb is usually found in low concentration in natural waters because Pb containing minerals are less soluble in water. Hence, the concentration of Pb in natural water increases mainly through anthropogenic activities as already quoted by Venugopal et al. (2009) from Adyar waters with values of 0.035–0.658 mg/L. But, in the present investigation, almost all stations (except Station V) showed that the concentration of Pb was found to be very high that it exceeded the maximal detectable limits (greater than 10 mg/L) during November–December 2010 period which may create health hazards.

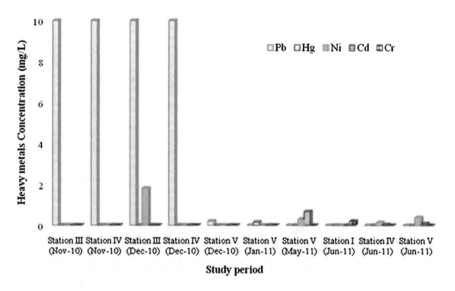

Fig. 5 Seasonal variation of heavy metals at various points across Adyar river

As a result of fresh rainwater input after December the concentrations of Pb became nil till the month of April 2011 at all the stations and thereafter a slow increase witnessed only at Station IV (0.33 mg/L) and V (0.71 mg/L) during the month of June 2011. In the same way, Venugopal et al. (2009) found that Pb Concentration during September 2005 was recorded to be ranged as 0.035–0.658 mg/L and for the month of February 2006, it was from 0.002–0.564 mg/L. Also, as a result of monsoon, the concentration of Pb was considerably decreased (0.05 mg/L) so that it came in the bracketed concentration as indicated in the prescribed permissible limits (Shanmugam et al. 2006) which is greatly in support of the current study. All other metals (Hg, Ni, Cd, and Cr) were found to be within the permissible limits of Tamil Nadu Pollution Control Board (TNPCB) at all the stations at all months. Similarly Kar et al. (2008) assessed heavy metal pollution in surface water of river Ganga and they reported significant seasonal changes in the concentration of Fe, Mn, Cd, and Cr. Whereas, the same study is also indicative of insignificant variations in the concentration of heavy metals like Pb, Ni, Cu and Zn with respect to seasons. According to a study by Shanmugam et al. (2006), it was evident that the concentration of Pb, Ni, Cd and Co appeared to be very high (0.5, 1, 0.04, and 0.09 mg/L) during September 2002 and exceeded the maximum permissible limit indicated by Moore (1991) and National Environmental Board Report (1994). Also in 1993, Joseph and Srivastava found that heavy metals (Pb, Ni, Cd, Cr, and Zn) in water and sediments in Adyar estuary was much higher (0.23, 0.08, 0.03, 0.15, and 0.11 mg/L) than the limits.

The wastewater coming from domestic and industrial sources is one of the favorable points where the microorganisms can develop resistance to heavy metals. The presence of small amount of heavy metals in the wastewater induces the emergence of heavy metal resistant microorganisms (Rajbanshi 2008). So we proposed during the commencement of our study, that it is logical to pick up bacteria from Station II, which is in the midst of the city and closer to an industrial estate, so that it might contain some potent metal resistant bacteria, if there could be a discharge of minute levels of heavy metals from anthropogenic sources. With this notion in mind, the present study involved isolation of more than 30 bacterial strains from Station II and the morphological characteristics of the same are indicated in Table 1. As we looked up for our previous experiments which indicated that Pb concentration was high at certain months, we mainly focused on bioremediation of Pb. However, during subsequent sample collection, we found that there are no significant levels of heavy metals including Pb. Hence, we amended the water with known quantities of heavy metals for bioremediation process. This is the first study to isolate the bacteria for bioremediation of heavy metals from Adyar river. The bioremediation studies showed that the bacterial isolates possess effective Pb tolerance capabilities within in a short period of time. Almost 15 bacterial isolates, a few to mention, B, C, L, M, Q, R, S, V, Y, etc., showed tolerances to Pb as observed in metal tolerance test. The growth of the bacteria during acclimation tests is seen in Fig. 6.

Table 1 Morphological
identification of bacterial
isolates

Strain name	Morphology
A	Small, more or less round, yellowish gray
B	Small, curve, gray
C	Small, curve, gray
D	Very small, white, round
E	Small, round, gray
F	Large, round, yellowish gray
G	Small, more or less round, gray
H	Small, more or less round, gray
I	Small, curve, gray
J	Very small, round, greay
K	Small, white, more or less round
L	Small, curve, gray
M	Large, round, white
N	Small, more or less round, gray
O	Very small, round, gray
P	Large, more or less round, gray
Q	Very small, round, white
R	Small, round, light gray
S	Large, round, white
T	Small, more or less round, white
U	Very small, round, yellow
V	Small, round, gray
W	Small, more or less, gray
X	Large, curve, white
Y	Very small, round, yellow
Z	Small, round, white
A1	Small, round, yellowish gray
B1	Large, more or less round, gray
C1	Small, round, white
D1	Very small, round, yellow

The order of removal of Pb from the waters by the isolates when grown at 50 mg/L is as follows, with highest being L (73.80%) followed by R (73.20%) and Y (70%) with the rest of the strains following. The results of the bioremediation experiments are indicated in Table 2. A variety of microbial metal resistance mechanisms exists, such as precipitation of metals as phosphates, carbonates, and sulfides; physical exclusion by electronegative components in membranes and exopolymers; energy-dependent metal efflux systems; and intracellular sequestration with low molecular weight cysteine-rich proteins (Gadd 1986; Hughes and Poole 1989; Silver 1998). Earlier reports suggested that some species of *Pseudomonas, Acinetobacter, Flavobacterium,* and *Aeromonas* can convert lead nitrate or trimethyl lead acetate to tetramethyl lead (Hughes and Poole 1991;

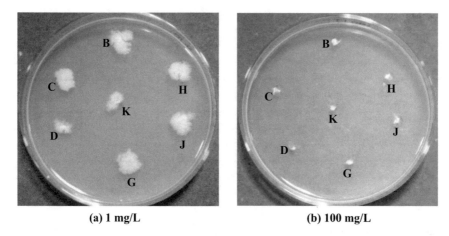

(a) 1 mg/L   (b) 100 mg/L

Fig. 6 The growth of the bacteria at 1 mg/L (**a**) and 100 mg/L (**b**) of lead acetate during acclimation tests

Table 2 Metal concentration in pre and post bioremediation process in culture supernatants

Strains	Lead concentration in mg/L (before bioremediation)	Lead concentration in mg/L (after bioremediation)	% Bioremediation
Control	50	50	0
B	50	17.83	64.34
C	50	17.50	65.00
D	50	16.81	66.38
G	50	15.70	68.60
H	50	24.11	51.78
K	50	20	60.00
L	50	13.10	73.80
M	50	16	68.00
Q	50	15.66	68.68
R	50	13.40	73.20
S	50	16.27	67.46
V	50	16.48	67.04
Y	50	15	70.00
C1	50	18.26	63.48
D1	50	18.58	62.84

Reisinger et al. 1981; Thayer and Brinckman 1982). In 2006, Jaysankar et al. isolated three Pb resistance bacteria from aquatic environment. The resistance mechanism to Pb of the three bacteria was reported to be different. Out of three, two strains precipitated Pb probably in the sulfide form, as was seen from blackening of the cell biomass upon exposure to air. In case of third strain, Pb seemed to be entrapped in the extracellular polysaccharides (EPS) as seen from the scanning

Table 3 Biochemical characterization of potent strains

Tests	G	L	R	Y
Gram staining	+	+	+	+
Indole	−	−	−	−
Methyl red	−	−	−	−
Voges Proskaeur	−	−	−	−
Citrate utilization	−	−	−	−
Triple sugar iron agar	−	−	−	−
Urease	+	−	+	−
Catalase	−	−	−	−
Motility test	Nonmotile	Nonmotile	Nonmotile	Nonmotile

electron microscopic (SEM) and energy dispersive X-ray spectrometric (EDX) studies. The present study will be extended to find out the exact mechanism of removal of Pb.

Potent Pb remediating bacterial isolates, in particular, L, is a Gram + cocci occurs in pairs, IMViC, Urease, TSI, and catalase negative, nonmotile isolate could be an Actinomyces as shown in Table 3. There has been an investigation by Valli et al. (2012) who reported the presence of Actinomycetes species from Adyar river. There have been many previous reports on the effects of heavy metals on a range of Actinomycetes species (Abbas and Edwards 1989; Amoroso et al. 1998; Lugauskas et al. 2005). However, with regard to the present study, a detailed identification of the strain L can be only determined with molecular characterization (16s rRNA) or pyrolytic method.

4 Conclusions

The levels of various physicochemical parameters in Adyar river of the Chennai city have been assessed and studied during November 2010–2011. During the study period, the observations revealed that temperature and pH were well within the specified limits and there was a uniform low dissolved oxygen level in all the stations across Adyar river. The Pb concentrations are found to be higher than the permissible limit of TNPCB at a few locations of the study area which may be due to anthropogenic activities. Out of 30 bacterial strains isolated from Adyar river, 15 strains were adapted for bioremediation of Pb as evidenced by luxuriant growth on Pb amended medium. Strains L, R, and Y were more potent to remediate Pb. Biochemical characterization studies revealed that the strain L could be a species of Actinomycetes. However, with regard to the present study, a detailed identification of the strain L can be only determined with molecular characterization (16s rRNA) or pyrolytic method. Hence we conclude that we have isolated potent Pb removing bacteria which may be a species of Actinomycetes from Adyar waters which can remove Pb up to 70% when grown even at 50 mg/L. This study may also help us acclimatize these potent strains for bioremediation of other metals and at higher concentration as well.

Acknowledgements The authors are grateful to Tamil Nadu State Council of Science and Technology/Department of Science and Technology for the financial assistance (Ref No: SSTP/ TN/09/21) and the Department of Biotechnology, Sri Ramachandra University, Porur, Chennai for providing the necessary infrastructure. We also thank Sargam Laboratories, Chennai for analyzing heavy metals and Biozone Research Technologies, Chennai for helping us with preliminary identification of bacterial strains.

References

Abbas A, Edwards C (1989) Effects of metals on a range of Streptomyces species. Appl Environ Microbiol 55:2030–2035

Amoroso MJ, Castro GR, Carlino FJ, Romero NC, Hill RT, Oliver G (1998) Screening of heavy metal-tolerant actinomycetes isolated from the Sali River. J Gen Appl Microbiol 44:129–132

APHA, AWWA, WPCF (1998) Standard methods for examination of water and waste water. American Public Health Organisation, Washington DC, USA

De Jaysankar, Sarkar A, Ramaiah N (2006) Bioremediation of toxic substances by mercury resistant marine bacteria. Ecotoxicology 15:385–389

Gadd GM (1986) Fungal responses towards heavy metals. In: Herbert RA, Codd GA (eds) Microbes in extreme environments. Academic Press, New York, pp 83–110

Hughes MN, Poole RK (1989) Metals and micro-organisms. Chapman and Hall, London

Hughes MN, Poole RK (1991) Metal speciation and microbial growth, the hard and soft facts. J Gen Microbiol 137:725–734

Joseph KO, Srivastava JP (1993) Pollution of Estuarine systems: heavy metal contamination in the sediments of Estuarine systems around Madras. J Indian Soc Soil Sci 41(1):79–83

Joshi-Tope G, Francis AJ (1995) Mechanisms of biodegradation of metal-citrate complexes by *Pseudomonas fluroescens*. J Bacteriol 177:1989–1993

Kar D, Sur P, Mandal SK, Saha T, Kole RK (2008) Assessment of heavy metal pollution in surface water. Int J Environ Sci Technol 5(1):119–124

Koshy M, Nayar TV (1999) Water quality aspects of river pamba. Pollut Res 18(4):501–510

Lokhande RS, Kelkar N (1999) Studies on heavy metals in water of Vasai Creek, Maharashtra. Ind J Environ Prot 19(9):664–668

Lugauskas A, Levinskaite L, Peeiulyte D, Repee kiene J, Motuzas A, Vaisvalavieius R, Prosyeevas I 2005. Effect of copper, zinc and lead acetates on microorganisms in soil. Ecology 1:61–69

Mergeay M (1991) Towards an understanding of the genetics of bacterial metal resistance. Trends Biotechnol 9:17–24

Moore JW (1991) Inorganic contaminants of surface water: research and monitoring priorities. Springer, New York

National Environmental Board Report (1994) Enhancement and conservation of national environmental quality. Act B.E.2537, Royal Government Gazette

Nies DH (1992) Resistance to cadmium, cobalt, zinc and nickel in microbes. Plasmid 27:17–28

Nurnberg HW (1982) Voltametric trace analysis in ecological chemistry of toxic metals. Pure Appl Chem 54(4):853–878

Prabhahar C, Saleshrani K, Dhanasekaran D, Tharmaraj K, Baskaran K (2011) Seasonal variations in physico-chemical parameters of Chennai, Coovum River, Tamil Nadu, India. Int J Curr Life Sci 1(2):33–35

Rajbanshi A (2008) Study on heavy metal resistant bacteria in Gueswori sewage treatment plant. Our Nat 6:52–57

Rajendran N, Sanjeevi SB, Khan SA, Balasubramanian T (2004) Ecology and biodiversity of Eastern Ghats—estuaries of India. ENVIS Newslett 10(3):2–7

Reisinger K, Stoeppler M, Nurnberg HW (1981) Evidence for the absence of biological methylation of lead in the environment. Nature 29:228–230

Shanmugam P, Neelamani S, Ahn Y-H, Philip L, Hong G-H (2006) Assessment of the levels of coastal marine pollution of Chennai city, Southern India. Water Resour Manag 21(7):1187–1206

Silver S (1998) Genes for all metals—a bacterial view of the periodic table. J Ind Microbiol Biotechnol 20:1–12

Thayer JS, Brinckman FE (1982) The biological methylation of metals and metalloids. Adv Organomet Chem 20:313–357

Thomas RL, Jaquet JM (1976) Mercury in the surficial sediments of Lake Erie. J Fish Res Board Can 33:404–412

Urrutia MM, Beveridge TJ (1993) Remobilization of heavy metals retained as oxyhydroxides or silicates by *Bacillus subtilis* cells. Appl Environ Microbiol 59:4323–4329

Uzel A, Ozdemir G (2009) Metal biosorption capacity of the organic solvent tolerant *Pseudomonas fluorescens* TEM08. Biores Technol 100:542–548

Valli S, Sugasini SS, Aysha OS, Nirmala P, Vinoth Kumar P, Reena A (2012). Antimicrobial potential of Actinomycetes species isolated from marine environment. Asian Pac J Trop Biomed 469–473

Venugopal T, Giridharan L, Jayaprakash M, Velmurugan PM (2009) A comprehensive geochemical evaluation of the water quality of river Adyar, India. Bull Environ Contam Toxicol 82:211–217

Winkler L (1888) The determination of the dissolved oxygen in water. J Am Chem Soc 21(2):2843–2855

Assessment of Water Quality Using Multivariate Statistical Analysis in the Gharaso River, Northern Iran

Yones Khaledian, Soheila Ebrahimi, Usha Natesan, Nabee Basatnia, Behroz Behtari Nejad, Hamed Bagmohammadi and Mojtaba Zeraatpisheh

Abstract The Gorgan Rod River, located in northern Iran in Golestan Province, is the largest watershed in the province, which its water quality is affected by natural and anthropogenic changes. Water samples were taken between 1984 and 2008 from 18 sampling stations, along the Gorgan Rod River. Determination of non-principal and principal monitoring stations was carried out for the Gorgan Rod watershed, south-east of the Caspian Sea. Water quality parameters including EC, TDS, bicarbonate, carbonate, chloride, total hardness, calcium, potassium, sodium, sodium adsorption ratio, sulfate, pH, and magnesium were measured. The graphic representations obtained underline that (i) PCA (principal component analysis) is associated with the natural and anthropogenic changes in the different stations; and (ii) the locations of the different stations studied are consistent with their apparent features. The results indicated that water quality in Basir Abad, Agh Ghala, Haji Ghochan, and Bagh Salian stations was of the poorest among other stations because of anthropogenic effects. The best water quality was observed in Ramian, Araz Kose, Nawda, Pas Poshtah, Lezore, Glikesh, and Tangrah stations because there were no changes in land uses.

S. Ebrahimi · B. B. Nejad · H. Bagmohammadi
Faculty of Soil Sciences, Gorgan University of Agricultural Sciences
and Natural Resources, Gorgan, Golestan, Iran

Y. Khaledian (✉)
Department of Agronomy, Iowa State University, Ames, IA, USA
e-mail: yones.khaledian@gmail.com

M. Zeraatpisheh
Department of Soil Science, College of Agriculture, Isfahan University
of Technology, 84156-83111 Isfahan, Iran

U. Natesan
Centre for Water Resources, Anna University, Chennai 600025, India

N. Basatnia
Faculty of Fisheries and Environmental Sciences, Gorgan University
of Agricultural Sciences and Natural Resources, Gorgan, Golestan, Iran

© Springer International Publishing AG 2018
A. K. Sarma et al. (eds.), *Urban Ecology, Water Quality and Climate Change*,
Water Science and Technology Library 84,
https://doi.org/10.1007/978-3-319-74494-0_18

Keywords CA · Land use change · Northern Iran · PCA · Water quality

Abbreviations

CA	Cluster analysis
Ca^{+2}	Calcium
Cl^{-1}	Chloride
EC	Electrical conductivity
HCO_3^{-1}	Bicarbonate
CO^{-3}	Carbonate
K^{+1}	Potassium
Mg^{+2}	Magnesium
Na^{+1}	Sodium
PCA	Principal component
SAR	Sodium adsorption ratio
SO_4^{-2}	Sulfate
T-Hard	Total hardness
Temp	Temperature
TDS	Total dissolved solids

1   Introduction

Totally, rivers play an utmost important role in a watershed for carrying off industrial and municipal wastewater and runoff from surrounding land (Shrestha and Kazam 2007) which are the most important water resources for human consumption, agricultural needs, and recreational and industrial purposes. Hence, it is important to have appropriate information on trends of water quality for efficient water management. This necessity is more pronounced in tourist regions such as the north of Iran, which experience water pollution problems in densely populated areas, with a rapid population growth due to the large influx of people seeking employment and pleasure. Furthermore, it is frequently determined, in river monitoring, whether related variables should be attributed to man-made pollution (spatial, anthropogenic) or natural changes (temporal, climatic) in the river hydrology (Razmkhah et al. 2010).

The impressive growth of human population and economic development has resulted in the current worldwide deterioration in water quality (Li and Zhang 2010). Natural processes and anthropogenic activities such as precipitation, soil erosion, industrial, and agricultural activities largely contribute to chemical pollutants in the fluvial systems (Li and Zhang 2010). Sources diffused by agricultural runoff have an increasingly great concern for nutrients due to widely overused fertilizer. Meanwhile, industrial developments, particularly electroplating, metallurgy, mining, and mineral processing, have been changing the biogeochemical cycles of heavy metals (Singh et al. 2005). Recent studies have also revealed that urban development and hydrological impacts by climate change greatly degraded water quality (Duh et al. 2008).

Creating a water quality monitoring system, with appropriate efficiency is one difficulty in evaluating surface water quality; and to measure variables that would express water quality changes as much as possible. To achieve this goal, a multivariate statistical method such as PCA can be utilized. Recently, PCA has been widely used to evaluate a variety of environmental issues. Multivariate statistic methods such as PCA and cluster analysis (CA) were used (Facchinelli et al. 2001; Azhar et al. 2015; Khaledian et al. 2016a, b; Noori et al. 2010) to predict potential nonpoint heavy metal sources in soil on a regional scale. Coming up with effective pollution control management for the surface water (Simeonov et al. 2003), using PCA and CA were able to interpret a large and complex data matrix of surface water variables in Northern Greece. Ouyang (2005) adapted PCA and PFA to identify important water quality variables in twenty-two stations, which were located at the main stem of the lower St. Johns River in Florida, USA. CA, PCA, PFA, and discriminated analysis techniques were applied (Shrestha and Kazama 2007; Khaledian et al. 2012) to evaluate temporal and spatial variations in northern Iran. Given the way it has been used, it has not yet provided answers to the questions this kind of study generally poses. Nevertheless, the originality of the lake system, in the study area, provided us with the opportunity to test the relevance of this tool.

Considering the above-cited studies, the research aims to investigate: (1) identification of the most informative water quality monitoring station in the Gorgan Rod watershed; (2) determination of the most important water quality variables in the river; (3) assessment of water quality changes in the Gharaso River in 25 years (1984–2008); (4) evaluation of the role of anthropogenic and natural influences on the river water quality; and (5) exploration of relationship between physical and chemical variables.

2 Materials and Methods

2.1 Sampling Stations

The Gorgan Rod River flowing in the Gorgan Rod watershed located at 53.63°–56.13°E longitude and 36.58°–37.8°N latitude. The maximum and minimum precipitation is in April and August, respectively. The river flows into the Caspian Sea.

Figure 1 illustrates the locations of 18 stations, which are identified by the Golestan Regional Water Company as appropriate sites to present water quality. Furthermore, Fig. 2 presents the map of land uses in the watershed. The stations

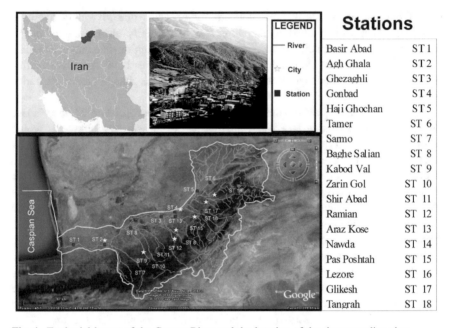

Fig. 1 Territorial layout of the Gorgan River and the location of the river sampling sites

were sampled monthly for 25 years on a rotational schedule. Some stations located in densely populated areas were sampled more often. Fifteen variables including; pH, electrical conductivity (EC), total dissolved solids (TDS), bicarbonate (HCO_3^{-1}), carbonate (CO_3^{-1}), sulfate (SO_4^{-2}), chloride (Cl^{-1}), temporal hardness (TemH), total hardness (TH), calcium (Ca^{+2}), magnesium (Mg^{+2}), potassium (K^{+1}), sodium (Na^{+1}), sodium percent (NaP), and sodium adsorption ratio (SAR) had been analyzed. The main reason to choose these parameters was to investigate the geochemical properties of the watershed. Also, this watershed plays an utmost important role in Golestan Province, as the river flows into the Caspian Sea. The area surrounding the Gorgan Rod river has seen an increased in land use change and urbanization (Fig. 2).

2.2 Analytical Procedures

Monitored variables were analyzed following APHA et al. (2005) and AOAC (1990) which are cited in parentheses. Measured variables include: calcium (3500-Ca B); magnesium (3500-Mg B); chloride (4500-Cl-B); bicarbonate (4500-HCO₃); and carbonate (4500-CO₃) measured with titration; potassium (3500-K B); sodium (3500-Na B) measured with flame photometry; sulfates (4500-SO₄ 2-D) measured with ICP (Source Spectrometry Techniques); temperature (2550 B, field measured);

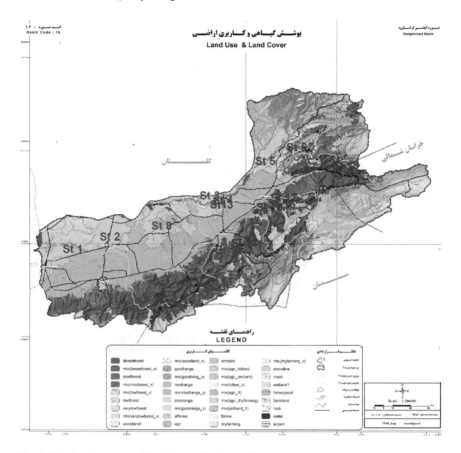

Fig. 2 The land use map of the Gorgan Rod watershed

pH (4500-Hþ B, field measured); total and temporary hardness (2340 C); total dissolved solids (2540 C); and conductivity (2510 B). All the analyses were run in duplicate and devices used were Jenway 430, Jenway Flame Photometer PFP7, Jenway Spectrophotometer 6715, and Hach 2100p (AOAC 1990; APHA, AWWA, WEF 2005).

2.3 Data Treatment

To test the suitability of these data for factor analysis, Kaiser–Meyer–Olkin (KMO) test was performed. KMO is a measure of sampling adequacy that indicates the proportion of variance which is common variance, i.e., might be caused by underlying factors. High value (close to 1) generally indicates that factor analysis may be useful, which is the case of this study: KMO = 0.85 (Table 1). If the KMO

Table 1 The destructive statistics of water parameters

	N	Range	Minimum	Maximum	Mean	Std. deviation	Variance
	Statistic	Statistic	Statistic	Statistic	Statistic	Statistic	Statistic
TDS	4530	2.25E4	0.00	22466.00	1.3474E3	2191.60527	4.803E6
EC	4530	3.03E4	0.00	30300.00	2.0659E3	3286.20463	1.080E7
pH	4530	2.84	6.06	8.90	7.4796	0.50287	0.253
CO_3	4530	2.90	0.00	2.90	0.0276	0.18677	0.035
HCO_3	4530	11.00	1.00	12.00	4.2651	1.06759	1.140
Cl	4530	243.80	0.20	244.00	9.9090	19.67844	387.241
SO_4	4530	169.22	0.08	169.30	8.0995	16.42658	269.833
Ca	4530	64.20	0.80	65.00	4.5139	4.63175	21.453
Mg	4530	87.94	0.06	88.00	6.3165	9.78512	95.748
Na	4530	232.94	0.06	233.00	11.4346	22.20554	493.086
K	4530	3.12	0.00	3.12	0.0919	0.11038	0.012
SAR	4530	31.55	0.04	31.59	3.6575	4.16997	17.389
NaP	4530	78.25	0.03	78.28	35.3667	19.89757	395.913
TemH	4530	598.60	1.40	600.00	2.0278E2	66.92150	4.478E3
TH	4530	6248.50	1.50	6250.00	4.9783E2	659.77944	4.353E5

test value is less than 0.5, factor analysis will not be useful. Bartlett's test of sphericity indicates whether correlation matrix is an identity matrix, which would indicate that variables are unrelated. The significance level which is 0 (Table 1) in this study (less than 0.05) indicate that there are significant relationships among variables. Finally, PCA was applied to normalized data, and so the covariance matrix coincides with the correlation matrix.

2.4 Multivariate Statistical Analysis

The exploratory data analysis was examined by PCA and CA on experimental data. Since the methods of classification used here are nonparametric, no assumptions about the underlying statistical distribution of the data were made; therefore, no evaluation of normal (Gaussian) distribution is necessary (Sharaf et al. 1986).

2.4.1 Cluster Analysis

The major goal of cluster analysis is to categorize objects (cases) into classes (clusters), where objects placed within a class are similar to each other, but different from those in other classes. In CA, the objects are grouped by linking inter-sample similarities and the outcome illustrates the overall similarity of variables in the data set (Li et al. 2009). Many applications of CA for water quality assessments have been reported (Singh et al. 2005; Mahbub et al. 2008; Li et al. 2009). CA allows the grouping of water samples on the basis of their similarities in chemical composition. Unlike PCA that normally uses only two or three PCs for display purposes, CA uses all the variance or information contained in the original data set. Hierarchical agglomerative CA was carried out by means of Ward's method, an extremely powerful grouping mechanism, which yields a larger proportion of correctly classified observations (Willet 1987), using squared Euclidean distances as a measure of similarity (Massart and Kaufman 1983).

2.4.2 Principal Component Analysis

PCA, as a technique for variable reduction, dispenses with nonhomogeneity in sampling data, missing value and periodic trends in data and identifies temporal variation in water quality. It also extracts the most important variable in polluted stations (Dillon and Goldstein 1984). In this way diagonal of the correlation matrix transforms the original principal correlated variables into principal uncorrelated (orthogonal) variables called principal components (PCs), which are weighed as linear combinations of the original variables (Dillon and Goldstein 1984). The eigenvalues of the PCs are a measure of associated variances and which the sum of the eigenvalues coincides with the total number of variables. Correlation of original variables and PCs is given by loadings, and scores are called individual transformed observations (Khaledian et al. 2012).

3   Results and Discussion

3.1   Comments on Some of Geochemical Variables Evolution

This part shows a synthesis of the measurements. Figure 3 depicts the mean values of studied variables, during period given. Table 1 presents the descriptive statistics of water parameters in the Gorgan Rod watershed.

3.1.1   Total Dissolved Solids, Electrical Conductivity, and Total Hardness

The significant spatial variation of TDS, EC, and TH was noted (Fig. 3). According to WHO specification TDS, the highest desirable and maximum permissible is up to 500 and 1,500 mg/L, respectively (Table 2). In the study area, the TDS value in all years varies between a minimum of 240 mg/L and a maximum of 5,167 mg/L (Fig. 3), indicating that some stations, including Basir Abad, Agh Ghala, Gonbad, Haji Ghochan, Bagh Salian, lie over the maximum permissible limit. In the last 4 years, the amount of TDS increased dramatically (Fig. 4).

Fig. 3 Average value of
some variables for each
station

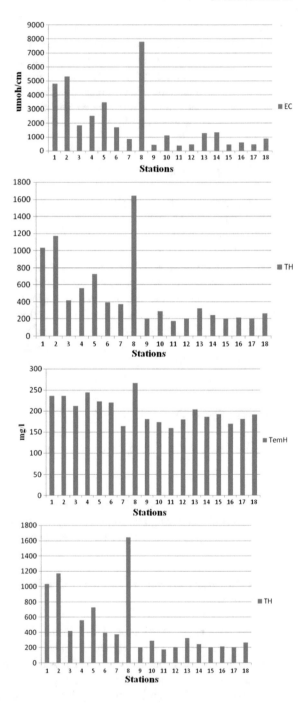

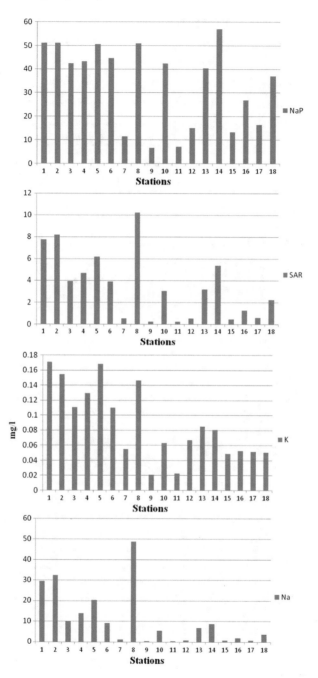

Fig. 3 (continued)

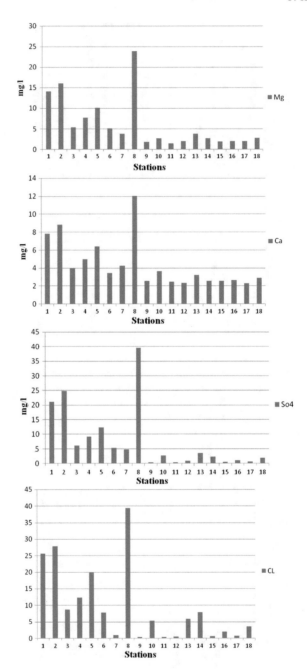

Fig. 3 (continued)

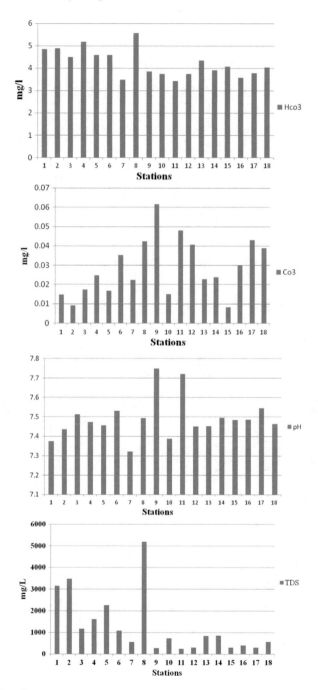

Fig. 3 (continued)

Table 2 Average of physicochemical characteristics of the Gorgan Rod watershed

Stations	TDS	EC	pH	CO$_3$	HCO$_3$	Cl	SO$_4$	Ca	Mg	Na	K	SAR	TH
St 1	3136	4803	7.37	0.0147	4.85	25.62	21.16	7.81	14.17	29.66	0.17	7.77	1032.69
St 2	3442	5256	7.43	0.0092	4.89	27.603	24.58	8.79	15.89	32.67	0.15	8.16	1158.35
St 3	1147	1799	7.5	0.0174	4.48	8.55	6.05	3.95	5.38	9.72	0.12	3.88	414.26
St 4	1577	2464	7.47	0.02	5.183	11.968	9.03	4.93	7.51	13.69	0.14	4.63	550.72
St 5	2264	3469	7.45	0.0167	4.606	20.024	12.39	6.43	10.18	20.49	0.16	6.18	725.09
St 6	1054	1642	7.52	0.0336	4.59	7.477	5.23	3.37	4.95	8.98	0.11	3.84	381.39
St 7	552	842	7.31	0.021	3.478	0.923	4.78	4.27	3.8	1.09	0.05	0.57	374.95
St 8	5167	7778	7.48	0.035	5.57	39.16	39.3	11.99	23.9	48.66	0.14	10.1	1634.61
St 9	266	410	7.74	0.0569	3.876	0.37	0.41	2.52	1.83	0.38	0.03	0.23	199.74
St 10	720	1115	7.3	0.0144	3.747	5.3	2.78	3.68	2.79	5.49	0.07	3.04	291.49
St 11	240	371	7.7	0.043	3.421	0.362	0.41	2.43	1.48	0.35	0.03	0.22	177.99
St 12	294	453	7.44	0.0352	3.75	0.465	0.93	2.39	1.91	0.79	0.06	0.52	205.1
St 13	824	1278	7.45	0.0185	4.33	5.812	3.57	3.27	3.72	6.9	0.09	3.15	320.98
St 14	848	1331	7.48	0.02	3.917	7.87	2.31	2.51	2.9	8.87	0.08	5.37	246.3
St 15	295	451	7.47	0.006	4.062	0.549	0.54	2.58	1.81	0.69	0.04	0.45	211.16
St 16	391	601	7.48	0.0262	3.57	1.88	1.16	2.61	2	1.99	0.06	1.26	217.26
St 17	293	450	7.53	0.04	3.778	0.682	0.65	2.22	1.96	0.81	0.05	0.58	200.81
St 18	571	875	7.46	0.0351	4.027	3.493	1.92	2.889	2.79	3.82	0.05	2.23	266.48
Desirable standards (WHO)	500	2000 (μS cm^{-1})	6.5–8.5	120 ppm	50 ppm	11 ppm	10 ppm	5 ppm	4 ppm	24 ppm	10 ppm	5	400

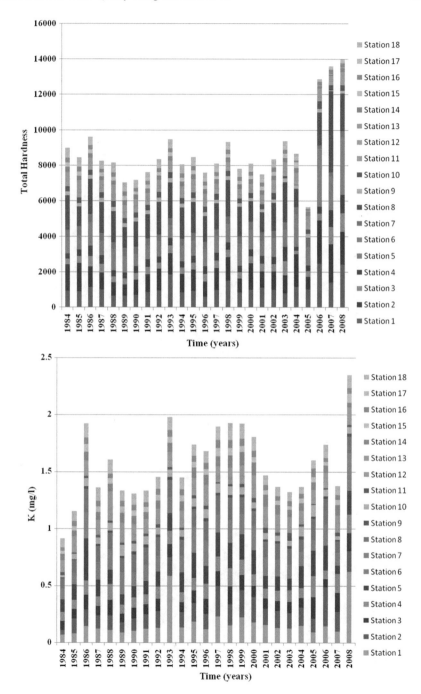

Fig. 4 The average of amount of each parameter for 18 stations during the 25-year period

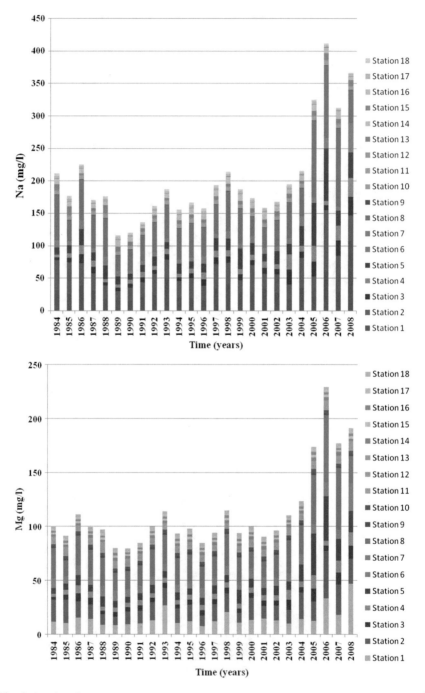

Fig. 4 (continued)

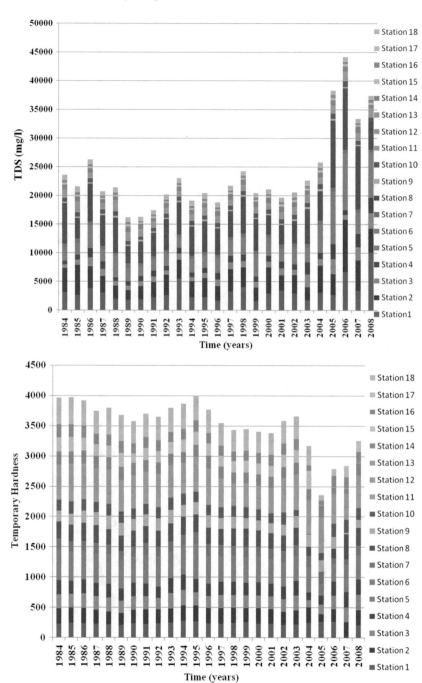

Fig. 4 (continued)

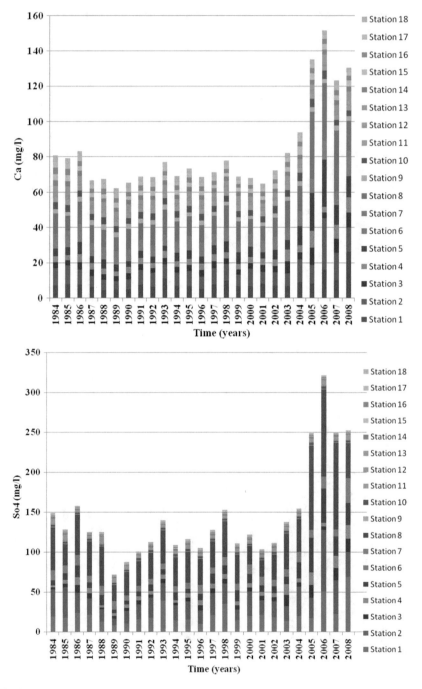

Fig. 4 (continued)

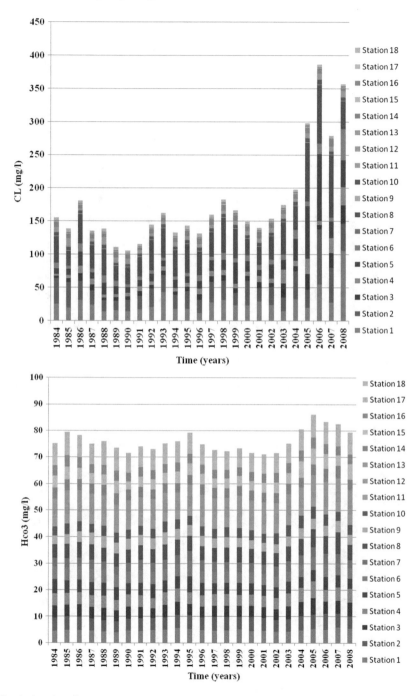

Fig. 4 (continued)

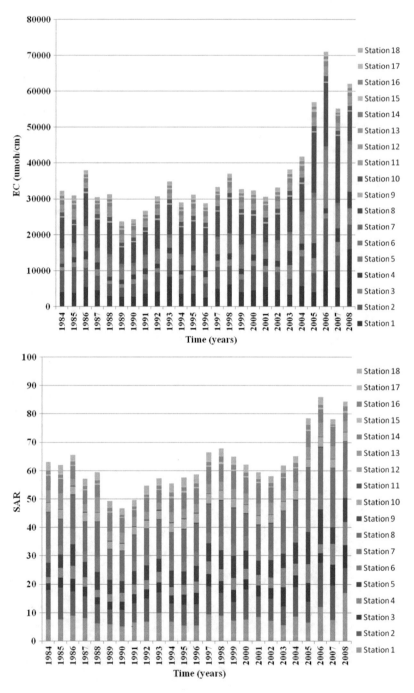

Fig. 4 (continued)

In terms of EC, the most desirable limit of EC is prescribed as 2,000 mhos/cm (WHO 2004) (Table 2). The EC is varying from 370 to 7,800 mhos/cm (Fig. 3). Higher EC in the study area indicates the enrichment of salts in the stations. The same stations reached over the permissible value those regarding TDS. In the last 4 years, the amount of EC reached its peak which it would be due to land use changes (Fig. 4).

Turning to total hardness, the total hardness is varying from 178 to 1643 mg/L (Fig. 3). Sawyer and McCarty (1967) classified water based on TH, as water with TH 75 (soft), 75–150 (moderately hard), 150–300 (hard) and more than 300 mg/L (very hard) (Table 2). According to the above categorization, the stations of Basir Abad, Agh Ghala, Ghezaghli, Gonbad, Haji Ghochan, Tamer, Sarmo, Bagh Salian, and Araz Kose belong to hard and the remaining comprises very hard water. Therefore, the analytical result presents that water quality ranges between hard and very hard. In the last 3 years, the amount of TH increased significantly rather than previous years (Fig. 4).

The TDS, EC, and TH recorded higher concentration than the other variables in the Bagh Salian, Basir Abad, Agh Ghala, Gonbad, and Haji Ghochan. This could be essentially explained here by the geochemical reactions due to intensive human activities, such as land use (Fig. 2) and natural changes. Comparison of these stations demonstrates that the high TDS, EC, and TH of water were as a result of land use changes (Fig. 2). On the other hand, the relatively higher TDS, EC, and TH in stations of Basir Abad and Agh Ghala is due to the location of station, which is in the output of the watershed joining the Caspian Sea; therefore, it received the highest amount of pollution. Such observations underline the fact that the environment has a strong feedback effect on the TDS, EC, and TH (Wellington and Dominic 2008; Xiao et al. 2014; Kilonzo et al. 2014).

3.1.2 pH

pH indicates the balance between the concentration of hydroxyl and hydrogen ions in water. The limit of pH value for drinking water ranges between 6.5 and 8.5 (WHO 2004). The pH value of most of the water samples in the study area varies from 7.2 to 7.8 (Fig. 3), which clearly shows that the water in the study area is slightly acidic in nature. The highest pH level is observed in the Kabod Val and Shir Abad stations; however, this does not show considerable difference with other stations. It would be due to the morphology of soils deposited by the river, which is consisted of mainly by limestone. This suggests that human activities (such as land use change) in the watershed are not modifying the pH, except for the cases where anomalous results were observed. Furthermore, this may be attributed to the anthropogenic activities, for example, sewage disposal and use of fertilizers, with natural phenomenon, such as intrusion of brackish water, which initiates the weathering process of underlain geology (Vega et al. 1998).

Table 3 Loadings of 12 experimental variables on three significant principal components for river water samples

Component matrix			
	Component		
	1	2	3
TDS	**0.977**	−0.157	−0.037
pH	0.070	−0.201	**0.969**
HCO$_3$	**0.585**	0.445	0.166
Cl	**0.961**	−0.171	−0.029
SO$_4$	**0.951**	−0.175	−0.039
Ca	**0.911**	−0.205	−0.029
Mg	**0.956**	−0.183	−0.028
Na	**0.979**	−0.139	−0.027
K	**0.615**	0.044	−0.058
SAR	**0.948**	0.067	0.002
TemH	0.511	**0.612**	0.003
TH	0.419	**0.793**	0.112

3.1.3   Calcium, Magnesium, Potassium, and Sodium

According to WHO specification Ca^{2+}, Mg^{2+}, and Na^+ up to 5, 4, and 24 mg/L is the highest desirable maximum permissible (Table 2). These parameters witnessed a dramatic increase in the last 4 years rather than previous years (Fig. 4). In the study area, these three parameters hit the permissible value in some stations, including Basir Abad, Agh Ghala, Haji Ghochan, and Bagh Salian. Sodium may be introduced into water from detergents, soaps, bleaches, and food. Water treatment may also introduce Na^+ and K^+, which are used as fertilizers. Furthermore, this would also be because of morphology of soils deposited in the river, primarily limestone. In the other stations, not much significant differences were noticed. Water hardness containing dissolved metals like Mg^{2+}, originating from industrial, agricultural, and domestic sewage runoff were in high levels in Basir Abad, Agh Ghala, Haji Ghochan, and Bagh Salian. It may be due to increasing soil weathering, erosion (seasonal effect), urbanization, and anthropogenic sources of pollution (Jinzhu et al. 2009).

The K^+ and sodium percentages in the studied stations of the Gorgan Rod River show significant differences in the stations of in Basir Abad, Agh Ghala, Haji Ghochan, and Bagh Salian, which lie over permissible value according to WHO for agricultural purposes. The highest content of K^{+1} observed in the Bagh Salian station (Fig. 3). The amount of K^+ fluctuation during the period (Fig. 4). This is probably due to the use of manure in cultivated areas, home construction, and agricultural and domestic sewage disposal (Zhang et al. 2007).

The results did not show any considerable changes in temporary hardness and HCO$_3$ (Fig. 4).

3.1.4 Sodium Adsorption Ratio (SAR)

SAR shows the suitability of water for agricultural irrigation, because sodium concentration can decrease the soil permeability and soil structure (Todd 1980). SAR illustrates alkali/sodium hazard to crops and it can be estimated by the following formula:

$$SAR = Na/((Ca + Mg)/2)0.5$$

where sodium, calcium, and magnesium are in meq/l.

The calculated values of SAR varies between 0.2 and 10.5 (Table 2). The SAR values of all the stations are found within the range of excellent to good category, except for Basir Abad, Agh Ghala, Haji Ghochan, and Bagh Salian, which is found to be unsuitable for irrigation purpose (Table 3). SAR showed fluctuated during the given period, while in the last 4 years reached its peak. The SAR value for irrigation purposes has a significant relationship with the amount of sodium is absorbed by the soils. Water with high SAR values for irrigation using may require soil amendments in order to prevent long-term damage to the soil, as the sodium in the water can displace the magnesium and calcium in the soil. This will also lead to decrease in infiltration and permeability of the soil to water leading to problems with crop production. This indicates that the water type in the stations of Basir Abad, Agh Ghala, Haji Ghochan, and Bagh Salian have no appropriate situation for irrigation on all type of soil (Lee et al. 2009).

3.1.5 Sulfate

The higher level of SO_4^{-2}, resulting from man-made, natural, and mixed source pollution, was measured in all stations. Even though in the first 3 years of the given period, the amount of SO_4^{-2} was high; it decreased slightly in following years and then increased again in the last 4 years (Fig. 4). The impermissible value of SO_4^{-2} was in Basir Abad, Agh Ghala, Haji Ghochan, and Bagh Salian. This can probably be an indication of weathering and erosion in this study area (Khaledian et al 2012; Jinzhu et al. 2009).

3.2 Spatial Variation of Water Quality

3.2.1 Principal Component Analysis

Because of the "feedback effect", which depends on the particular characteristics of each station, the above comments pointed out that the intrinsic values of analytical data are not sufficient to make a correct assessment of changes in the land uses. The

Table 4 Rotation if component matrix for 12 variables

Rotated component matrix

	Component		
	1	2	3
TDS	**0.965**	0.218	0.026
pH	0.042	−0.015	**0.991**
HCO₃	0.363	**0.651**	0.108
Cl	**0.955**	0.201	0.037
SO₄	**0.948**	0.192	0.028
Ca	**0.921**	0.151	0.041
Mg	**0.955**	0.188	0.040
Na	**0.960**	0.237	0.034
K	**0.559**	0.263	−0.042
SAR	**0.853**	0.418	0.026
TemH	0.249	**0.753**	−0.084
TH	0.088	**0.900**	−0.012

land use changes will indirectly affect the relationship between analytical variables. To maintain consistency and to avoid any biased approach, the number of PCs in order to comprehend the underlying data structure (Jackson 1991) was determined by the built-in criteria of the software throughout the analysis. The Scree plots showed a pronounced change of slope after the first eigenvalue; even though it has been suggested (Cattell and Jaspers 1967) to use all the PCs up to and including the first one after the break, again using the default software criteria of eigenvalues greater than one used for determining the number of PCs.

Loading is the projection of the original variables on the subspace of the PCs, which coincides with the correlation coefficients between PCs and variables (Vega et al. 1998). The loading of the three retained PCs is presented in Table 4. PC1 explains 62.4% of the variance and is highly contributed by most variables, EC, TDS, Ca^{+2}, Mg^{+2}, Cl^{-1}, Na^+, SO_4^{2-}, SAR, and total hardness. PC2 explains 9.7% of the variance and includes Na percent, and temporary hardness and finally PC3 (8.4% of variance) is positively contributed to by CO_3 and pH. As shown in Table 4, PC1 shows the effects of high concentration of mineral compounds, but PC2 and PC3 indicate anthropogenic and mineral pollution, respectively. Therefore, these components could not be explained only in terms of organic or mineral pollution. The variables identified in the first two principle components were plotted in the large and small ellipses in Fig. 3a, b, showing the loadings of the third component on the first with the three main variables plotted in the ellipse.

A rotation of PCs (Upper Loads for each component are in bold format) can achieve a simpler and more meaningful representation of the underlying factors by decreasing contributions to PCs by variables with minor significance and increasing the more significant ones. Rotation produces a new set of factors, each involving primarily a subset of original variables with as little overlap as possible, so that the original variables are divided into groups somewhat independent of each other (Sharaf et al. 1986). Although rotation does not affect quality of the principal components solution, the variance explained by each factor is modified. Rotated PCs is redounded a varimax rotation of principal components, which is called henceforth varifactors (Table 4). Therefore, three varifactors were extracted, with VFs explaining 56% of the variance. It should be noted that rotations lead to increase the number of factors necessary to identify the same amount of variance in an original data set. Therefore, the first two varifactors used for graphical representation, explained less variance than that shown before rotation. However, smaller groups of variables can be now linked to individual rotated factors with a hydrochemical meaning (Vega et al. 1998; Viswanath et al. 2015). VF1 with high and positive scores on EC, TDS, Ca^{+2}, Mg^{+2}, Cl^{+}, $Na^{+,}$ SO_4^{2-}, and SAR explains 56% of the total variance. This variable cluster points to a common origin for these minerals, such as dissolution of limestone, marl, and gypsum in water. VF2, containing 15% of the total variance, includes HCO_3, Na percent, and temporary hardness where pH has a negative contribution to this varifactor. The same can be explained taking into account that hydrolysis of HCO_3^{-}. VF3 (8.6% of variance) has a high and positive load of CO_3^{-2} and hydrolysis of calcium carbonate compounds cause an increase in the water pH value.

3.2.2 Cluster Analysis

The dendrogram of the stations obtained by Ward's method in different stations in Fig. 5.

Three well-differentiated clusters can be seen in Fig. 5. The first group from the top is formed by two subgroups, with water quality decreasing from top to bottom. Basir Abad, Agh Ghala, and Haji Ghochan stations are in the same group and the Bagh Salian record the least similarity to another group. It shows the largest decrease of water quality in these groups of stations in comparison with the other stations. In general, physical and chemical properties in stream are primarily controlled by forest cover and vegetated land (including forest and shrub) within the rainy season and dry season, respectively (Mahbub et al. 2008; Li et al. 2009).

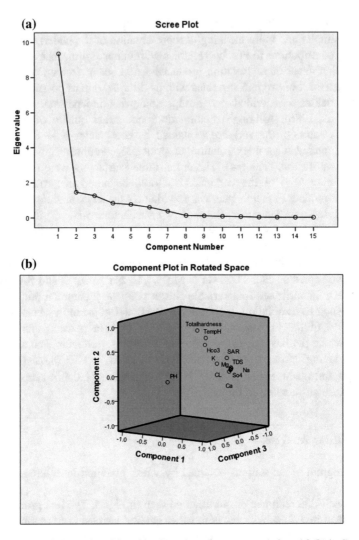

Fig. 5 Scree plot of eigenvalues (**a**) and loading plots of component 1, 2, and 3 (**b**) in Gorgan Rod watershed

Also, it depicts that there was the largest decrease of water quality in these groups compared to the other stations which are attributable to increase in agricultural runoff in high flows and urban domestics, and lower forest coverage (Li and Zhang 2010) (Fig. 6).

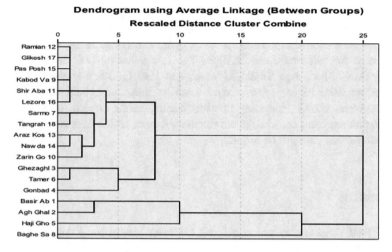

Fig. 6 Dendrogram based on agglomerative hierarchical clustering (Ward's method) for river stations

4 Conclusions

From the observational study of these stations, it is obvious that the "feedback effect" can be applied not only to land use and natural processes, but also to other geochemical ones. This "feedback effect" could be extended to every station. Therefore, it is observed that a change in every relationship linking analytical variables would alter water quality. PCA made with correlation coefficients, takes into account these changes, and becomes an easy and appropriate descriptive tool. Surface water quality data for the Gorgan Rod River (in northern Iran) were evaluated for spatial variations and the effects of rapid growth of land use changes and urbanization. The PCA was applied to determine important monitoring stations and water quality variables. Environmental analytical chemistry generates multi-dimensional data that requires multivariate statistics to analyze and interpret underlying information. PCA reduced the number of varifactors to three that explain high percentage of the experimental data set variance. By explaining mean 80.5% of spatial variations with three varifactors, PCA in combination with CA has nearly allowed identification and assessment of spatial sources of variation affecting the quality and hydrochemistry of river water. CA was not able to define details but guided us through classification. Performing PCA for each cluster formed by CA could help identify pollution sources of stations in each cluster. The worst water quality was in Basir Abad, Agh Ghal, and Bagh Salian stations in comparison with other stations. The Basir Abad, Agh Ghal stations located in the output of the river and watershed, where it joins to the Caspian Sea, received the more amounts of pollution and; therefore, the greatest threat to the health of aquaculture. Whilst Haji Ghochan was not at end of the river, this station was polluted as much as the stations at end of the river, such as Basir Abad, Agh Ghal. It indicates Haji Ghochan, the land use changes, therefore soil degradation was remarkable. The Bagh Salian station had the lowest water quality in comparison to other stations because of intensive land use changes. The low value of TDS and EC in Bagh Salian, Basir Abad, Agh Ghala, Gonbad, and Haji Ghochan stations confirm it. Human activities as important factors such as intensive urbanization, land use changes were caused decreasing in water quality in the Gorgan Rod River. Soil degradation and land use change are the main factors which lead to increase solute concentration in the surface runoff.

References

AOAC (1990) Official methods of analysis, vol 1, 15th edn. Association of Analytical Chemists, Arlington, VI, USA, p 312

APHA, AWWA, WEF (2005) Standard methods for the examination of water and wastewater, 21st edn. American Public Health Association/American Water Works Association/Water Environment Federation, USA

Azhar SC, Aris AZ, Yusoff MK, Ramli MF, Juahir H (2015) Classification of river water quality using multivariate analysis. Procedia Environ Sci 30:79–84

Cattell RB, Jaspers J (1967) A general plasmode (No. 30-10-5-2) for factor analytic exercises and research. Multivar Behav Res Monogr 67:1–212

Dillon WR, Goldstein M (1984) Multivariate analysis methods and application. Wiley

Duh VJ, Shandas H, Chang L (2008). George, rates of urbanization and the resilience of air and water quality. Sci Total Environ 400:238–256

Facchinelli A, Sacchi E, Mallen L (2001) Multivariate statistical and GIS-based approach to identify heavy metals sources in soils. J Environ Poll 114:313–324

Jackson JE (1991) A user's guide to principal components. Wiley, New York

Jinzhu M, Zhenyu D, Guoxiao W et al (2009) Sources of water pollution and evolution of water quality in the Wuwei basin of Shiyang river, Northwest China. J Environ Manag 90:1168–1177

Khaledian Y, Ebrahimi S, Bag Mohamadi H (2012) Effects of urbanization and changes land use on water geochemical properties. A case study: Gharaso River in Golestan Province, North of Iran. Terr Aquat Environ Toxicol 6(2):77–83

Khaledian Y, Kiani F, Ebrahimi S, Brevik EC, Aitkenhead-Peterson JA (2016a) Assessment of urbanization on soil quality using multivariate statistical, Northern Iran. Lad Degradation and Development. https://doi.org/10.1002/ldr.2541

Khaledian Y, Pereira P, Brevik EC, Pundyte N, Paliulis D (2016b) The influence of organic carbon and pH on heavy metals levels in lithuanian soils. https://doi.org/10.1002/ldr.2638

Kilonzo F, Masese FO, Van Griensven A, Bauwens W, Obando J, Lens PN (2014) Spatial–temporal variability in water quality and macro-invertebrate assemblages in the Upper Mara River basin, Kenya. Phys Chem Earth, Parts A/B/C 67:93–104 (Chicago)

Lee JH, Hamma SY, Cheong JY et al (2009) Characterizing riverbank-filtered water and river water qualities at a site in the lower Nakdong River basin, Republic of Korea. J Hydrol 376:209–220

Li S, Zhang Q (2010) Risk assessment and seasonal variations of dissolved trace elements and heavy metals in the Upper Han River, China. J Hazard Mater 181:1051–1058

Li S, Liu W, Gu S, Cheng X, Xu Z, Zhang Q (2009) Spatial-temporal dynamics of nutrients in the upper Han River basin, China. J Hazard Mat 162:1340–1346

Mahbub H, Syed Munaf A, Walid A (2008) Cluster analysis and quality assessment of logged water at an irrigation project, eastern Saudi Arabia. J Environ Manag 86(1):297–307

Massart DL, Kaufman L (1983) The interpretation of analytical chemical data by the use of cluster analysis. Wiley, New York

Noori R, Sabahi MS, Karbassi AR, Baghvand A, Taati Zadeh H (2010) Multivariate statistical analysis of surface water quality based on correlations and variations in the data set. Desalination. https://doi.org/10.1016/j.desal.04.053

Ouyang Y (2005) Evaluation of river water quality monitoring stations by principal component analysis. J Water Res 39:2621–2635

Razmkhah H, Abrishamchi A, Torkian A (2010) Evaluation of spatial and temporal variation in water quality by pattern recognition techniques: a case study on Jajrood River (Tehran, Iran). J Environ Manag 91:852–860

Sawyer GN, McCarthy DL (1967) Chemistry of sanitary engineers, 2nd edn. Mc Graw Hill, New York, p 518

Sharaf MA, Illman DL, Kowalski BR (1986) Chemometrics. Wiley, New York

Shrestha S, Kazama F (2007) Assessment of surface water quality using multivariate statistical techniques: a case study of the Fuji river basin, Japan. J Environ Modell Softw 22:464–475

Simeonov V, Stratis JA, Samara C, Zachariadis G, Voutsa D, Anthemidis A, Sofoniou M, Kouimtzis T (2003) Assessment of the surface water quality in Northern Greece. J Water Res 37:4119–4124

Singh KP, Malik A, Sinha S (2005) Water quality assessment and apportionment of pollution sources of Gomti river (India) using multivariate statistical techniques—a case study. Anal Chim Acta 538:355–374

Todd DK (1980) Groundwater hydrology, 2nd edn. Wiley, New York, p 535

Vega M, Pardo R, Barrado E, Deban L (1998) Assesment of seasonal and polluting effects on the quality of river water by exploratory data analysis. J Water Res 32(12):3581–3592

Viswanath NC, Kumar PD, Ammad KK (2015) Statistical analysis of quality of water in various water shed for Kozhikode City, Kerala, India. Aquat Procedia 4:1078–1085

Wellington RLM, Dominic M (2008) Impact on water quality of land uses along Thamalakane-Boteti River: an outlet of the Okavango Delta. Phys Chem Earth 33:687–694

WHO (2004) Guidelines for drinking water quality. World Health Organisation, Geneva

Willet P (1987) Similarity and clustering in chemical information systems. Research Studies Press, Wiley, New York

Xiao J, Jin Z, Wang J (2014) Geochemistry of trace elements and water quality assessment of natural water within the Tarim River Basin in the extreme arid region, NW China. J Geochem Explor 136:118–126

Zhang CL, Zou XY, Yang P et al (2007) Wind tunnel test and 137Cs tracing study on wind erosion of several soils in Tibet. J Soil Tillage Res 94(2):269–282

Stability Analysis of Earthen Embankment of Kollong River, Near Raha, Nagaon, Assam

Rituraj Buragohain, Dimpi Das, Rubia Sultana Choudhury and Rajib K. Bhattacharjya

Abstract Stability of an earthen slope can be analyzed using circular slip circle method. In this method, an optimal search is carried out to find the most critical slip circle corresponding to the minimum factor of safety. For obtaining the factor of safety of the critical slip circle, information about the subsurface strata is necessary. Most of the time, the slip circle is considered as homogeneous in nature. In this study, heterogeneity of the bank material is considered in the stability analysis of a river bank. A field survey is initially conducted using Ground Penetrating Radar (GPR) to acquire information about the nature of the subsurface soil strata of the embankment. Information extracted from the GPR survey was verified using actual bore log data. Stability analysis of the embankment is then carried out using circular slip circle method. The methodology is applied for the river bank analysis of the river Kollong at Raha, Assam. The evaluation shows that the methodology is effective in studying the stability analysis of river banks.

Keywords Slope stability · Slip circle method · GPR

R. Buragohain (✉)
Department of Civil Engineering, Indian Institute of Technology Guwahati, Guwahati, Assam, India
e-mail: b.rituraj@iitg.ernet.in

D. Das · R. S. Choudhury · R. K. Bhattacharjya
Department of Civil Engineering, National Institute of Technology Silchar, Silchar, Assam, India
e-mail: dimpidas13@gmail.com

R. S. Choudhury
e-mail: rubia.choudhury@yahoo.com

R. K. Bhattacharjya
e-mail: rkbc@iitg.ernet.in

© Springer International Publishing AG 2018
A. K. Sarma et al. (eds.), *Urban Ecology, Water Quality and Climate Change*,
Water Science and Technology Library 84,
https://doi.org/10.1007/978-3-319-74494-0_19

1 Introduction

Embankment failure is a very common problem for the rivers in Assam. Stability of the slope of an earthen embankment can be evaluated using the circular slip circle method. For obtaining the factor of safety of the most critical slip circle, information about the subsurface strata is necessary. Most of the time, this information is either not available or scanty information is available. The information about the subsurface strata can be obtained using the Ground Penetrating Radar (GPR). It is a geophysical method which uses radar pulses to image the subsurface.

In the present study, GPR survey is conducted initially for acquiring the subsurface information of the earthen embankment. The information obtained by the GPR is used to obtain the factor of safety of the most critical slip circle. The methodology is applied to study the stability of the earthen embankment of river Kollong at Raha, Assam. The two sites near Raha, of river Kollong, have been suffering from frequent embankment failure. The water resources department of Government of Assam has taken some initiative to protect the bank of the river. As protection measures, concrete piles were applied on the banks. However, it could not prevent the sliding of the slope of the embankment. As a result of newly driven piles, the weight of the sliding mass of the slope has increased and thus, the slope mass slided along with the piles.

2 Theoretical Background

Stability of a slope can be investigated by a number of methods. Some of them are as follows.

2.1 Culmann's Method of Planer Failure Surface

Culmann (1866) considered a simple failure mechanism of a slope of homogeneous soil with plane failure surface passing through the toe of the slope. Culmann's method is suitable for very steep slopes.

2.2 The Swedish Circle Method or the Slip Circle Method

This is a general method developed by Swedish engineers, which is equally applicable to homogeneous soils, stratified soils, fully or partly submerged soils, nonuniform slopes, and also when seepage and pore pressure exist within the soil.

2.3 The Friction Circle Method

In this method, it is assumed that the resultant reaction is tangential to the friction circle.

2.4 Bishop's Method

Bishop (1955) took into consideration the forces acting on the sides of the slices, which were neglected in the Swedish method.

3 Study Area and Methods

3.1 Overview

At first, the area was surveyed using GPR. Bore log data were also collected from the site. For analyzing the stability of the surveyed area, slip circle method was used to find out the most critical slip circle corresponding to the minimum factor of safety. Different field conditions have been tried to evaluate the most critical condition of the sites.

3.2 Study Area

The area under study is described schematically in Fig. 1. A number of boreholes were dug in the area for investigating soil conditions. The area is located at the bank of Kollong River.

3.3 Methodology

Slip circle method: Fig. 2 shows the section of a slope with AB as the trial slip surface. The analysis of stability in the case of $c-\varnothing$ soil is carried out by the method of slices which was first introduced by Fellenius (1926). In this method, the soil mass above the assumed slip circle is divided into a number of vertical slices of equal width.

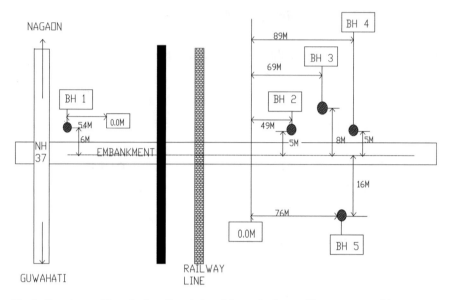

Fig. 1 Bore log positions in the affected site of the embankment (figure not to scale)

Fig. 2 $c-\varnothing$ analysis,
method of slices

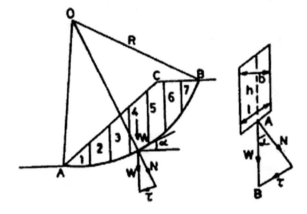

Considering the whole slip surface AB of length L, the total driving and resisting forces are

$$\text{Driving forces} = \sum T.$$
$$\text{Driving moment} = \sum TR.$$
$$\text{Resisting forces} = \sum c'L + \sum N \tan \varnothing'.$$
$$\text{Resisting moment} = \sum c'LR + \sum NR \tan \varnothing'.$$

Table 1 .

Fellenius's criteria for locating center circle of a slope in a $\Phi_u = 0$ soil

Slope angle (degree)	Slope ratio	Angle α* (degree)	Angle ψ* (degree)
60	1:0.58	29	40
45	1:1	28	37
	1:1.5	26	35
	1:2	25	35
	1:3	25	35
	1:5	25	37

Therefore, the factor of safety against sliding, F can be written as

$$F = \frac{\sum c'L + \tan \varnothing' \sum N}{\sum T},$$

where

N $W \cos \alpha$ and
T $W \sin \alpha$

The above equation can be written in the form

$$F = \frac{\sum c'L + \tan \varnothing' \sum W \cos \alpha}{\sum W \sin \alpha}$$

Fellenius (1936) proposed an empirical procedure to find the center of the most critical circle in a $\varnothing_u = 0$ soil.

The center can be located at the intersection of the two lines drawn from the ends A and B of the slope at angles α and ψ. The angles α and ψ vary with respect to slope B as shown in Table 1.

Junikis (1962) further extended this method to the case of a homogeneous $c' - \varnothing$, soil. After obtaining the center, point P is located at a distance $4.5H$ horizontally from the toe of the slope and H below the toe of the slope. The center of the critical circle is then assumed to lie on the extension of the line PO and the factors of safety obtained are plotted to obtain a locus from which the minimum factor of safety can be obtained.

3.4   GPR Data Collection

3.4.1   Background

GPR is a technology used for subsurface exploration which works on the principle of reflection of radio waves from underground features. The time of receiving the

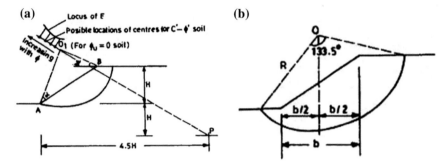

Fig. 3 **a** Toe failure, **b** base failure

reflections is used to compute the depth of the cause of reflection. The reflections are caused by changes encountered in dielectric constant within the medium of propagation.

3.4.2 GPR Survey

GPR survey was conducted at the two problematic locations of the river embankment of river Kollong near Raha. SIR-3000 GPR system was used in the survey with a MODEL 3200 MLF bistatic antenna of 80 and 35 MHz frequencies. For the first site, both the antennas were used. The depth of penetration for 80 MHz was 18 m and that of 35 MHz was 39 m. For the second and third sites, only 80 MHz antenna was used and the depth of penetration was 18 and 24 m, respectively. The depth of penetration is related to the dielectric constant that was set by default as 6.25. The dielectric constant was later reassessed using the available borehole information (Fig. 3).

4   Results and Discussions

4.1   Bore Log Data

The bore log data of the affected area has been collected for the classification of soil so that the various properties of the soil are known to us. The soil samples collected from the boreholes were taken to the lab for sieve analysis. The results of sieve analysis at each meter depth show that the soil from the borehole locations #3 and #4 is a bit coarse-grained in the top couple of meters but below that it becomes fairly fine-grained and cohesive in nature. The following two charts show an example of the variation of particle-size distribution with increasing depth. The results from 1 and 4 m depths, respectively, from borehole #3 are shown in Fig. 4a, b.

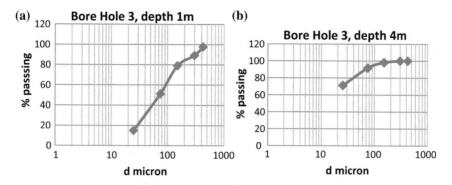

Fig. 4 **a** Sieve analysis result of borehole 3 at 1 m depth, **b** sieve analysis at 4 m depth

4.2 GPR Analysis

The data collected from the GPR was analyzed in RADAN software. RADAN is GSSI's GPR processing software. RADAN is an acronym that stands for **Ra**dar **D**ata **AN**alyzer.

The GPR plot acquired in the field is in Fig. 5. After processing, the quality of the plot was improved and it is shown in Fig. 6.

The GPR plot shows a dark region near 40 to 70 m zone in the horizontal axis, which is the approximate location of the boreholes 3 and 4. The dark region represents the clay deposit from where little reflections are coming thus, verifying the sieve analysis data (Fig. 7).

Fig. 5 GPR data (after stretching and removing air gap). The dielectric constant is 25

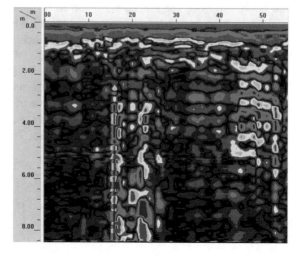

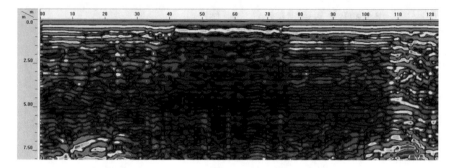

Fig. 6 GPR data showing the region of clay deposit (darker region)

Fig. 7 Slip circle method

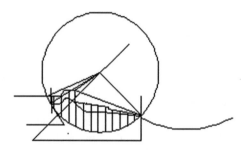

4.3 Stability Analysis

Stability analysis of the embankment is carried out using the slip circle method discussed above. The bulk density for clay is considered as 1.56, for sand as 1.976, and for sandy silt as 1.82. The Ø value for the clay soil is considered as 9° and for the sandy soil as 35°. The cohesion c for the clay soil is considered as 25 kPa, for the sandy soil as 0, and for the sandy silt as 0.

Various trials have been carried out to obtain the most critical factor of safety of the most critical slip circle. Table 2 shows the factor of safety obtained in various trials. It may be observed that the factor of safety of the most critical slip circle is 2.29. It seems that the slope is quite stable and the slope should not fail.

In addition to the above trials, one more trial is carried out in order to find out the stability of the slope by filling up the uneven area of the slope with a layer of sand of bulk density 1.976. In this case, the critical factor of safety of the slope comes up as 2.01. As such the slope is quite stable in nature. Therefore, the probable cause of

Table 2 .

Trial	Radius	θ°	L	FOS
I	15.92	103	28.628	2.29
II	20.14	47	16.522	2.39
III	17.00	79	23.439	2.35

the failure is the seepage pressure developed in the sliding mass due to the clay nature of the soil in the problematic zone. The detailed seepage analysis of the area has to be carried out to find out the exact cause of the failure.

5 Conclusion

After the complete analysis of the affected area, we came up with an approximate idea about the reason of the slope failure. Since the factor of safety is much greater than 1 in each trial, one can come to the conclusion that the slope failure was not due to the instability of the slope, but the failure may be due to the development of excessive pore pressure on the sliding mass. The presence of the clay deposit causes obstruction to the flow of groundwater and pore pressure in the region and in its neighborhood increases which gets released only through the failure of the slope.

References

Abramson LW, Lee TS, Sharma S, Boyce GM (2002) Slope stability and stabilization methods, 2nd edn, pp 329–367

Annan AP, Davis JL. Ground penetrating radar—coming of age at last!! Electrical and Electromagnetic Methods, Paper 66

Berilgen MM (2007) Investigation of stability of slopes under drawdown conditions. Comput Geotech 34(2007):81–91

Mello UT, Pratson LF (1999) Regional slope stability and slope-failure mechanics from the two-dimensional state of stress in an infinite slope. Mar Geol 154(1999):339–356

US Army Corps of Engineers Manual on Slope Stability. EM 1110-2-1902; 31 Oct 2003

Part IV
Urban Waste: Concerns and Remediation

Elimination of Chromium(VI) from Waste Water Using Various Biosorbents

Shubhrima Ghosh and Dipannita Mitra

Abstract In recent years, pollution of water due to the presence of heavy metal ions is a serious socio-environmental problem caused by the discharge of industrial effluents. Their toxicity, non-biodegradable nature and accumulation in food chains are a matter of concern and thus their removal deserves urgent attention for a cleaner biosphere. Chromium, a major heavy metal pollutant, in its hexavalent form is highly toxic and persistent. Conventional methods like precipitation, evaporation, ion-exchange, electrodialysis and membrane processes incur high costs, need high energy, require chemical reagents and are often inadequate for large-scale industrial applications. Biosorption represents a biotechnological innovation and an economical strategy for elimination of chromium from industrial discharge. The objective of this review study is to contribute in the search for promising cost-effective adsorbents and their utilization possibilities, from various materials of agricultural and biological origin such as sugarcane bagasse, sawdust, rice hull, etc., for the elimination of chromium from polluted water. Removal of chromium(VI) from wastewaters by algal, fungal and bacterial biomass is also discussed.

Keywords Biosorbents · Heavy metal · Hexavalent · Biomass

1 Introduction

Pollution of aquatic environments by heavy metals due to urbanization and industrialization is a serious problem worldwide. Their toxicity, non-biodegradable nature and persistency in the food chain pose great threat to living organisms. Lead,

S. Ghosh (✉)
Enzyme and Microbial Biochemistry Laboratory, Department of Chemistry,
Indian Institute of Technology Delhi, Hauz Khas, New Delhi 110016, India
e-mail: shubhrima.ghosh@gmail.com; cyz138123@chemistry.iitd.ac.in

D. Mitra
Department of Molecular Signal Processing, Leibniz Institute for Plant Biochemistry,
Weinberg 3, 06120 Halle (Saale), Germany

© Springer International Publishing AG 2018
A. K. Sarma et al. (eds.), *Urban Ecology, Water Quality and Climate Change*,
Water Science and Technology Library 84,
https://doi.org/10.1007/978-3-319-74494-0_20

cadmium and mercury are examples of heavy metals that have been classified as priority pollutants by the U.S. Environmental Protection Agency (Sun and Shi 1998). Chromium, one of the major pollutants of the environment exists in nature in stable hexavalent and trivalent forms. The hexavalent form of chromium is more toxic than trivalent chromium and is often present in wastewater as chromate (CrO_4^{2-}) and dichromate ($Cr_2O_7^{2-}$). Chromium gets introduced into natural waters from a variety of industrial wastewaters including those from the textiles, leather tanning, electroplating and metal finishing industries. The hexavalent form of chromium is considered a class 'A' carcinogen due to its mutagenic and carcinogenic nature (Helmer and Bartley 1971). Therefore, removal of Cr(VI) from wastewater prior to its discharge into natural water systems requires serious and immediate attention. Commonly used methods for the removal of chromium(VI) from waste waters include precipitation, evaporation, ion-exchange, electrodialysis and membrane processes. But disadvantages like incomplete metal removal, high reagent and energy requirements, generation of toxic sludge or other waste products that require careful disposal has made it imperative for a cost-effective treatment method that is capable of removing chromium from aqueous effluents. Biosorption of Cr(VI) using fungal, algal or bacterial biomass (growing, resting and dead cells) and agricultural waste materials has been recognized as a promising alternative to the existing conventional methods for detoxification of industrial wastewater (Sen and Dastidar 2010). These processes are advantageous over conventional methods because of their low-cost, increased metal removal, regeneration of biosorbent and possibility of metal recovery.

1.1 Biosorption Mechanisms

Biosorption can be defined as the removal of metallic ions by means of passive adsorption or complexation by living biomass or solid organic waste (Davis et al. 2003). The process involves a solid phase (sorbent; biological material) and a liquid phase (solvent; normally water) containing a dissolved species to be sorbed (sorbate; metal ions). Owing to the higher affinity of the sorbate for the sorbent, they are attracted to it and are subsequently removed. This process continues till equilibrium is established between the amount of sorbate deposited on the sorbent and its portion left in the solution. The binding mechanisms of heavy metals by biosorption could be explained by the physical and chemical interactions between cell wall ligands and adsorbents by ion-exchange, complexation, coordination, chelation, physical adsorption and micro-precipitation (Nouri et al. 2006). The diffusion of the metal from the bulk solution to active sites of biosorbents predominantly occurs by passive transport mechanisms and various functional groups such as carboxyl, hydroxyl, amino and phosphate existing on the cell wall of biosorbents which can bind to the heavy metals. A number of factors such as pH, initial metal ion concentration and biomass concentration in solution affect biosorption.

1.2 Natural Products as Biosorbents

1.2.1 Sugarcane Bagasse

Bagasse pitch is a waste product from sugar refining industry. It is the residual cane pulp remaining after sugar has been extracted. Bagasse pitch is composed largely of cellulose, hemicellulose, and lignin (Mohan and Singh 2002). It has been reported that an adsorbent dose of 0.8 g/50 ml is sufficient to remove 80–100% Cr(VI) from aqueous solution having an initial metal concentration of 50 mg/l at a pH value of 1 but the efficiency reduced sharply to 15% at pH 3 (Khan et al. 2001). The effect of solution pH, Cr(VI) concentration, adsorbent dosage and contact time were studied in a batch experiment. The removal was in general most effective at low pH values and low Cr(VI) concentration. The maximum removal obtained was around 99.8% at pH 2. Gupta and Ali (2004) studied the utilization of sugarcane bagasse ash in Cr (VI) removal in batch and column operations. In batch mode, the effect of contact time, concentration, pH, adsorbent dose, temperature, effect of interfering ions and particle size were optimized. They obtained 96–98% removal with 20 mg/l of chromium concentration with 10 g/l adsorbent dose at pH 5 in batch and 95–96% in column operations with a flow rate of 0.5 ml/min. The utilization types of sugar-cane bagasse as adsorbent is summarized in Table 1.

1.2.2 Sawdust

Sawdust is a widely available waste by-product of the timber industry that is either used as a cooking fuel or a packing material. Studies have shown that phosphate-treated sawdust (PSD) has greater sorption capacity of Cr(VI) as compared to untreated sawdust (Ajmal et al. 1996). Adsorption of Cr(VI) on PSD is highly pH dependent and the maximum adsorption of Cr(VI) is observed at pH 2. A general decrease in adsorption densities is observed as the adsorbent dose is increased from 0.2 to 3 g. 100% removal of Cr(VI) from synthetic wastewater as well as from actual electroplating waste containing up to 50 mg/l Cr(VI) was achieved by batch as well as column process. Removal of Cr(VI) from aqueous solution by adsorption onto activated carbon prepared from coconut tree sawdust has also been studied (Selvi et al. 2001). Batch mode adsorption studies were carried out by varying agitation time, initial Cr(VI) concentration, carbon concentration and pH. The adsorption of Cr(VI) was pH dependent and maximum removal was observed in the acidic pH range. Sawdust types and its efficiency are summarized in Table 1.

1.2.3 Rice Hull

Rice husk is an agricultural waste generated in rice-producing countries. Dry rice husk contains 70–85% of organic matter (cellulose, lignin, sugars, etc.)

Table 1 Types of agricultural wastes as adsorbent for Cr(VI)

Adsorbent	Initial concentration of Cr(VI) (mg/l)	Adsorbent dose (g/l)	% of Cr(VI) removed	Cr(VI) removed mg/g of biosorbent	pH	Reference
Activated bagasse carbon	5	2	99.97	2.5	2.5	Chand et al. (1994)
Raw sugarcane bagasse	50	16	80–100	–	1	Khan et al. (2001)
Bagasse ash	20	10	96–98	4.35	5	Gupta and Ali (2004)
Phosphate-treated sawdust	40	4	100	10	2	Ajmal et al. (1996)
Untreated sawdust	40	4	2	0.2	7	Ajmal et al. (1996)
Coconut tree sawdust carbon	5	4–15	98.84	3.46	3	Selvi et al. (2001)
Rice husk carbon	10	10	93.28	0.93	2	Bishnoi et al. (2004)
Raw rice bran	5	10	40	0.0048	1.5–2	Oliveira et al. (2005)
Raw rice husk	50	70	66.6	2.2	2	Subramaniam and Khan (1970)
Rice husk ash	10	40	59.12	0.49	2	Panjai Saueprasearsit (2010)

(Vempati et al. 1995). Studies on chromium removal by rice husk carbon showed greater than 93% removal of hexavalent chromium. Studies on adsorption of Cr(VI) on rice husk-based activated carbon have concluded that rice husk carbon is a good sorbent for the removal of Cr(VI) from aqueous solution with metal concentration ranging from 5 to 10 mg/l with adsorbent dose of 10 g/l at pH <5 under the minimum equilibration time of 1.5 h. Maximum reported adsorption resulted in >93% removal of Cr(VI). Maximum removal (66%) of Cr(VI) has been obtained for raw rice husk at pH 2, when it is given an adsorbent dose of 70 g/l for 2 h (Subramaniam and Khan 1970). Table 1 summarizes the usage of types of rice husk as an adsorbent.

1.2.4 Algal Biomass

Algae are photosynthetic organisms found abundantly along the coastline and beaches in both tropical and temperate regions of the world. They also occur in still waterbodies such as lakes and ponds. They are attractive biosorbents as they are ubiquitous in natural environment, have large surface area to volume ratio and high binding affinity to pollutants. The cell surface binding is the first step involved in the binding of Cr(VI) ions to algal species which is a rapid process and is metabolism independent. This is followed by the second step of intracellular accumulation of metal due to the simultaneous effect of growth and surface biosorption. This step is metabolism dependent and is a much slower process. Both growing and non-living cells of algae such as Chlorella, Cladophora and Spirogyra are capable of removing Cr(VI) and are shown in Table 2 (Deng et al. 2009; Gupta et al. 2001). Pena-Castro et al. (2004) studied Cr(VI) in continuous cultures of the microalga *Scenedesmus incrassatulus*. The effect of various factors such as pH, algal dosage, initial Cr(VI) concentration, temperature and coexisting ions are generally studied for optimal adsorption.

1.2.5 Fungal Biomass

Fungi and yeasts are the eukaryotic organisms which are used as biosorbents for the removal of heavy metals due to high percentage of biomass with excellent metal-binding properties. The biosorption mechanism of Cr(VI) ions by fungal biomass has been studied largely in relation to chitin, its deacylated derivatives, chitosan and cellulose. Metal ion uptake by fungal biomass similarly takes place as discussed in algal cells. The fungal cells can be killed for biosorption by physical and chemical methods. A wide range of fungal species under non-living condition have been studied by different researchers for the removal of Cr(VI) from the wastewaters under different conditions. They are able to adsorb metal through a combination of surface reactions, intracellular and extracellular precipitation and extracellular complexation reactions (Sen and Dastidar 2011). A summary of the usage of different types of fungi as adsorbent is given in Table 2.

Table 2 Types of algal, fungal and bacterial biomass as adsorbent for Cr(VI)

Adsorbent	Initial concentration of Cr(VI) (mg/l)	Adsorbent dose (g/l)	% of Cr(VI) removed	Cr(VI) removed mg/g of biosorbent	pH	Reference
Cladophora albida (green algae)	154	2	95	47.1	0.5–3	Deng et al. (2009)
Scenedesmus incrassalulus (green microalgae)	1.2	–	52.7	–	8	Pena-Castro et al. (2004)
Spirogyra species (green filamentous algae)	5	15	90	14.7	2	Gupta et al. (2001)
Aspergillus niger (filamentous fungi)	25	5	90	9.23	2	Issac and Prabha (2011)
Fusarium solani (filamentous fungi)	500	4.5	63.9	60	4	Sen and Ghosh (2011
Candida lipolytica (fungi)	30	1	96	6.35	1	Ye et al. (2010)
Bacillus sp. QC1-2 (bacteria)	58.2	–	100	–	7	Campos et al. (1995)
Staphylococcus saprophyticus (bacteria)	193.66	20	45	4.34	2	Ilhan et al. (2004)
Microbacterium liquefaciens MP30 (bacteria)	16.2	75	90–95	–	7	Pattanapipitpaisal et al. (2001)
Agrobacterium radiobacter EPS-916 (bacteria)	97	2.5×10^{13} cells/ml	100	–	7	LLovera et al. (1993)

1.2.6 Bacterial Biomass

Because of their small size, ubiquity and ability to grow under a wide range of environmental situations, numerous studies have identified potential bacterial species for Cr(VI) accumulation. *Bacillus* sp. has been reported to have a very high potential for Cr(VI) removal from waste water (Campos et al. 1995). The complexity of microorganisms' structure implies that there are many ways for the metal to be captured by the cell. Various mechanisms involved in metal removal by bacteria include cell surface binding, extracellular precipitation, intracellular accumulation, oxidation and reduction and methylation/demethylation. The ability of bacterial cells to bind metal ions is associated with its cell envelope, which is formed of many layers and separates the cell protoplasm from the surrounding area. The metal binding to the surface occurs through a passive mechanism which involves stoichiometric interaction of the metal with the reactive chemical groups, followed by intracellular accumulation due to the simultaneous effects of growth and surface biosorption. The chromium adsorption efficiency of various bacterial species has been shown in Table 2.

2 Conclusions

Biosorption is being demonstrated as a useful alternative to conventional systems for the removal of toxic metals such as chromium from industrial effluents. Research over the past few decades has provided a better understanding of metal biosorption by potential biosorbents. The group of cheap biosorbents of natural and waste origin constitutes the basis for new cost-effective technology for removal of chromium from waste water. The sorption capacity is dependent on the type of the adsorbent investigated and the nature of the wastewater treated. The development of such processes requires further investigation in the direction of mathematical modelling for process optimization and for providing predictions of the biosorption process under different operating conditions. For commercial exploitation, strategies need to be developed where further processing of the biosorbent can regenerate the biomass as well convert the recovered metal into a usable form.

References

Ajmal M, Khan Rao RA, Siddiqui BA (1996) Studies on removal and recovery of Cr(VI) from electroplating wastes. Water Res 30(6):1478–1482

Bishnoi NR, Bajaj M, Sharma N, Gupta A (2004) Adsorption of Cr(VI) on activated rice husk carbon and activated alumina. Biores Technol 91(3):305–307

Campos J, Martinez-Pacheco M, Cervantes C (1995) Hexavalent-chromium reduction by a chromate-resistant *Bacillus* sp. strain. Antonie Van Leeuwenhoek 68(3):203–208

Chand SH, Agarwal VK, Kumar P (1994) Removal of hexavalent chromium from wastewater by adsorption. Indian J Environ Health 36(3):151–158

Davis TA, Volesky B, Mucci A (2003) A review of the biochemistry of heavy metal biosorption by brown algae. Water Res 37(18):4311–4330

Deng L, Zhang Y, Qin J, Wang X, Zhu X (2009) Biosorption of Cr(VI) from aqueous solutions by nonliving green algae *Cladophora albida*. Miner Eng 22(4):372–377

Gupta V, Shrivastava A, Jain N (2001) Biosorption of chromium(VI) from aqueous solutions by green algae Spirogyra species. Water Res 35(17):4079–4085

Gupta VK, Ali I (2004) Removal of lead and chromium from wastewater using bagasse fly ash—a sugar industry waste. J Colloid Interface Sci 271(2):321–328

Helmer LG, Bartley EE (1971) Progress in the utilization of urea as a protein replacer for ruminants. A Review1. J Dairy Sci 54(1):25–51

Ilhan S, Nourbakhsh MN, Kiliçarslan S, Ozdag H (2004) Removal of chromium, lead and copper ions from industrial waste waters by *Staphylococcus saprophyticus*. Turk Electron J Biotechnol 2:50–57

Issac R, Prabha L (2011) Equilibrium and Kinetic studies on biosorption of Cr(VI) by non-living mycelial suspensions of *Aspergillus niger*. Int J Pharma Biosci 2(3):1–7

Khan NA, Ali SI, Ayub S (2001) Effect of pH on the removal of Chromium (Cr)(VI) by Sugar Cane Bagasse. Sci Technol 6:13–19

LLovera S, Bonet R, Simon-Pujol MD, Congregado F (1993) Effect of culture medium ions on chromate reduction by resting cells of *Agrobacterium radiobacter*. Appl Microbiol Biotechnol 39(3):424–426

Mohan D, Singh KP (2002) Single- and multi-component adsorption of cadmium and zinc using activated carbon derived from bagasse—an agricultural waste. Water Res 36(9):2304–2318

Nouri J, Mahvi A, Babaei A, Jahed G, Ahmadpour E (2006) Investigation of heavy metals in groundwater. Pak J Biol Sci 9(3):377–384

Oliveira EA, Montanher SF, Andrade AD, Nóbrega JA, Rollemberg MC (2005) Equilibrium studies for the sorption of chromium and nickel from aqueous solutions using raw rice bran. Process Biochem 40(11):3485–3490

Panjai Saueprasearsit JW (2010) Biosorption of chromium(VI) using rice husk ash and modified rice husk ash. Environ Res J 4(3):244–250

Pattanapipitpaisal P, Brown NL, Macaskie LE (2001) Chromate reduction by Microbacterium liquefaciens immobilised in polyvinyl alcohol. Biotech Lett 23(1):61–65

Pena-Castro JM, Martinez-Jeronimo F, Esparza-Garcia F, Canizares-Villanueva RO (2004) Heavy metals removal by the microalga *Scenedesmus incrassatulus* in continuous cultures. Bioresour Technol 94(2):219–222

Selvi K, Pattabhi S, Kadirvelu K (2001) Removal of Cr(VI) from aqueous solution by adsorption onto activated carbon. Biores Technol 80(1):87–89

Sen M, Dastidar MG (2010) Chromium removal using various biosorbents. Iran J Environ Health Sci Eng 7(3):189

Sen M, Dastidar MG (2011) Biosorption of Cr(VI) by resting cells of *Fusarium solani*. Iran J Environ Health Sci Eng 8(2):153

Subramaniam P, Khan N (1970) The effects of various parameters on Cr(VI) adsorption by raw rice husk. Malays J Sci 25(1)

Sun G, Shi W (1998) Sunflower stalks as adsorbents for the removal of metal ions from wastewater. Ind Eng Chem Res 37(4):1324–1328

Vempati RK, Musthyala SC, Mollah MYA, Cocke DL (1995) Surface analyses of pyrolysed rice husk using scanning force microscopy. Fuel 74(11):1722–1725

Ye J, Yin H, Mai B, Peng H, Qin H, He B, Zhang N (2010) Biosorption of chromium from aqueous solution and electroplating wastewater using mixture of *Candida lipolytica* and dewatered sewage sludge. Biores Technol 101(11):3893–3902

Biodegradation Kinetics of Toluene, Ethylbenzene, and Xylene as a Mixture of VOCs

Aviraj Datta and Ligy Philip

Abstract Toluene, xylene and ethylbenzene constitutes more than 50% of the total VOC emissions from manufacturing and application processes of surface coatings. Although biodegradability of these compounds has been reported by many researchers, the biodegradation characteristics are mostly measured for their biodegradation as a single substrate. In industrial VOC emissions, these compounds are mostly emitted as mixture. Hence, it is important to study the biodegradation of these compounds as a mixture with different relative concentrations of individual VOCs. The present work focused on the biodegradation of toluene, ethylbenzene, and xylene by acclimatized mixed culture under aerobic condition. The experimental data were fitted into various biokinetic models to predict the individual biokinetic parameters, viz., maximum specific growth rate (μ_{max}), half saturation constant (K_s), inhibition constant (K_i), and yield coefficient (Y_T). Monod's kinetic model, Monod's inhibition, and Haldane model were used for this purpose. The results indicated that for the same initial concentrations of the three mono-aromatic compounds, xylene degradation took longer time than ethylbenzene and toluene though its presence in the mixture did not have a significant inhibitory effect on the biodegradation of this two substrates. Four inhibition models, viz., no-interaction, competitive, uncompetitive, and noncompetitive models, were applied to study the nature of inhibition for different combinations of these compounds. The biodegradation of these VOCs was also studied in presence of easily biodegradable hydrocarbons like glucose.

Keywords Biokinetic parameters · Yield coefficient · Acclimatized mixed culture

A. Datta · L. Philip (✉)
Department of Civil Engineering, Indian Institute of Technology Madras,
Chennai, India
e-mail: ligy@iitm.ac.in

© Springer International Publishing AG 2018
A. K. Sarma et al. (eds.), *Urban Ecology, Water Quality and Climate Change*,
Water Science and Technology Library 84,
https://doi.org/10.1007/978-3-319-74494-0_21

1 Introduction

Volatile organic compounds (VOCs) include most solvent thinners, degreasers, cleaners, lubricants, and liquid fuels. VOCs include most solvent thinners, degreasers, cleaners, lubricants, and liquid fuels. Control techniques for anthropogenic VOC emission have gained importance after understanding its role in inducing high ground level ozone concentration through photochemical reactions. Particularly in the urban atmosphere in presence of nitrogen oxides (NO_x) and other photoactive precursors, VOCs catalyze different chain reactions, resulting in buildup of high ground level ozone concentration (Sillman 1999). High localized ambient ozone concentration in the troposphere poses a health threat, by causing respiratory problems. In addition, high concentrations of tropospheric ozone can damage crops and buildings because of its oxidizing nature (WHO 2003). The aromatic compounds benzene, toluene, and xylene are suspected carcinogens and may lead to leukemia through prolonged exposure (Blair et al. 1998). The wide industrial application of toluene, benzene, and ethylbenzene as chemical intermediates, solvents or precursors for paints and lacquers, thinners, rubber products, adhesives, inks, cosmetics, and pharmaceutical products (NTP 2005) is the reason that these three compounds are among the most abundantly produced chemicals, with worldwide annual production of 5–10 million tons of toluene (ATSDR 2000), 5–10 million tons of ethylbenzene (IPCS 1996), and 10–15 million tons of xylenes (IPCS 1997). The paint industry is one of largest anthropogenic sources of toluene, ethylbenzene, and o-xylene mixtures into the atmosphere. Indian paints industry is estimated at $6.25 billion out of which organized sector accounts for 65% ($\sim$\$ 4.1 billion) of the total market (Web reference 1). Top five organized players control 80% of the organized market in India. In the unorganized segment, there are about 2000 units having small- and medium-sized paint manufacturing plants. Demand for paints comes from two broad categories: decorative and industrial. Decorative and industrial coatings accounted for 72 and 28% of the total Indian paints and coatings market, respectively (Web reference 2). The paint industry can easily grow at 12–13% annually over the next few years from its current size. The per capita paint consumption in India which is a little over 4 kg is still very low as compared to the developed western nations. Therefore, as the country develops and modernizes, the per capita paint consumption is bound to increase. Industry players expect close to 12% growth in business volume and 10–12% rise in sales in the financial year 2016. Industrial coatings are expected to witness a higher growth rate compared to decorative coatings, owing to the increasing number of industrial facilities in India. Major paint application industries that include car, ship, and aircraft industrial coatings are expected to witness a higher growth rate compared to decorative coatings, owing to the increasing number of industrial facilities in India. This rise in paint production and application processes triggers more and more volumes of VOCs being emitted into the atmosphere. Hence, there is a necessity to develop sustainable and economical VOC treatment techniques to curb this pollution. Unlike petrochemical industries, the heat value of the emitted VOC streams from paint manufacturing and application processes being significantly lower high energy intensive thermal oxidation processes becomes unviable. The adsorption coupled thermal processes result in an increase of carbon emission. Biological VOC treatment techniques surely are the most sustainable solution to VOC control technique. The process essentially utilizes the "microbial energy" to oxidize the VOCs unlike any external energy input apart from that needed to maintain optimal microbial activity. However, to enhance the reliability and predictability of biological VOC treatment processes deeper, understanding of the biodegradation kinetics of these pollutants is necessary. Though there are previous works by numerous researchers (Deshusses 1994; Babu et al. 2004; Wu et al. 2006) on biodegradation of these compounds, there is a need to find out biodegradation pattern of these VOCs as a mixture. The study used an acclimatized mixed culture for the treatment of toluene, ethylbenzene, and o-xylene alone and when present together.

2   Literature Review

Various techniques for off-gas treatment include adsorption on activated carbon, physicochemical methods, thermal incineration, scrubbing, (electro) filtration, catalytic oxidation, dry chemical treatment, and biological methods such as biofiltration, bioscrubbing, and biotrickling filtration (Lee et al. 2002). Oxidation of VOCs at elevated temperature of 1300–1800 °C gained popularity owing to its high and reliable removal efficiency. Because of the high energy cost, however, only in industries such as petrochemical refineries with a flare burner already in place, VOC incineration seems justifiable. When coupled with partial heat recovery system, the process looks attractive particularly for the treatment of exhaust stream consisting of xenobiotics or recalcitrant compounds (Nacken et al. 2007). Catalytic oxidation systems operate at a lower temperature typically about 350–500 °C. Destruction efficiencies in excess of 90% are common for thermal oxidation processes (Khan and Ghoshal 2000). Use of catalysts reduces the combustion energy requirements. The high operating cost associated with high-temperature requirement for oxidation process can thus be significantly reduced with the use of catalysts such as thermally aged Ce/Zr mixed oxides (Rivas et al. 2009), TiO_2–V_2O_5–WO_3, or cobalt oxide (Lojewska et al. 2009). An overview of catalytic combustion of chlorinated VOCs is presented nicely in the article published by Everaert and Baeyens (2004). The work of Li et al. (2009) gives a thorough overview of the application of different non-noble metal catalysts in VOC abatement processes (Mitsui et al. 2008). Application of nonthermal plasma for catalytic and non-catalytic VOC removal is an emerging technique in this field (Vandenbroucke et al. 2010). Catalytic combustion, however, may not be economically feasible for the treatment of low heat value VOC emissions with very low concentration VOCs (Baek et al. 2004). For the control of the low concentration VOCs, a good combined adsorbent/catalyst dual functional system was reported by Baek and Ihm (2004). Adsorption process which was initially applied for odor removal from exhaust gas with straw has come a long way, and researchers over the years have successfully removed VOCs such as acetone, BTEX, and trichloroethylene (Engleman et al. 2000; Hamdi et al. 2004). Unlike oxidation, adsorption of VOCs merely transfers the pollutants from air phase to a solid adsorbent matrix (Qu et al. 2009). Adsorbent materials commonly used are zeolites (Tao et al. 2004), activated carbon, activated alumina (Zaitan et al. 2008), silica gel (Dou et al. 2011), etc. These collection/capture techniques merely generate a concentrated pollutant stream in water or solid matrix. Adsorption aided with thermal processes find application where the VOC concentration in the gas is too dilute. Thermal processes such as incineration, thermal oxidation, and catalytic oxidation are limited in their application because of higher cost and energy required. All these processes actually result in more carbon in the atmosphere than original load in the inlet point because of additional heat and fuel requirement.

The first evidence of application of bioreactors for the treatment of odorous gas appeared in the 1950s (Groenestijn and Kraakman 2005). These processes utilize the ability of microorganisms to degrade a variety of inorganic and organic compounds. These techniques were initially applied to remove odorous compounds, such as ammonia, H_2S, and other organo-sulfur compounds from wastewater treatment plants (Schlegelmilch et al. 2005). The biological VOC removal mechanism involves absorption of volatile contaminants in an aqueous phase or biofilm followed by oxidation through the action of microorganisms (Brauer 1986). Under appropriate environmental conditions, microorganisms can oxidize numerous compounds (substrates) into mineral end products and new cell material (Aizpuru et al. 2001). The batch biodegradation works reported can be broadly divided into two types: pure culture biodegradation studies and mixed culture biodegradation studies. A certain VOC present in a mixture may show toxicity to the microbes degrading another VOC (Okpokwasili et al. 2006). Barker et al. (1987) commented that the biodegradation rates of the BTEX compounds in a series of microcosm experiments were very similar to each other. The BTEX compounds, when arranged by median rate constant, show the order: toluene > ethylbenzene > benzene > m-xylene, o-xylene, p-xylene. In general, however, the differences in rate constant values between the BTEX compounds are not very high. However, acetone as sole carbon source is reported to be easily biodegradable with the acclimatized microorganism (Shim and Yang 1999). Novak et al. (1985) have reported the degradation of 1000 mg/L of methanol under natural environment without any toxicity. Most of the liquid batch studies that have been carried out in the past focused primarily on the degradation of individual substrates with a microbial population acclimatized to the same substrate. A few studies have assessed the possibility of degrading a variety of compounds by one particular strain of microbes that was previously acclimatized with a single substrate. Shim et al. (2002) have studied the potential of toluene degrading microorganisms to degrade B,

T, E, and X at high concentrations, ≤ 1000 mg/L. Singh et al. (2011) isolated strains of solvent tolerant *Pseudomonas putida* which exhibited solvent stability of its psychro-thermoalkalistable protease. Gibson et al. (1970) isolated strains of *P. putida* F1 that was able to grow on two or more aromatic compounds, and most of these studies have been carried out at low concentrations (<70 mg/L) of pollutants. This particular strain used toluene, benzene, ethylbenzene, phenol, and other aromatics as the sole carbon source. Several studies conducted during the last two decades demonstrated that aerobic mixed populations, grown on hydrocarbons such as methane, propane, butane, and phenol, can effectively degrade most chlorinated aliphatic hydrocarbons by means of co-metabolism (Priya and Philip 2013). These chlorinated compounds may be co-metabolized in the presence of a more easily biodegradable non-chlorinated hydrocarbon serving as the primary carbon and energy source. Aerobic co-metabolism has been exploited extensively for bioremediation of chlorinated aliphatic compounds such as trichloroethylene in groundwater and in some cases for chlorinated benzenes and phenols (Balasubramanian et al. 2011). Bielefeldt and Stensel (1999) studied the degradation of a multi-substrate system consisting of benzene, toluene, ethylbenzene, o-xylene, and p-xylene, using mixed cultures from a manufactured gas site and modeled the system with competitive inhibition kinetics. Field studies investigating the fate of acetone in a shallow stream reported a biodegradation rate constant of 1.2 per day. The rate constant was estimated by subtracting the volatilization rate constant from the overall loss rate constant. Pugh et al. (1996) reported that though acetone was not biodegraded over 30 days in a non-amended microcosm using soil and groundwater from a contaminated solvent storage site. However, when nitrogen and phosphorus were added, however, half-lives reduced from 49.5 to 20 days. A second study by the same authors using pre-acclimated soil (over 30 days), reported a half-life of 3.6 days for acetone. Application of biofiltration techniques, such as biofilters, bioscrubbers, and biotrickling filters, has been a matter of extensive research over the last few decades. Though most of these were conceived as odor control techniques, their potential in VOC control is increasingly being recognized. Industrial application of these techniques is attractive, for their low operation and maintenance costs, potential for high contaminant removal efficiency, safety, environmental acceptability, and non-generation of harmful by-products, and they are highly preferred to the conventional physical and chemical methods. Condensation and absorption too when followed by a biological treatment can greatly enhance the efficiency of the later, and lead to complete mineralization of VOCs.

3 Materials and Methods

3.1 Materials

Toluene, ethylbenzene, and xylene used in the present study were of HPLC grade procured from Rankem (India). The chemicals used for the preparation of minimal salt media were of analytical grade procured from Rankem (India).

3.2 Minimal Salt Media (MSM)

All bacterial strains were grown at 30 °C in minimal salt medium (MSM). MSM has the following composition (in g/L): K_2HPO_4—0.8, KH_2PO_4—0.2, $CaSO_4 \cdot 2H_2O$—0.05, $MgSO_4$–$7H_2O$—0.5, $(NH_4)_2SO_4$—1.0, and $FeSO_4$—0.01 in one liter of distilled water. The value of pH of MSM obtained was 6.7 after the addition of all the salts (Raghuvanshi 2009). MSM were autoclaved for 15 min at 121 °C and 15 psi.

3.3 Microorganism Culture Conditions

The development of enriched culture was carried out in MSM with target chemical/s as the sole carbon source. The source for the microbes was the activated sludge collected from the aeration tank of the nearby sewage treatment plant (Nesapakkam, Chennai, Tamil Nadu). The activated sludge obtained was allowed to settle for 4 h. 10 g of settled sludge was taken and thoroughly mixed with 100 mL of physiological saline water in a beaker. Shaking was carried out gently, and then, the sludge was allowed to settle for 1 min in order to screen out the inorganic solid particles. Supernatant was then taken in a 50 mL centrifuge tube. The centrifugation was carried out for 2 min at 10,000 rpm in a centrifuge (REMI R-23 and 24, India). The supernatant was removed carefully from the top of the centrifuge tube without disturbing the pellet. The shake flask study was carried out by taking out the pellet with the help of a loop and transferred into a 500 mL flask containing 200 mL MSM in an aseptic environment with respective carbon sources (Raghuvanshi and Babu 2009).

3.4 Enrichment Procedure

Cultures were grown in 500 mL reagent bottles (Borosil), equipped with silicon septum at the mouth, containing 200 mL of MSM. The cultures were kept in a BOD-Orbital Shaker (REMI), maintained at 30 °C temperature and 100 rpm speed.

3.5 Inoculation Procedure

One hundred milliliter of the stock culture was taken in two 50 mL capped tubes and was centrifuged (REMI R-23 and 24) at 10,000 rpm for 2 min. The pellets were re-suspended in 100 mL of fresh MSM. The centrifugation and subsequent re-suspension steps were repeated one more time to ensure no residual VOC is present in the inoculum. Inoculation was done by adding 2 mL of this inoculum to 200 mL of MSM. The initial OD (540 nm) was around 0.05.

3.6 Measurement of Cell Density

Optical densities (OD) of bacterial suspension at different dilutions were monitored at 540 nm using UV–Visible spectrophotometer (Techcomp, UK). Known volumes, of different dilutions of bacterial suspension, were filtered through 0.45 μm (Whatman) filter papers. The difference between initial and final weight of the dried filter papers gave the corresponding mixed liquor suspended solid (MLSS) concentrations of these dilutions. A plot between OD (540) and MLSS was prepared as standard graph. During the different batch studies, the OD reading was taken at discrete time intervals, and corresponding MLSS concentrations were calculated using this OD versus MLSS standard graph.

3.7 Gas Chromatographic Analyses

Perkin Elmer Clarus 500 gas chromatograph with flame ionization detector (GC-FID) was used for analyzing residual VOC concentrations in liquid samples. GC was equipped with an autosampler, an on-column, split/splitless capillary injection system, and a capillary column (Perkin Elmer Elite (PE)-624, 30 m × 0.53 mm × 0.5 mm film thickness). During the analysis, the column was held initially at an oven temperature of 90 °C for 10 min. Temperatures of injector and detector were maintained at 150 and 300 °C, respectively. Nitrogen was used as the make-up and carrier gas at flow rates of 60 and 1.0 mL/min, respectively. Standard graphs for respective solvents were prepared individually by injecting known amounts of respective compound into a sealed reagent bottle equipped with Teflon septum as per the standard method. Liquid samples were then transferred to GC vials and analyzed by GC-FID (USEPA 2001).

3.8   Sample Preparation

At a particular time interval, 2 mL of samples was collected in a 2-mL air-tight centrifuge tube (Eppendorf, Germany). Then the samples were centrifuged at 10,000 rpm for 2 min. The supernatants thus collected were then transferred to 1.8 mL screw-cap GC vials with micropipettes.

3.9   Batch Degradation Studies

Batch studies were performed in 100 mL reagent bottles, sealed with silicone septum and kept in orbital shaker (REMI) at 100 rpm at 30 °C for the study. Fifty milliliters of MSM was taken in each reagent bottle and autoclaved properly. Known volumes of VOCs were injected in, sealed serum bottles containing 50 mL of sterilized MSM, with gas-tight syringes (SGE, Australia). Initial concentrations were noted after an equilibration time of four hours. Inoculation was done to attain an initial OD_{540} of approximately 0.05 to initiate the biodegradation. Liquid samples (5 mL each) were withdrawn at discrete time intervals, and two milliliters from each sample was subjected to close centrifuging in order to remove microbes. The supernatant was then utilized for GC-FID analysis. Remaining three milliliters of each sample was used for measuring optical density. Both biotic and abiotic controls were employed and all samples were analyzed.

3.10   Mathematical Models

3.10.1   Single Substrate Biodegradation Kinetics

Several kinetic models are available in the literature to find out the biokinetic parameters of the microbial system (Monod 1949; Haldane 1968; Lourn et al. 2006). Batch kinetic data were fitted to these models to determine the biokinetic parameters such as maximum specific growth rate (μ_{max}), yield coefficient (Y_T), half saturation concentration (K_s), and inhibition concentration (k_i) for the growth of microbes in corresponding substrates.

3.10.2 Mixed Substrate Biodegradation Kinetics

The most common model used for describing microbial growth in substrate mixtures is one in which the specific growth rate is the sum of the specific growth rates on each substrate i (μ_i). In its simplest form, Monod expressions are used for each μ_i, yielding a model in which the presence of one substrate does not affect the biodegradation rate of the other. For a mixture with n substrates (say $i = 1$ to n), general governing equations for this no-interaction sum kinetics model are as follows:

$$\frac{dM}{dt} = (M) * \sum_{i=1}^{n} \left(\frac{\mu_{maxi} * S_i}{K_{S_i} + S_i} \right), \tag{1}$$

$$\frac{dS_i}{dt} = -\left(\frac{M}{Y_{Ti}} \right) * \sum_{i=1}^{n} \left(\frac{\mu_{maxi} * S_i}{K_{S_i} + S_i} \right), \tag{2}$$

in which M is the biomass concentration in mg/L, S_i is concentration of ith substrate in mg/L, μ_{maxi} is the maximum specific growth rate of bacteria when only ith substrate is present in 1/h, K_{S_i} is the half saturation constant for the ith substrate in mg/L when only that substrate is present, Y_{Ti} is the yield coefficient when only ith substrate is present, and t is the time from the start of the process.

As the metabolic pathway used in the catabolism of both substrates is same, it may be possible that the two substrates compete for the active site on any of the enzyme in the pathway. A sum kinetics model incorporating purely competitive substrate kinetics was proposed by Yoon et al. (1977). For a mixture of substrates, the general governing equation for microbial growth under competitive substrate biodegradation kinetics is

$$\frac{dM}{dt} = (M) * \sum_{i=1}^{n} \left(\frac{\mu_{maxi} * S_i}{K_{S_i} + S_i + \sum_{j=1}^{n} \left(\frac{K_{S_i}}{K_{S_j}} * S_j \right)} \right). \tag{3}$$

Another form of interaction between an enzyme and two substrates is non-competitive inhibition. In this type of inhibition, a nonreactive complex is formed when both substrates are simultaneously bound to the enzyme (Balasubramanian et al. 2010). The general governing equation for cell growth under noncompetitive interaction for a mixture of substrates is

$$\frac{dM}{dt} = (M) * \sum_{i=1}^{n} \left(\frac{\mu_{maxi} * S_i}{(K_{S_i} + S_i) * \left(1 + \sum_{j=1}^{n} \frac{S_j}{K_{S_j}} \right)} \right). \tag{4}$$

In case of uncompetitive enzyme inhibition, one of the compounds (the inhibitor) can bind only to the enzyme–substrate complex and not to the free enzyme. The general governing equation for cell growth for a mixture of substrates under uncompetitive substrate interaction conditions is

$$\frac{dM}{dt} = (M) * \sum_{i=1}^{n} \left(\frac{\mu_{maxi} * S_i}{K_{S_i} + S_i * \left(1 + \sum_{j=1}^{n} \frac{S_j}{K_{S_j}}\right)} \right).$$

(5)

In each of the above cases, the substrate depletion follows the same equation as that for biomass growth, modified with a factor $-(1/Y_t)$, as given in Eq. (2). Different model performances were statistically evaluated using the dimensionless modified coefficient of efficiency (Balasubramanian et al. 2010),

$$E = 1 - \frac{\sum_{i=1}^{n} \left[|E(t_i) - O(t_i)|\right]}{\sum_{i=1}^{n} \left[|O(t_i) - \overline{O}|\right]},$$

(6)

where $E(t_i)$ is the numerically simulated value of a variable at time t_i, $O(t_i)$ is the experimentally observed value of the same variable at time t_i, and $\overline{O}$ is the mean value of the observed variable ranges between $-\infty$ and 1.0, higher values indicating better model prediction. Positive values of E represent an acceptable simulation, whereas $E > 0.5$ represents a good simulation. E equal to one indicates a perfect simulation.

4 Results and Discussion

For all the three mono-aromatic compounds, biodegradation patterns showed distinct lag phase, logarithmic phase as well as stationary phase. Toluene biodegradation pattern was studied for five different set of concentrations. About 80–100% removal of toluene was observed within 15 h irrespective of the initial concentration, implying good acclimatization of the culture. The complete biodegradation of Toluene even up to a concentration of 500 mg/L was achieved within 20 h (Fig. 1). Ortho-xylene was the last compound to degrade in all the cases. Even for an acclimatized culture, the long lag phase observed during the biodegradation of both ethylbenzene as well as o-xylene as single substrate is remarkable. During acclimatization, mixture of mono-aromatic compounds was used. An alternate approach could have been acclimatization of three separate cultures with toluene, ethylbenzene, and o-xylene, respectively, and utilizing the composite enriched culture for the biodegradation study. However, such a route could not ensure enrichment of microbes tolerant to all three target mono-aromatic compounds. Moreover, toluene dehydrogenase is a broad spectrum and ubiquitous enzyme widely (Alvarez and Vogel 1991) reported for its capacity to degrade common

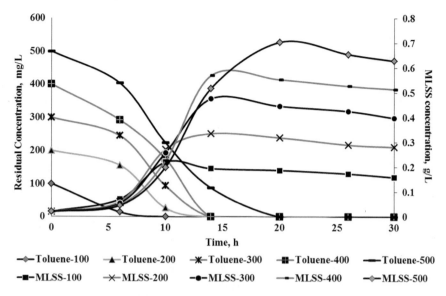

Fig. 1 Kinetics of toluene biodegradation as a single substrate for various initial concentrations

mono-aromatic compounds such as toluene, ethylbenzene, as well as o-xylene. The experimental data of only the logarithmic phase of biodegradation were fitted against different biokinetic models to find out the individual biokinetic parameter of these compounds. Three models were studied; they are Haldane, Monod growth, and Monod inhibition model. Monod growth model gave the best predictions with a higher E value. The biokinetic parameters given by Monod growth model for these compounds are given in Table 1 with corresponding E values. The single substrate biodegradation kinetics for both ethylbenzene and o-xylene was also carried out for an initial concentration of 100 mg/L. Unlike toluene and ethylbenzene, biodegradation of o-Xylene (Fig. 2) exhibited a longer lag phase compared to ethylbenzene (Fig. 3) followed by rapid degradation. The overall biodegradation took 48 h an initial concentration of 100 mg/L of o-xylene. Previous researchers also reported slower biodegradation of ethylbenzene compared to toluene when present as a sole

Table 1 Different biokinetic parameters of toluene, ethylbenzene, and o-xylene by Monod growth model

Solvents	μ^m (h^{-1})	K_s (mg/L)	Y_t	E values obtained for different initial substrate concentrations (So) (mg/L)				
				Initial substrate concentrations				
				100	200	300	400	500
Toluene	0.3412	1.0045	0.9607	0.6564	0.7533	0.7639	0.7372	0.8039
Ethylbenzene	0.0789	0.0539	0.8797	0.9610	–	–	–	–
o-Xylene	0.081	11.731	0.4067	0.7081	–	–	–	–

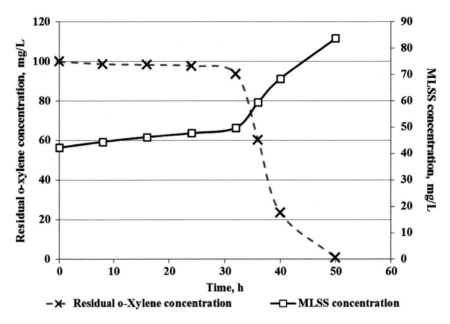

Fig. 2 Kinetics of xylene biodegradation as a single substrate

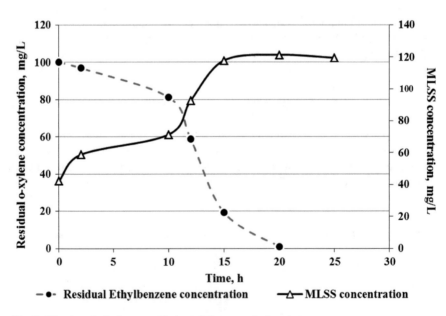

Fig. 3 Kinetics of ethylbenzene biodegradation as a single substrate

carbon source (Chang et al. 1993). The yield coefficient obtained for o-xylene biodegradation using Monod growth model was lower than the other two mono-aromatic compounds. Previous researchers also reported xylene as one of the least biodegradable of the common mono-aromatic compounds or even non-biodegradable especially under oxygen-limited conditions (Corseuil and Weber 1994). The low specific growth rate observed for both o-xylene and ethylbenzene (approximately 0.08 for both) was in agreement with the specific growth of 0.011/h reported by Shim et al. (2006) for these two compounds. The value of Ks was found to be 1 mg/L, which is higher than the values (0.23 mg/L) reported by previous researchers (Bielefeldt and Stensel 1999). The biodegradation pattern of toluene, ethylbenzene, and o-xylene as a single substrate by the acclimatized open mixed bacterial culture clearly exhibited a significant difference in the length of the lag phase observed for the three compounds for same initial concentrations of 100 mg/ L. Moreover, the long lag phase was always followed by rapid biodegradation. As reported by previous researchers (Alvarez and Vogel 1991), biodegradation of these mono-aromatics often takes place by enzymes of broad substrate specificity such as toluene dehydrogenase. These types of broad-spectrum enzymes enable microbes to utilize a wider range of carbon sources as substrates while also increasing their tolerance to these compounds. The induction efficiency of these broad substrate enzymes may vary from substrate to substrate. However in absence of the key substrate which induces the broad spectrum enzyme most effectively, e.g., toluene in case of toluene dehydrogenase, such prolong lag phases may be observed for other co-substrates. The biodegradation pattern thus observed may be explained by a relatively higher induction capacity of toluene, followed by ethylbenzene and o-xylene for a common broad substrate enzyme responsible for their biodegrada-tion. The effect of these mono-aromatic on the biodegradation pattern of each other was further investigated through biodegradation as tertiary VOC mixtures. The biodegradation kinetics of ethylbenzene (E), toluene (T), and o-xylene (X) as ter-tiary mono-aromatic mixtures (termed as ETX) were carried, out and the biodegradation kinetics is given in Fig. 4. All the tertiary mixtures studied con-tained the 100 mg/L each of ethylbenzene and o-xylene concentrations. Irrespective of their actual concentration in an emission, low water solubility of ethylbenzene and o-xylene restricts their highest possible bioavailable concentration. The initial toluene concentration was varied from 100 to 500 mg/L, and the mixtures were termed as ETX-a, where "a" is the initial toluene concentration in that particular tertiary mixture. The presence of toluene evidently shortened the lag phase sig-nificantly in the biodegradation study involving different ETX mixtures (Fig. 4). This further strengthens the argument of higher induction capacity of toluene for the broad substrate enzymes involved in the biodegradation of mono-aromatic VOCs. Presence of 100 mg/L of toluene was found to eliminate the lag phase observed for ethylbenzene biodegradation in Fig. 2 and facilitate faster biodegradation. Presence of toluene and ethylbenzene was found to reduce the lag phase observed for o-xylene biodegradation in Fig. 3. Moreover, with increasing toluene concentra-tions, the lag phase got reduced incrementally. The biokinetic parameters given in Table 1 were applied in all the four multiple substrate growth kinetic models, and

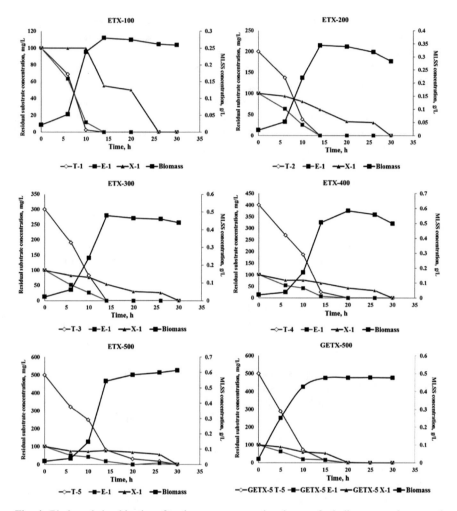

Fig. 4 Biodegradation kinetics of tertiary mono-aromatic mixture of ethylbenzene, toluene, and o-xylene where only the initial toluene concentration was varied, and the mixtures were termed as ETX-*a*, where *a* denotes the initial toluene concentration

no-interaction and competitive inhibition models seem to be predicting better than uncompetitive and noncompetitive models. Competitive inhibition of one compound on another is common in biodegradation involving broad substrate enzymes, in multiple substrates mixtures. Presence of glucose (termed as GETX-500) triggered a rapid increase of the MLSS concentration which facilitated a lower food to microorganism (F/M) ratio. The high F/M ratio in turn facilitated faster biodegradation of the three mono-aromatics present in the mixture (Fig. 4).

5 Conclusions

The acclimatized open mixed aerobic culture used in this present biodegradation study could effectively remove all the three mono-aromatic compounds. Increase in initial concentration increased the time taken for complete biodegradation of toluene as a single substrate. Single substrate biodegradation of o-xylene took long time even for a concentration of 100 mg/L. Increasing concentration of toluene was found to reduce the lag phase observed for ethylbenzene and o-xylene during their biodegradation as a single substrate. The shortening of the lag phase in presence of toluene for both of these compounds suggests biodegradation of ethylbenzene and o-xylene through broad-spectrum enzymes, induced by the presence of toluene. The biodegradation data fitted well with zero-interaction model for low cumulative concentrations and with competitive inhibition model at higher cumulative concentrations of these mono-aromatics. The presence of glucose was found to increase the MLSS concentration and enabled faster biodegradation through a higher food to microorganism ratio.

References

Aizpuru A, Malhautier L, Roux JC, Fanlo JL (2001) Biofiltration of a mixture of volatile organic emissions. J Air Waste Manage Assoc 51:1662–1670

Álvarez PJJ, Vogel TM (1991) Substrate interactions of benzene, toluene and para-xylene during biodegradation by pure cultures and mixed culture aquifer slurries. Appl Environ Microbiol 57:2981–2985

ATSDR (2000) Toxicological profile for toluene. US Department of Health and Human Services, Agency for Toxic Substances and Disease Registry, USA

Babu BV, Raghuvanshi S (2004) Biofiltration for VOC removal: a state-of-the-art review. In: Proceedings of international symposium & 57th annual session of IIChE in association with AIChE (CHEMCON-2004), Mumbai, 27–30 Dec 2004

Baek S, Ihm S (2004) Design of metal loaded zeolites as dual functional adsorbent/catalyst system for VOC control. Stud Surf Sci Catal 154(C):2458–2466

Baek S, Kim J, Ihm S (2004) Design of dual functional adsorbent/catalyst system for the control of VOC's by using metal-loaded hydrophobic Y-zeolites. Catal Today 93–95:575–581

Balasubramanian P, Philip L, Murty BS (2010) Biodegradation of chlorinated and non-chlorinated VOCs from pharmaceutical industries. Appl Biochem Biotechnol 163:497–518

Balasubramanian P, Philip L, Bhallamudi SM (2011) Biodegradation of chlorinated and non-chlorinated VOCs from pharmaceutical industries. Appl Biochem Biotechnol 163(4):497–518

Barker JF, Patrick GC, Major D (1987) Natural attenuation of aromatic Hydrocarbons in shallow sand aquifer. Winter, GWMR, pp 64–71

Bielefeldt A, Stensel HD (1999) Modeling competitive inhibition effects during biodegradation of BTEX mixtures. Water Res 33:707–714

Blair A, Hartge P, Stewart PA, McAdams M, Lubin J (1998) Mortality and cancer incidence of aircraft maintenance workers exposed to trichloroethylene and other organic solvents and chemicals: extended follow up. Occup Environ Medicine 55:161–171

Brauer H (1986) Biological purification of waste gases. Int Chem Eng 26(3):387–395

Chang MK, Voice T, Criddle CS (1993) Kinetics of competitive Inhibition and cometabolism in the biodegradation of benzene toluene, and p-Xylene by two Pseudomonas isolates. Biotechnol Bioeng 41:1057–1065

Corseuil HX, Weber WJ (1994) Potential biomass limitations on rates of degradation of monoaromatic hydrocarbons by indigenous microbes in subsurface soils. Wat Res 28: 1415–1423

Deshusses MA (1994) Biodegradation of mixtures of ketone vapours in biofilters for the treatment of waste air. Dissertation submitted to Swiss Ferederal Institute of Technology Zurich, 94–98

Dou B, Li J, Wang Y, Wang H, Ma C, Hao Z (2011) Adsorption and desorption performance of benzene over hierarchically structured carbon–silica aerogel composites. J Hazard Mater 196:194–200

Engleman VS (2000) Updates on choices of appropriate technology for control of voc emissions. Met Finish 98(6):433–445

Everaert K, Baeyens J (2004) Catalytic combustion of volatile organic compounds. J Hazard Mater 109:113–139

Gibson DT, Hensley M, Yoshioka H, Mabry TJ (1970) Formation of (+)-cis-2,3-dihydroxy-1-methylcyclohexa-4,6-diene from toluene by *Pseudomonas putida*. Biochemistry 9:1626–1630

Haldane (Andrews) JF (1968) A mathematical model for the continuous culture of microorganisms utilizing inhibitory substance. Biotechnol Bioeng 10(6):707–723

Hamdi B, Houari M, Hamoudi SA, Kessaïssia Z (2004) Adsorption of some volatile organic compounds on geomaterials. Desalination 166:449–455

IPCS (1996) Environmental health criteria 186: Ethylbenzene. International Programme on Chemical Safety, World Health Organization, Geneva, Switzerland

IPCS (1997) Environmental health criteria 190: Xylenes. International Programme on Chemical Safety, World Health Organization, Genevsa, Switzerland

Khan FI, Ghoshal AK (2000) Removal of volatile organic compounds from polluted air. J Loss Prev Process Ind 13:527–545

Lee EY, Jun YS, Cho KS, Ryu HW (2002) Degradation characteristics of toluene, benzene, ethylbenzene, and xylene by *Stenotrophomonas maltophilia* T3-C. J Air Waste Manag Assoc 52:400–406

Li N, Gaillard F (2009) Catalytic combustion of toluene over electrochemically promoted Ag catalyst. Appl Catal B: Environ 88:152–159

Łojewska J, Kołodziej A, Łojewski T, Kapica R, Tyczkowski J (2009) Structured cobalt oxide catalyst for VOC combustion. Part I: catalytic and engineering correlations. Appl Catal A: General 366:206–211

Louarn E, Aulenta F, Levantesi C, Majone M, Tandoi V (2006) Modeling substrate interactions during aerobic biodegradation of mixtures of vinyl chloride and ethane. J Environ Eng 132 (8):940–948

Mitsui T, Tsutsui K, Matsui T, Kikuchi R, Eguchi K (2008) Catalytic abatement of acetaldehyde over oxide-supported precious metal catalysts. Appl Catal B: Environ 78:158–165

Monod J (1949) The growth of bacterial cultures. Annu Revision Microbiol 3:371

Nacken M, Heidenreich S, Hackel M, Schaub G (2007) Catalytic activation of ceramic filter 415 elements for combined particle separation, NO_x removal and VOC total oxidation. Appl Catal B 70:370–376

Novak JT, Goldsmith CD, Benoit RE, O'Brien JH (1985) Biodegradation of methanol and tertiary butyl alcohol in subsurface systems. Water Sci Technol 17:71–85

NTP (2005) Report on carcinogens, Eleventh edn. US Department of Health and Human Services, Public Health Service, National Toxicology Program, USA

Okpokwasili GC (2006) Microbes and the environmental challenge. Inaugural lecture series No 53. University of Port Harcourt Press, Port Harcourt, pp 1–77

Priya VS, Philip L (2013) Biodegradation of dichloromethane along with other VOCs from pharmaceutical wastewater. Appl Biochem Biotechnol 169(4):1197–1218

Pugh ALV, Norton SA, Schauffler M, Jacobson GL, Kahl JS, Brutsaert WF, Mason CF (1996) Interactions between peat and salt-contaminated runoff in Alton Bog, Maine, USA. J Hydrol 182:83–104

Qu F, Zhu L, Yang K (2009) Adsorption behaviors of volatile organic compounds (VOCs) on porous clays hetero-structures (PCH). J Hazard Mater 170:7–12

Raghuvanshi S, Babu BV (2009) Biodegradation kinetics of methyl iso-butyl ketone by acclimated mixed culture. Biodegradation 21(1):31–42

Rivas B, López-Fonseca R, Sampedro C, Gutiérrez-Ortiz JI (2009) Catalytic behaviour of thermally aged Ce/Zr mixed oxides for the purification of chlorinated VOC-containing gas streams. Appl Catal B Environ 90:545–555. https://doi.org/10.1016/j.apcatb.2009.04.017

Schlegelmilch M, Streese J, Stegmann R (2005) Odour management and treatment technologies: an overview. Waste Manag 25(9):928–939

Sene L, Converti A, Felipe MGA, Zilli M (2002) Sugarcane bagasse as alternative packing material for biofiltration of benzene polluted gaseous streams: a preliminary study. Biores Technol 83:153–157

Shim H, Yang S (1999) Biodegradation of benzene, toluene, ethylbenzene, and o-xylene by a coculture of Pseudomonas putida and Pseudomonas fluorescens immobilized in a fibrous-bed bioreactor. J Biotechnol 67:99–112

Shim H, Shin E, Yang ST (2002) Biodegradation of benzene, toluene, ethylbenzene and o-xylene by a coculture of Pseudomonas putida and Pseudomonas fluorescens in a fibrous-bed bioreactor. Adv Environ Res 7(1):203–216

Shim EH, Kim J, Cho KS, Ryu HW (2006) Biofiltration and inhibitory interactions of gaseous benzene, toluene, xylene, and methyl-butyl ether. Environ Sci Technol 40(9):3089–3094

Sillman S (1999) The relation between ozone, NO_x and hydrocarbons in urban and polluted rural environments. Atmos Environ 33:1821–1845

Singh SK, Tripathi VR, Khare SK, Garg SK (2011) A novel psychrotrophic, solvent tolerant Pseudomonas putida SKG-1 and solvent stability of its psychro-thermoalkalistable protease. Proc Biochem 46:1430–1435

Tao WH, Yang TCK, Chung TW (2004) Effect of moisture on the adsorption of volatile organic compounds by Zeolite 13x. J Environ Eng 130(10):1210–1216

USEPA (U.S. Environmental Protection Agency) (2001) Method 8015 C. Non-halogenated organics by gas chromatography

van Groenestijn JW, Kraakman NJR (2005) Recent developments in biological waste gas purification in Europe. Chem Eng J 113:85–91

Vandenbroucke A, Morent R, Geyter ND, Dinh MTN, Giraudon JM, Lamonier J, Leys FC (2010) Non-thermal plasma technic for air pollution control. Int J Plasma Environ Sci Technol 42

Web reference 1: file:///H:/IIT%20G_1/IIT%20G_Dec%202016/Indian%20Paints%20Industry%20Report%20-%20Paints%20Sector%20Research%20&%20Analysis%20in%20India%20-%20Equitymaster.html

Web reference 2: http://cmrindia.com/external/reports/CMR_Paints_Study_IMG%20Report%202016.pdf

WHO (2003) Health aspects of air pollution with particulate matter, ozone and nitrogen dioxide, Report on a WHO working group, Bonn, Germany, 13–15 Jan 2003

Wu D, Quan X, Zhao Y, Chen S (2006) Removal of p-xylene from an air stream in a hybrid biofilter. J Hazard Mater 136(2):288–295

Yoon H, Klinzing G, Blanch HW (1977) Competition for mixed substrates by microbial populations. Biotechnol Bioeng 19:1193–1210

Zaitan H, Bianchi D, Achak O, Chafik T (2008) A comparative study of the adsorption and desorption of o-xylene onto bentonite clay and alumina. J Hazard Mater 153(1–2):852–859

Aerobic Degradation of Complex Organic Compounds and Cyanides in Coke Oven Wastewater in Presence of Glucose

Naresh Kumar Sharma, Ligy Philip and B. S. Murty

Abstract This study aims at determining the degradation aspect of pollutants from coke oven effluents such as phenols, aromatic hydrocarbons, and cyanide by aerobic mixed culture. Enriched mixed culture was developed from the sludge collected from aeration tank of a sewage treatment plant by serial enrichment technique. The acclimatized culture was able to degrade phenol, cresol, xylenol, quinoline, indole, and cyanide individually, for their concentrations usually found in coke oven wastewater. Xylenol and indole with concentrations above 250 mg/L were highly recalcitrant for biodegradation. A co-substrate such as glucose (1000 mg/L) had an adverse effect on the biodegradation of all the above pollutants; the degradation time was extended with the increase in the concentration of pollutant, although glucose was completely oxidized independent of the pollutant concentration. Biodegradation during mixed pollutant (100 mg/L of each organic compound) conditions were tested in presence of glucose (1000 mg/L) and glucose (1000 mg/L) with cyanide (2.5 mg/L). Aerobic microbes showed increased substrate affinity in the following order phenol > cresol > quinoline > indole > xylenol. The COD at the end of the experiment was found to be less than 0.5 mg/L showing no accumulation of intermediates. In the presence of glucose (1000 mg/L) and cyanide (2.5 mg/L), the lag phase for microbial growth was increased by several days and cyanide was found to be oxidized before the organic pollutants. Xylenol was highly recalcitrant during this experiment and was not degraded even after 20 days. The experimental results highlight the effect of high concentration of co-substrate (1000 mg/L) and the combined toxic influence of cyanide and organics on the microbes treating coke oven wastewater. These results and the ongoing work are aimed at developing

N. K. Sharma (✉)
Department of Biotechnology, Kalasalingam Academy of Research and Education, Krishnankoil 626126, Tamil Nadu, India
e-mail: naresh@klu.ac.in

L. Philip · B. S. Murty
Department of Civil Engineering, Indian Institute of Technology, Madras, India
e-mail: ligy@iitm.ac.in

B. S. Murty
e-mail: bsm@iitm.ac.in

© Springer International Publishing AG 2018
A. K. Sarma et al. (eds.), *Urban Ecology, Water Quality and Climate Change*,
Water Science and Technology Library 84,
https://doi.org/10.1007/978-3-319-74494-0_22

high-rate bioreactors for efficient treatment of phenolics, aromatic hydrocarbons, and cyanide-containing wastes emanating from industrial activities.

Keywords Activated sludge · Glucose · Cyanide · Phenolics · Heterocyclic aromatics

1 Introduction

A mixture of phenolics, heterocyclic aromatics along with cyanide moieties represents one of the toxic combinations of wastewater known to create havoc not only to microbes used for their treatment but also to the ecosystem and to the human race at large. The most common sources of such a combination of these fatal chemicals at lethal doses are from the coal processing and coke manufacturing industries, which satiate our increasing demands for the steels used inconsiderately in this age. Effluents from industries such as coke oven, coal gasification, and coal processing contain myriad of organic and inorganic pollutants (Lai et al. 2008).

The composition of coke oven wastewater is complex and varies from one plant to another, depending upon the quality of raw coal, carbonation temperature, and method used for byproduct recovery (Pal and Kumar 2014). The compounds could be grouped into phenolics, polyaromatic hydrocarbons, nitrogenous heterocyclic compounds, oxygen or sulfur-containing compounds, inorganic compounds, and oil and grease along with suspended solids. Phenol, cresol, xylenol, alkyl phenols are the major contributors in the phenolics category as are quinoline, pyridine, indole in the nitrogen heterocyclic compounds and ammonia, thiocyanate, cyanide, and sulfide in the inorganic group. There are many published reports on the eco-toxicity of these chemical compounds and on the biodegradation of these compounds individually, under different redox conditions (Fetzner 1998). Phenolic compounds can migrate in different aqueous environments and contaminate groundwater (Li and Zheng 2004). Most of the PAHs and NHCs have been reported to be mutagenic and even carcinogenic. Cyanides are toxic to aquatic species and mammals even at low concentrations. Oil and tar form oil slicks over water bodies and influences the distribution of O_2 and CO_2 in water affecting the fishes and aquatic organisms.

Coke oven wastewater (CWW) contains a wide range of PAHs and nitrogen, oxygen and sulfur containing heterocyclic compounds (Zhang et al. 1998; Jianlong et al. 2002). Some of these compounds are known to have long-term environmental impacts and are reported to be mutative and carcinogenic (Qi et al. 2007). Most of the COD that is present after the biological treatment of coke oven wastewater have found to contain these refractory compounds (Lai et al. 2008) because of their nonbiodegradable nature.

The treatment techniques reported in the literature are numerous, including the use of conventional aerobic lagoons, constructed wetlands, photo-bioreactors

(Tamer et al. 2006), bioreactors (aerobic, anoxic, and anaerobic) with different configurations (Zhao et al. 2009; Lai et al. 2008; Maranon et al. 2008; Chakraborti and Veeramani 2005), physicochemical methods (Ghose 2002), membrane bioreactors (Zhao et al. 2009), and the newly enriched bioreactors like anaerobic ammonium oxidation processes. Nevertheless, an economical and effective way of treating coke oven wastewater (CWW) is still a challenge; mainly because the pollutants in CWW are highly varied and it is generally not feasible to remove all the pollutants under similar conditions to the standard discharge levels.

There has been a number of reports on full-scale biological treatment of CWW, but only a few of them follow environmental discharge regulation at a high level of sophistication and price. The bioremediation of contaminant and the rate at which it is achieved depends on the conditions of treatment (pH, temperature, substrate loading, biomass concentration, nutrients, inhibitors concentration), microbial diversity, nature, and chemical structure of the compound being degraded (Haritash and Kaushik 2009). Thus, to devise a biological treatment system, several factors are responsible which must be addressed and explored.

The challenge is to develop a treatment system which could overcome the toxic effects of these chemical mixtures and still maintain the stability in operation and discharge levels. A synthetic wastewater containing the major organic compounds representative of each group corresponds well with the actual wastewater. Henceforth, phenol, cresol, xylenol, quinoline, indole, and cyanide are the target compounds which are selected to represent the actual wastewater from coke oven and coal processing industries. The biodegradation of these compounds with a co-substrate and in presence of an inhibitor contributes to an in-depth study of biodegradation of these pollutants in order to optimize the treatment processes. This knowledge helps in the development of a stable and economically viable treatment plant.

2   Materials and Methods

2.1   Aerobic-Activated Sludge

The activated sludge used in the experiments was obtained from a domestic sewage treatment plant, Chennai and was maintained in the laboratory by a medium containing dextrose as the carbon source. Then the composition of the medium was gradually replaced with phenol, cresol, xylenol, quinoline, indole, and cyanide. Eventually, the activated sludge was cultured in a medium containing these compounds as carbon source over a period of 8 months in a 2 Lx g laboratory unit. The composition of the ultimate medium containing carbon source (in mg/L) was as follows: phenol (500), cresol (100), xylenol (100), quinoline (100), indole (100), cyanide (2.5) along with mineral media having the composition (in mg/L):

NH$_4$SO$_4$(230),   CaCl$_2$(8),   FeCl$_3$(1),   MnSO$_4$.H$_2$O(100),   MgSO$_4$.7H$_2$O(100), K$_2$HPO$_4$(500), KH$_2$PO$_4$(250), and pH 7.5 ± 0.5 (Saravanan et al. 2008) under agitation condition at room temperature. This medium was supplied to the culture at three days intervals by replacement of 1 L. This acclimated sludge formed light brown flocs and had good settleability.

2.2 Biomass Estimation

To estimate the cell concentration, a known volume of cell suspension was filtered through 0.45 μm filter paper followed by weighing the dried cell mass retained on the filter paper. Protein standard curve was plotted using bovine serum albumin. Protein contents of the acclimated cultures, with known bacterial concentrations (dry weight in mg/L), were determined using Lowry's method (Lowry et al. 1951) and the correlation between protein content of the cells and dry cell weight was established. The modified Lowry's method used was as follows. 2 mL of sample was taken and centrifuged at 8000×g for 8 min. The pellets were then resuspended in 2 mL of phosphate buffer (pH 7) and sonicated at 100 Hz at 15 s intervals (15 s on and 15 s off) for 3 min. The solution was centrifuged again at 8000×g for 8 min. 2 mL of alkaline copper reagent was added to 0.5 mL of supernatant or diluted supernatant of a suitable concentration, incubated for 10 min, followed by addition of 0.2 mL of Folin phenol reagent, and incubated again for 30 min. Reagent blank, containing 0.5 mL of distilled water instead of bacterial suspension, was treated in a similar way. The optical density was measured at 600 nm using a UV–Vis spectrophotometer (Techcom, UK) against the reagent blank. Samples with known bacterial concentrations were used for preparing the calibration curve.

2.3 Individual Pollutant Analysis

2.3.1 High-Performance Liquid Chromatography

Phenol, o-cresol, m-xylenol, quinoline, and indole concentrations were quantified by high-performance liquid chromatography (HPLC) (Dionex, Ultimate 3000). The aqueous samples of the suspended culture were centrifuged at 8000×g for 10 min and filtered through a 0.45 μm filter paper. Then the cell-free supernatants were determined for residual chemical concentrations in the solution. HPLC was performed on a reverse phase C-18 column with acetonitrile/water (50/50, v/v) mobile phase at a flow rate of 1.0 mL/min, and detected using UV at 275 nm.

2.3.2 UV-Spectrophotometry

Free cyanide ion concentrations were analysed using standard methods (APHA 1985) using (Chloramine-T) pyridine-barbituric acid reagent and the light pink color developed was measured colorimetrically at 578 nm using UV-spectrophotometry (Techcom, UK).

2.4 Chemical Oxygen Demand

COD of liquid samples was estimated from the centrifuged ($8000\times g$ for 10 min) and filtered samples and the supernatant was analysed by the closed reflex method as suggested in standard methods (APHA 1985). Closed reflex digestion was conducted in HACH COD digester (Model No 45600, USA) and the remaining $K_2Cr_2O_7$ was titrated with FAS to calculate the amount of organics.

2.5 Aerobic Biodegradation Studies

All biodegradation experiments using the acclimated mixed culture were performed in 500 mL Erlenmeyer flask containing 100 mL of MSM containing glucose and pollutants at different concentrations. The initial pH of the reaction mixture was 7.5 ± 0.5. The initial dissolved oxygen concentration was between 6 and 6.5 mg/L in the entire batch degradation studies and it never dropped below 3 mg/L during the biodegradation studies. Upon incubation of the flasks at 30 °C under agitation condition ($150\times g$), samples were withdrawn at regular time intervals, centrifuged ($10,000\times g$ for 5 min), filtered, and analysed for the residual pollutant, COD, and protein concentration. For each concentration, duplicate experiments were performed under the same conditions and average values are reported. Each experiment was carried out for a period until the residual concentration of the pollutants and the amount of biomass in the flask had reached asymptotic values with time. Abiotic controls were also monitored during the study.

3 Results and Discussion

3.1 Effect of Co-substrate on Individual Pollutant Biodegradation

An attempt was made to study the degradation of different pollutants, with varying concentrations, along with 1000 mg/L of glucose. Such a high COD of glucose was chosen to represent the COD of cww. However, 1000 mg/L of glucose had an adverse effect on the rate of biodegradation of all pollutants. The degradation time was extended with the increase in the concentration of the pollutant, although glucose was completely oxidized independent of pollutant concentration; as corroborated by the biomass growth and COD at the end of the experiment. Similarly, Lob and Tar (2000) observed decreased degradation rate of phenol when glucose concentration exceeded 1 g/L and attributed it to the catabolite repression by glucose, which has been observed by other researchers also (Papanastasiou 1982; Satsangee and Ghosh 1990). Phenol (200 mg/L) which degraded within 10 h took a maximum degradation time of more than 14 h in presence of 1000 mg/L of glucose as shown in Fig. 1.

Optimum concentration of glucose during which degradation rate of the pollutant observed to be higher was around 350 mg/L reported by Lob and Tar (2000) and concentrations above 780 mg/L increased the pollutant removal time. The COD at the end of the batch was less than 0.1 mg/L which indicated the complete oxidation of glucose and metabolic intermediates.

Xylenol, the most toxic organic compound did not degrade to a concentration higher than 200 mg/L even in the presence of glucose (Fig. 2), although complete

Fig. 1 Phenol biodegradation and biomass growth in presence of glucose (1000 mg/L)

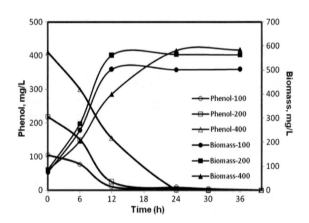

Fig. 2 Xylenol biodegradation and biomass growth in presence of glucose (1000 mg/L)

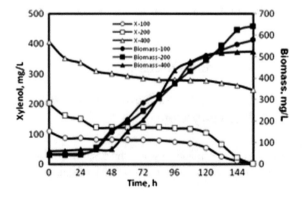

oxidation of glucose and biomass growth was observed as shown in Fig. 2. Xylenol has already been reported to be 32-fold more toxic than phenol by Acuna-Arguelles et al. (2003). Other organic pollutants such as cresol (Fig. 3), quinoline (Fig. 4), and indole (Fig. 5) showed increased removal time period in presence of glucose as co-substrate. The abiotic losses of the organic compounds were always found to be less than 1%.

Fig. 3 Cresol biodegradation and biomass growth in presence of glucose (1000 mg/L)

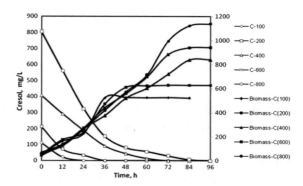

Fig. 4 Quinoline biodegradation and biomass growth in presence of glucose (1000 mg/L)

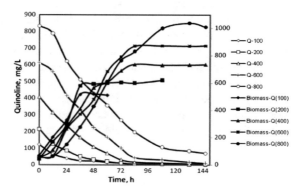

Fig. 5 Indole biodegradation and biomass growth in presence of glucose (1000 mg/L)

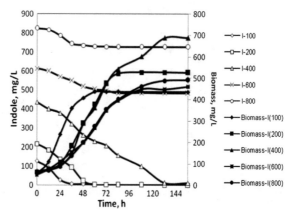

There are no reports available on the effect of cyanide on the biodegradation of cresol, xylenol, quinoline, and indole, with some exceptions on phenol. Therefore, experiments have been carried out to understand the effect of varying concentrations of cyanide on the biodegradation of phenolics and aromatic hydrocarbons under aerobic conditions. Initial concentrations of phenol, quinoline, cresol, indole, and xylenol were kept at 250 mg/L, while the cyanide concentration was varied from 0 to 20 mg/L. Phenol (250 mg/L) degraded completely within 24 h. Phenol degradation was insignificant as long as cyanide was present in the system, and the degradation of phenol was very fast once the cyanide disappeared from the system. A similar trend was observed in case of cresol and other organics (Sharma et al. 2012).

3.2 Effect of Co-substrate on Mixed Organic Pollutant Biodegradation

Effect of glucose (1000 mg/L) in mixed pollutants (100 mg/L of each organic compound) was studied of inorganic inhibitors. The lag phase for degradation of pollutant was increased mainly because of the combined influence of each pollutant on the aerobic consortia. Inhibition due to the presence of multiple pollutants has been reported by several authors (Lai et al. 2008; Zhang et al. 1998). Phenol and cresol were degraded in less than 4d in presence of other organic pollutants and were most preferable substrates, xylenol on the other hand with only one methyl group excess to cresol was the most toxic organic compound even compared to the heterocyclics like quinoline and indole. The substrate affinity for aerobic consortia was found to be phenol > cresol > quinoline > indole > xylenol as shown in Fig. 6. The degradation of xylenol was found to start only after the oxidation of

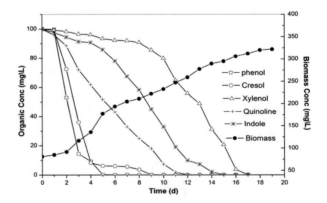

Fig. 6 Mixed organic substrate biodegradation and biomass growth in presence of glucose (1000 mg/L)

other organic compounds. It has been seen from the single substrate studies that xylenol was found to be the most recalcitrant compound due to excess methyl group present in its structure. Also, the degradation time was longest for xylenol than for other single-target compounds. Hence, it was found that the acclimated biomass had a high proclivity toward other organic substrates compared to xylenol. Other studies have shown that the recalcitrant nature of the pollutant increases with increase in the methyl group along the benzene ring of the compound (O'Connor and Young 1996).

Mixed bacterial culture was able to biodegrade the mixed pollutants in absence of cyanide without any accumulation of organic intermediates as evident from the TOC analysis. The initial biomass concentration was maintained at 50–80 mg/L while the concentrations of target organic pollutants were at 100 mg/L each along with 2.5 mg/L of cyanide. Coke oven wastewater contains many inorganic toxic inhibitors such as ammonia, sulfide, thiocyanate, and cyanide. Effect of the presence of cyanide (2.5 mg/L) on mixed organic pollutant degradation was studied. Cyanide had a detrimental effect on the aerobic consortia even at very less concentration (2.5 mg/L). Cyanide was oxidized before other organic compounds, whose concentrations were 40 times more than cyanide concentration. Thus, the time period for organic degradation increased, without any change in the organic substrate affinity.

3.3 Effect of Cyanide on Mixed Pollutant Biodegradation in Presence of Glucose

Coke oven wastewater contains many inorganic toxic inhibitors such as ammonia, sulfide, thiocyanate, and cyanide. Effect of presence of cyanide (2.5 mg/L) on mixed organic pollutant degradation in presence of glucose as co-substrate (1000 mg/L) was experimented. Abiotic loss of compound was very minimal. In case of cyanide, which is highly volatile below pH 8.4, the abiotic loss was less than 1%. Cyanide ion hydrolyzes in water to form molecular hydrogen cyanide (HCN) and hydroxyl ions ($OH-$), with a corresponding increase in pH. Though HCN has a high vapor pressure, the rate of volatilization depends on the HCN concentration (a function of total cyanide concentration and pH); the surface area and depth of the liquid; temperature and transport phenomena associated with mixing (Huiatt et al. 1983). Moreover, at a pH of 7.1–8.4, it was reported that at low cyanide concentrations (<50 mg/L), the volatilization was less than 1%, while at high concentrations of cyanide (100–150 mg/L) the removal due to volatilization was about 14%, under similar conditions of batch experiments (Haghighi-Podeh and Siyahati-Ardakani 2000). Many researchers who conducted experiments at pH of 7–8.5 and at room temperature (25–30 °C) observed cyanide volatilization loss to be less than 1%, for different initial KCN concentrations (Chen et al. 2008). Cyanide had a detrimental effect on the aerobic consortia, although very less in concentration (2.5 mg/L). Cyanide was oxidized before other organic compounds whose concentrations were 40 times more than cyanide concentration. Thus, the time period for organic degradation was increased, without any change in the organic substrate affinity. Xylenol has been reported to be 32-fold more toxic than phenol (Acuna-Arguelles et al. 2003). Xylenol was found to be very recalcitrant in presence of cyanide and mostly remained in the medium even after 20d as shown in Fig. 7; this shows the combined toxicity of cyanide and organic substrates on the aerobic consortium.

Semple and Cain (1997) observed that xylenol was significantly oxidized only after the consumption of phenols by a pure culture of *O. dancia*. After 12 h, the synthetic wastewater which was colorless initially turned into dark pinkish brown color, representative of actual coke oven wastewater. This color development was mainly due to the oxidation of aromatics like quinoline and indole, as observed by other researchers (Bai et al. 2009, 2010). The mixture of the pollutants without microbial consortia remained visibly colorless for several days and the abiotic loss of each of the compounds was always less than 10%.

In absence of glucose, Phenol and cresol were degraded in less than 96 h in presence of other organic pollutants and were most preferred substrates. Xylenol, on the other hand, with only one methyl group excess to cresol was the most toxic organic compound even compared to the heterocyclic compounds like quinoline and indole. The biomass concentration was found to increase up to 330 mg/L.

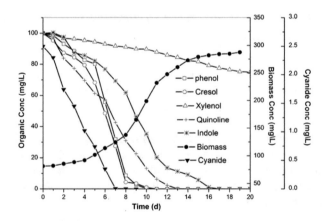

Fig. 7 Mixed substrate biodegradation and biomass growth in presence of glucose (1000 mg/L) and cyanide (2.5 mg/L)

4 Conclusion

Our experiments indicated that in presence of an external carbon source (1000 mg/L) the degradation time for all the pollutant tested was increased, most probably due to additional organic loading and catabolic repression. Xylenol which has been showed to be 32-fold more toxic than phenols was also observed to be the most toxic pollutant to inhibit microbial growth above 200 mg/L, similarly indole beyond 400 mg/L proved inhibitory to aerobic microbes. In the mixed organic study, substrate affinity of aerobic microbes was found to be phenol > cresol > quinoline > indole > xyelnol. Nevertheless, the presence of co-substrate increased the biomass concentration. In presence of cyanide, xylenol was not removed. Thus, the combination of xylenol and cyanide was proved to be fatal to the aerobic microbes by completely arresting the microbial growth and substrate degradation.

References

Acuna-Arguelles ME, Olguin-lora P, Razo-Flores E (2003) Toxicity and kinetic parameters of the aerobic biodegrdation of the phenol and alkyl phenols by a mixed culture. Biotech Lett 25:559–564

American Public Health Association (APHA) (1985) Standards methods for examination of water and wastewater. American water works association, water pollution control federation (15), Washington DC

Bai Y, Sun Q, Zhao C, Wen D, Tang X (2009) Simultaneous biodegradation of pyridine and quinoline by two mixed bacterial strains. Appl Microbiol Biotechnol 82(5):963–973

Bai Y, Sun Q, Zhao C, Wen D (2010) Bioaugmentation treatment for coking wastewater containing pyridine and quinoline in a sequencing batch reactor, 1943–1951

Chakraborti S, Veeramani H (2005) Anerobic-anoxic-aerobic sequential degradation of synthetic wastewaters. Appl Biochem Biotechnol 102–103:443–451

Chen CY, Kao CM, Chen SC (2008) Application of Klebsiella oxytoca immobilized cells on the treatment of cyanide wastewater. Chemosphere 71:133–139

Fetzner S (1998) Bacterial degradation of pyridine, indole, quinoline and their derivatives under different redox conditions. Appl Microbial biotechnol 49:237–250

Ghose MK (2002) Phyisco chemical treatment as an suitable option for treatment of coke plant effluents. IE (I) J-CH 84:1–6

Haritash AK, Kaushik CP (2009) Biodegradation aspects of polycyclic aromatic hydrocarbons (pahs). J Hazard Mater 169(1–3):1–15

Haghighi-Podeh MR, Siyahati-Ardakani G (2000) Fate and toxic effects of cyanide on aerobic treatment systems. Water Sci Technol 42:125–129

Huiatt JL, Kerrigan JE, Olson FA, Potter GL (1983) Cyanide from mineral processing. In: Proceedings of Workshop Sponsored by National Science Foundation, USBM & Industry. Salt Lake City, UT Utah Mining and Mineral Resources, Research Institute

Jianlong W, Quian X, Libo W, Yi Q, Werner H (2002) Bioaugumentation as a tool to enhance removal of refractory compounds in coke plant wastewater. Process Biochem 38:777–781

Lai P, Zhao H, Ye Z, Ni J (2008) Assessing the effectiveness of treating coking effluents using anerobic and aerobic biofilms. Process Biochem 43:229–237

Li W, Zheng SK (2004) A combination of anaerobic and aerobic treatment for ammonia-laden coke plant effluent: the pilot study. Environ Inform Arch 2:602–610

Lowry OH, Rosenbrough NJ, Farr AL, Randall RJ (1951) Protein measurement with the Folin phenol reagent. J Biol Chem 193:265–275

Maranon E, Vazquez I, Rodriguez R, Castrillon L, Fernandez Y (2008) Coke wastewater treatment by a three step activated sludge system. Water Air Soil Pollut 192:155–164

O'Connor OA, Young LY (1996) Effects of six different functional groups and their position on the bacterial metabolism of monosubstituted phenols under anaerobic conditions. Environ Sci Technol 30(5):1419–1428

Pal P, Kumar R (2014) Treatment of coke wastewater: a critical review for developing sustainable management strategies. Sep Purif Rev 43(2):89–123

Papanastasiou AC (1982) Kinetics of biodegradation of 2,4-Dichlorophenoxyacetate in presence of glucose. Biotechnol Bioeng 24:2001–2011

Qi R, Yang K, Yu Z (2007) Treatment of cokes plant wastewater by SND fixed biofilm hybrid system. J Environ Sci 19:153–159

Saravanan P, Pakshirajan K, Saha P (2008) Growth kinetics of an indigenous mixed microbial consortium during phenol degradation in a batch reactor. Biores Technol 99:205–209

Satsangee R, Ghosh P (1990) Anaerobic degradation of phenol using an acclimated mixed culture. Appl Microbial Biotechnol 34:127–131

Semple KT, Cain RB (1997) Degradation of phenol and its methylated homolouges by Ochromonas dancia. FEMS Microbiol Lett 152:133–159

Sharma NK, Philip L, Murty Bhallamudi S (2012) Aerobic degradation of phenolics and aromatic hydrocarbons in presence of cyanide. Biores Technol 121:263–273

Tamer E, Amin MA, Ossam ET, Bo M, Benoit G (2006) Biological treatment of industrial wastes in photobioreactor. Water Sci Technol 53(1):117–125

Zhao WT, Xia Z, Lee DJ (2009) Enhanced treatment of coke plant wastewater using anaerobic anoxic-aerobic membrane bioreactor system. Sep Purif Technol 66:279–286

Zhang M, Tay JH, Quian Y, Gu YS (1998) Coke plant wastewater treatment by fixed biofilm system for COD and NH3-N removal. Water Res 32(2):519–527

Benzene Biodegradation During Growth by *Aerococcus* sp. Isolated from Oil Sludge

Priyadarshini Dey, Ranjit Das and Sufia K. Kazy

Abstract Monoaromatic hydrocarbons such as benzene have been found in large quantities of sludge created by oil production facilities and industries. Benzene is highly volatile and soluble in water, it seeps into the ground and encounters water, sediment and soil layers of the environment. Benzene is considered as a carcinogenic substrate and biodegradation, principally under aerobic conditions, is an important environmental fate process for water and soil associated benzene. *Aerococcus* sp. strain BPD-6 was characterized in terms of its growth in presence of benzene (50–1000 mg L^{-1}) as the sole carbon and energy source in the mineral salt medium. The maximum biomass growth was achieved at a pH of 7.0, temperature of 37 °C, with an initial benzene concentration of 50 mg L^{-1} and incubation period of 27 h. Various other monoaromatic compounds, such as toluene, ethyl benzene, and xylene were also tested as substrates for the growth of this bacterial isolate. The growth kinetics of the strain was analysed and the Haldane model was found to be a good fit for the experimental data with kinetic constants maximum specific growth rate $(\mu_{max}) = 0.02157$ h^{-1}, half saturation constant $(K_s) = 19.56$ mg L^{-1}, and substrate inhibition constant $(K_i) = 1584$ mg L^{-1}. The biodegradation rate kinetic parameter suggested that the strain can potentially be utilized in bioremediation of benzene and other monoaromatic compounds.

Keywords Benzene · *Aerococcus* sp. · Bioremediation · HPLC

1 Introduction

Volatile organic compounds (VOCs) like benzene, toluene, ethyl benzene, and xylenes originating from petrochemical and allied industries are of great public health concern due to toxic, carcinogenic, and mutagenic nature (Rahul et al. 2013;

P. Dey · R. Das · S. K. Kazy (✉)
Department of Biotechnology, National Institute of Technology Durgapur,
Mahatma Gandhi Avenue, Durgapur, West Bengal 713209, India
e-mail: sufia_kazy@yahoo.com

© Springer International Publishing AG 2018
A. K. Sarma et al. (eds.), *Urban Ecology, Water Quality and Climate Change*,
Water Science and Technology Library 84,
https://doi.org/10.1007/978-3-319-74494-0_23

305

Luo et al. 2016). The chemical mixtures of VOCs are present in wastewaters from industrial and municipal sources as well as in contaminated groundwater (Chen et al. 2014; Luo et al. 2016). Benzene is also the transformation product of gasoline and other petroleum fuels, pesticides, and wood-treating substances that are released into the environment during manufacture, transportation, usage, and disposal, leakage in underground storage tanks and pipelines, and through leachate from landfills (Chen et al. 2014; Luo et al. 2016). Benzene is widely used in plastic, detergents, and pesticide production (Chakraborty and Coates 2005). It has been considered as a priority pollutant worldwide only because of their large migration abilities, solubility in water, toxicity, and volatility (Farhadian et al. 2009). Such environmental relevant measures have facilitated the benzene to the environments and commonly found in soils, aquifers, and in the atmosphere (Mathur and Majumder 2010; Li et al. 2017). According to the US Environmental Protection Agency (1996) and the Agency for Toxic Substances and Disease Registry (2005), the maximum permissible level of benzene in water is 0.5 ppm (Mukherjee et al. 2012). The liquid as well as gaseous states of this compound in the environment pose a significant threat to human. After inhalation or absorption of benzene is effected the organs,viz., liver, kidney, lung, heart, and brain etc. Benzene causes haematotoxicity through its phenolic metabolites that act in concert to produce DNA strand breaks and chromosomal damage along with it also causes apneas, lung cancer, or leukemia and other adverse effects (Munoz et al. 2007; Singh et al. 2010; Rahul et al. 2013; Aburto-Medina et al. 2015). The Clean Air Act Amendments of 1990 (CAAA 90) proposed by the US Environmental Protection Agency (EPA) places special emphasis on the handling, usage, and treatment of monoaromatic compounds such as benzene. Hence, there is a pressing need for the development of efficient and cost-effective technologies for the treatment of such pollutant (Singh et al. 2010; Rahul et al. 2013).

Bioremediation of such organic solvents has been considered as an adjunct/ alternative to the traditional physical and chemical treatment methods that are costly, energy intensive, and produce hazardous by-products (Cerqueira et al. 2011; Gillespie and Philp 2013). Biodegradation employing growing cells is an excellent alternative as viable cells provide rich metabolic flux in the form of enzymes capable of degrading organic compound (Choudhary and Sar 2011). Degradation of benzene component by pure bacterial strains has been well reported by Gram-negative bacteria including *Pseudomonas fluorescens*, *Pseudomonas aeruginosa*, and *Pseudomonas putida* (Maliyekkal et al. 2004; Kim et al. 2005; Mathur and Majumder 2010). However, benzene degradation by Gram-positive bacteria is scarce and particularly by *Aerococcus* sp. has not been reported so far. It was earlier considered that the outer membrane present in Gram-negative bacteria shields the organisms from the attack of the organic lipophilic compounds. Fahy et al. (2008) demonstrated that Gram-positive bacteria belonging to the genera *Bacillus, Rhodococcus, Staphylococcus,* and *Arthrobacter* have good tolerance to various concentrations of benzene. In order to design a proper biodegradation process, it is necessary to determine the degradation kinetics parameters of the benzene compound by bacterial isolate (Mayer et al. 2016). The kinetic parameters of K_s and K_i

are distributed over a wide range, depending on cell type and culture environments (Loh and Wang 1997). Moreover, large variations also derive from culture history, the method used to determine the parameters, the correlation between parameters, and nature of nonlinearity of the model. According to both Monod's and Haldane's equations, the kinetic parameters μ_{max}, K_s, and K_i can describe the growth rate of microorganisms (Yang and Humphrey 1975). Therefore, by knowing bacterial growth kinetics physical dimensions of treatment units can be designed and removal efficiency can be improved.

In the present study, a new indigenous bacterial isolate of *Aerococcus* sp. strain BPD-6 obtained from a petroleum hydrocarbon containing oily sludge of a storage reservoir of the Bharat Petroleum Corporation Limited, Rajbandh, Durgapur, West Bengal, India. The primary investigation was to examine the benzene degradation during cells growth and estimation of different growth kinetic parameters such as specific growth rate (μ_{max}) and saturation constant (K_s) of the strain. The effects of pH, temperature, and salinity on growth of the strain were also studied.

2 Materials and Methods

2.1 Materials

Benzene used in the study is of analytical grade, glucose and other inorganic salts used in preparing microbial growth media were of reagent grade. All the chemicals and other reagents were purchased from Merck (India).

2.2 Preparation of Media

Luria-Broth (LB) and mineral salt medium (MSM) of culture media were used for the growth of this strain. The composition of MSM is as follows (in g L^{-1} of deionized water): K_2HPO_4, 0.348; KH_2PO_4, 0.272; NaCl, 4.68; NH_4Cl, 1.07; KCl, 1.49; Na_2SO_4, 0.43; $MgCl_2 \cdot 6H_2O$, 0.2; $CaCl_2 \cdot 2H_2O$, 0.03. The pH of the medium was adjusted to 7.2 and sterilized by autoclaving (121 °C and 15 lb/seq. in. for 15 min) prior to the addition of the trace element solution (filter sterilized) and the organic substrate of benzene. Stock glucose solution is prepared by dissolving 10 g of glucose in 100 mL of distilled water.

2.3 Batch Growth and Biodegradation Study

All growth and biodegradation experiments using the indigenous *Aerococcus* sp. strain BPD-6 (JN377811.1) were examined by acclimatization of the strain to the increasing concentration of benzene 10–50 mg L^{-1} as the sole carbon source. The experiments were conducted in 100 mL serum bottles containing 25 mL of MSM supplemented with 50–1000 mg L^{-1} of benzene as the sole source of carbon and energy. The culture serum bottles were closed with butyl rubber caps and sealed with aluminum crimps to prevent volatilization of benzene and incubated in an orbital shaker maintained in dark at 37 °C and 150 rpm. Before sealing 2% of the acclimatized culture was added to each serum bottles directly under aseptic conditions. All the experiment was carried out in duplicate. Uninoculated bottles were maintained under the same conditions to determine abiotic loss during the experiment. Samples were withdrawn at a regular time interval in polytetrafluoroethylene (PTFE) tubes using airtight disposable syringe and analyzed for biomass concentration. Subsequently, the samples were centrifuged (12,000×g for 10 min) and the resulting supernatant was analyzed for residual benzene concentration. The results obtained from the same set of duplicate experiments were averaged and are plotted.

Different physicochemical factors such as pH, temperature, and salinity were assayed in presence of benzene as a substrate during growth of the strain BPD-6 and examined the optimum pH, temperature, and salinity. The pH was maintained in the medium ranging from 2 to 12 using 1 M HCl or 1 M NaOH and the NaCl salt concentration and temperature were maintained at 1.0–6.0 g L^{-1} (w/v) and 20–60 °C, respectively. In all the cases, 2% v/v of acclimatized inoculum was taken and inoculated in 25 mL MSM containing 100 mg L^{-1} of benzene. All the experimented serum bottles were incubated in dark at 37 °C under shaking conditions (150 rpm) till the growth reached the late log phase. Samples were collected at regular intervals of 4 h up to 36 h of time and the bacterial cell concentration was determined by UV–visible spectrophotometer (U–2800, Hitachi) by measuring the absorbance or optical density (OD) of the cell suspensions at 600 nm.

2.4 Software Used

Regression analysis was performed with the data analysis tool pack of Microsoft Excel®. The model equations were solved using nonlinear regression method using Prism® 5.0.

2.5 Analytical Methods

The benzene content in the biomass free samples was determined quantitatively using high performance liquid chromatography (HPLC) equipped with a UV–Visible detector and C18 column with acetonitrile and water (75:25) as mobile phase at a flow and the detection wavelength of 208 nm. Biomass concentration in the samples was monitored by measuring its absorbance at 600 nm wavelength using a UV–Visible Spectrophotometer 2310.

2.6 Scanning Electron Microscopy

Aerococcus sp. strain BPD-6 cells were growth in MSM in presence of glucose (0.5%) as well as benzene (50 mg L^{-1}). The log phase culture was fixed with 2.5% (v/v) glutaraldehyde in phosphate buffer for 4 h and rinsed in the same buffer for three times. Then the samples were dehydrated by sequential immersion in increasing concentration of ethanol (30, 50, 70, 90, and 100%) each for 10 min. Subsequently, the specimens were coated with gold and observed using Hitachi S-530 scanning electron microscope at an acceleration voltage of 25 kV.

3 Results and Discussion

3.1 Effect of Initial Benzene Concentration on the Growth of the Culture

The growth profile of the strain BPD-6 at different initial concentrations of benzene is shown in (Fig. 1). It was observed that a lag phase of around 6 h during the growth of strain BPD-6 was evident at all concentrations. The lag phase observed in its utilization (degradation) and therefore the culture growth could be attributed to the highly toxic nature of the compound. The experimental specific growth rate (μ) data were plotted against the initial concentration of benzene in order to show the variation in the experimental specific growth rate against the initial substrate concentrations (Fig. 2). Higher concentration of the substrate enhanced the reaction rate and is responsible to render the driving force to subdue the mass transfer resistance of benzene between the two phases of aqueous and gaseous (Ayed et al. 2009). Thus in this plot, a typical trend has been observed and the specific growth rate increases with the increase of initial benzene concentrations and then it started decreasing with the increasing of benzene concentrations. The result showed that

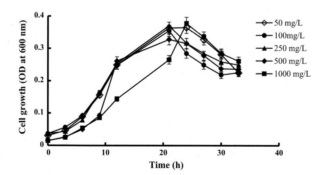

Fig. 1 Growth profile of the *Aerococcus* sp. strain BPD-6 at different benzene concentrations (50–1000 mg L^{-1}) as sole carbon and energy sources and growth was observed at 600 nm

Fig. 2 Specific growth rate versus initial benzene concentration using free cells of *Aerococcus* sp. strain BPD-6

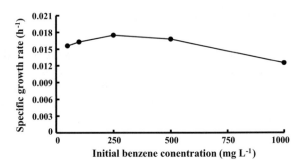

the maximum specific growth rate μ_{max} was reached at a benzene concentration of 250 mg L^{-1} and then the specific growth rate was gradually decreased. It was obvious that beyond 250 mg L^{-1} the inhibition became prominent. Thus the parameter of μ has been found to be a strong function of initial benzene concentration of S.

3.2 Modeling the Kinetics of the Culture Growth

The degradation of a substrate by an organism proceeds through cell mass growth, the kinetic parameters were evaluated on the basis of various growth models such as Haldane, Yano, Aiba, and Tessier at the optimum growth temperature of 37 °C and pH 7.0 of the strain. The value of μ is determined at the exponential phase of the growth curve.

$$\mu = (1/X).(dX/dt)$$

where μ in the above expression represents the specific growth rate of biomass (X) on sole substrate, which is a function of the concentration of the resource. There are two broad opinions regarding the use of equations relating to specific biodegradation growth rates to benzene concentration (Arutchelvan et al. 2006). One view was that benzene could be consider as non-inhibitory compound for an adapted populations and was represented by Monod's non-inhibitory kinetics equation as given shown below $\mu = (\mu_{max} S)/(K_s + S)$. Where μ is the specific growth rate (h^{-1}), S the substrate concentration (mg L^{-1}), μ_{max} the maximum specific growth rate (h^{-1}), K_s the half saturation coefficient or the substrate affinity constant (mg L^{-1}), this constant is defined as the substrate concentration at which μ is equal to half μ_{max} and K_i is the substrate inhibition constant (mg L^{-1}). A low apparent K_s value is indicating the bacterium can efficiently remove the pollution down to low concentration. A large K_i value indicates that the culture is less sensitive to substrate inhibition (Onysko et al. 2000), so low value of K_i showed that the inhibition effect could be observe at low benzene concentration. The other view considered benzene as a growth inhibitory compound, especially at high concentration. Various single substrate growth kinetic models are available in literature to represent the growth kinetics data of an inhibitory compound. In the present study, four growth kinetic models were considered for all substrate concentration values and fitted to the experimental data. The values of kinetic parameters and correlation coefficients (R^2) of four substrate inhibition models were (Haldane, Aiba, Teissier, and Yano) mentioned in Table 2. The results of all four kinetic models and the experimental run were shown in Fig. 3. This figure showed that all models were close to the experimental values under the substrate concentrations ranging from 50 to 1000 mg L^{-1} and Haldane model showed the best result. But at high substrate concentration, the prediction of models of Haldane, Aiba, Teissier, and Yano were diverged and so at high substrate concentration there should be a distinction among these models (Aiba et al. 1968). Monod model did not fit the experimental data at all. The calculations for standard squares of deviation of predicted values of μ from experimental values showed that Yano and Koga and Aiba models with R^2 values (0.9683 and 0.9028), respectively followed by Haldane model with R^2 value (0.8680) were more suitable model for full substrate concentrations of 50–1000 mg L^{-1}. Experimental values showed that Yano and Koga, Aiba and Haldane kinetic models provided a comparable prediction with high R^2 values and showing a good fitted plot. Since the improvement is not significant to justify the use of a four parameter/complex model, it is suggested that Haldane and other kinetic models can also be successfully used for benzene due to its mathematical simplicity for representing the growth kinetics of inhibitory substrates. Table 2 provided the Haldane kinetics model parameters for various microorganisms/benzene waste system. In the present study on benzene using bacterium *Aerococcus* sp. strain BPD-6, the value of μ_{max} was equal to 0.02157. Based on this study, specific growth rate of *Aerococcus* sp. BPD-6 was assumed to be inhibited by the production of metabolites or products and higher initial substrate concentrations. All the models except Monod, adopted in this study have generally been used to describe substrate inhibition on growth of the isolate. Therefore, it was more likely that these models fitted the experimental data obtained in the study reasonably well. It was evident that the maximum specific growth rate (μ_{max}) of the bacterium *Aerococcus* sp. strain BPD-6 was 0.02157 h^{-1}, which was below the range of literature results 0.096–0.62 h^{-1} (Table 1). However, some models showed slight deviation in the values of biokinetic constants, such as μ_{max}, K_s, and K_i, probably due to their differences in origin of development. For example,

Fig. 3 Comparison of growth kinetic models. The software Prism 5.0 is employed to process the experimental data

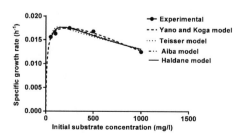

Table 1 Estimated values of biomass growth kinetic model parameters for benzene degradation

Compounds	Models	Equations	μ_{max} (h^{-1})	K_S (mg L^{-1})	K_i (mg L^{-1})	R^2	RMSE values
Benzene	Haldane	$\mu = \frac{\mu_{max} S}{K_S + S + \frac{S^2}{K_i}}$	0.02157	19.56	1584	0.8680	0.00099
	Yano and Koga	$\mu = \frac{\mu_{max} S}{K_S + S + \frac{S^3}{K_i^2}}$	0.0175	11.08	1456	0.9683	0.00048
	Aiba	$\mu = \frac{\mu_{max} S \exp(-\frac{S}{K_i})}{K_S + S}$	0.02115	18.04	2091	0.9028	0.00085
	Teisser	$\mu = \mu_{max}\left[\exp\left(-\frac{S}{K_i}\right) - \exp\left(-\frac{S}{K_s}\right)\right]$	0.01874	28.44	2818	0.7707	0.00131

Table 2 Comparison of benzene Haldane kinetic parameters from published studies

Microorganism	μ_{max} (h^{-1})	K_s (mg L^{-1})	K_i (mg L^{-1})	Reference
Aerococcus sp. strain BPD-6	0.0216	19.56	1584.0	This study
Ralstonia sp. strain YABE411	0.096	3.589	1235.0	Lin and Cheng (2007)
Pseudomonas fluorescens	0.0973	64.3	170.2	Maliyekkal et al. (2004)
Pseudomonas putida F1	0.62	1.65	180.0	Abuhamed et al. (2004)
Pseudomonas putida MTCC 1194	0.1631	71.18	340.15	Mathur and Majumder (2010)
Pseudomonas aeruginosa	0.3	30	130.0	Kim et al. (2005)

Haldane model is based only on the effect of substrate on the growth of a culture (Monteiro et al. 2000). Aiba model is based on that substrate inhibition could be due to the formation of intermediates or products; changed activity of one or more enzymes; dissociation of one or more enzymes; or formation of metabolic aggregates (Raghuvanshi and Babu 2010). The value of K_s obtained from Haldane model was 19.56 mg L^{-1} and slightly low but was within the limits of literature ranges (1.65–71.18 mg L^{-1}). The low K_s showed that the maximum reaction rate maintained until a low concentration was obtained (Gomes et al. 2006). Thus the substrate was oxidized at maximum rate until a low concentration value was reached and thus the strain could be grown in such low concentration of substrate. The inhibition constant K_i was 1584 mg L^{-1} and on the higher range when compared to the values reported in literature (130.0–1235.0 mg L^{-1}). A large K_i value indicated that the culture was less sensitive to substrate inhibition, so high value of K_i showed that the inhibition effect could be observed at high benzene concentration (Abuhamed et al. 2004). When the organic contaminant surpassed the K_i value, the rate of degradation was reduced leading to accumulation of substrate thus affecting the efficacy of the treatment process. Thus these values of kinetic parameters showed a high efficiency of microbial culture *Aerococcus* sp. strain BPD-6 to grow in presence of benzene, and so a complete degradation of the compound in the process.

3.3 Degradation of Benzene by Aerococcus sp.

Figure 4 showed that the time profile of benzene biodegradation for concentration ranging from 50 to 1000 mg L^{-1} using the acclimatized culture of *Aerococcus* sp. strain BPD-6. The removal pattern of benzene was a function of the initial concentration of benzene. Within the first few hours of the experiment, a rapid decline in the concentrations of benzene was noticed with increase in the cell

Fig. 4 Potential of
Aerococcus sp. strain BPD-6
to utilize benzene at different
initial benzene concentrations
$(50\text{--}1000 \text{ mg L}^{-1})$

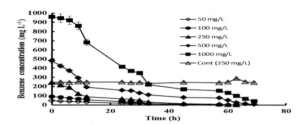

growth. Benzene was removed 28.8% at 50 mg L^{-1} during the first 3 h. Similarly, maximum biomass growth was observed in higher concentrations of benzene of 100, 250, 500 mg L^{-1} in 24 h and during this time benzene was removed 97.5, 73.1, and 89.9%, respectively. At the high concentration of 1000 mg L^{-1}, a marginal removal of benzene was observed about 64.91%. It was observed that the increase in the concentrations of benzene not only increased the benzene removal but also stimulated better biomass production during growth. The strain BPD-6 was also able to utilize BTEX compound as a sole carbon source in the MSM medium (Das and Kazy 2014).

3.4 Scanning Electron Microscopy

The Scanning Electron Microscopy was used to study and understand the morphological changes on LB solid plate in the presence of benzene at a concentration of 100 mg L^{-1} using cells from exponentially growing cultures. There was no marked distortion but little shrinkage of the cells in the presence of benzene as observed under the microscope (Fig. 5a, b). This indicates that *Aerococcus* sp. strain BPD-6 was quite resistant to the compound benzene.

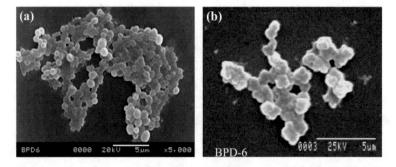

Fig. 5 SEM micrograph of *Aerococcus* sp. strain BPD-6 **a** Control cell grown in presence of glucose. **b** Cells grown in presence of benzene as a carbon source

3.5 Effect of Initial pH on the Growth of the Strain

It was known that availability of macronutrients and the enzyme activity are highly pH-dependent (Skladany and Baker 1994) and specific growth rate should be optimized with respect to medium pH during growth. High acidic and basic environmental pH increased the toxicity toward the cell which directly inhibited the biomass formation (Shing et al. 2008). Therefore, to investigate the effect of pH on the growth of *Aerococcus* sp. strain BPD-6 was studied experimentally at different initial pH values, varying from 2.0 to 12.0 under identical environmental conditions like temperature at 37 °C and the initial concentration of benzene at 100 mg L^{-1}. Most of the organisms cannot tolerate pH levels above 9.5 or below 4.0. However, the optimum pH for different microorganisms observed in different ranges (Singh et al. 2008). Consequently, the effect of pH on growth of *Aerococcus* sp. strain BPD-6 was in agreement with most microorganisms, which favored growth at pH levels ranging from 6.0 to 8.0 (Skladany and Baker 1994). Subsequently, growth profile study at varying pH was carried out and maximum growth was obtained at pH value 9.0 (Fig. 6). The important thing observed in the present study was that the isolated strain could grow at the pH values 8.0–10.0 and 12.0 as indicated by high growth at these pH values which is in context with the properties of *Aerococcus* sp. is a high pH tolerant. However, the growth of the strain was significantly declined when the pH was less than 5.0. The activity of *Aerococcus* sp. strain BPD-6 at wide pH indicated that it can be applied to different climatic condition on wide range of pH. In the present work, the pH of the medium was considered to be 7.0 for further degradation studies.

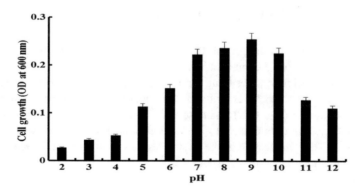

Fig. 6 Growth of *Aerococcus* sp. strain BPD-6 in MSM at 100 mg L^{-1} of benzene as sole source of carbon at different pH range (2–12)

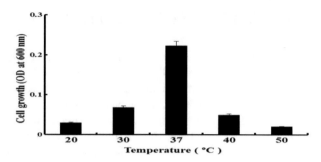

Fig. 7 Growth of *Aerococcus* sp. strain BPD-6 in MSM at 100 mg L^{-1} of benzene as sole source of carbon growth at different temperature (20–60 °C)

3.6   Effect of Temperature on the Growth of the Strain

Comparisons of the growth in presence of benzene by *Aerococcus* sp. strain BPD-6 was incubated at various temperatures to determine the observed growth rate at elevated temperature. The strain was tested for growth on 100 mg L^{-1} benzene and at an optimum pH of 7.0. A wide range of temperatures from 20 to 60 °C was tested. As shown in Fig. 7 the cell growth was increased at the temperatures 30, 37, and slightly at 40 °C while it decreased sharply at incubation of 20 and 50 °C and inhibited entirely at 60 °C. Results thus indicated that at 37 °C incubated strain exhibited shorter benzene lag time relative to when incubated at 30 °C.

3.7   Effect of Salinity on the Growth of the Strain

The effect of salinity on the growth of the strain was observed in the batch set up containing different NaCl concentrations (1.0–6.0 g L^{-1}). High salinity effects cell numbers and distribution, resulting in reduced microbial metabolic rates. The maximum growth obtained at different salt concentrations in the mineral salt medium is shown in (Fig. 8). It was observed that the bacteria grows exceptionally well up to the concentration of 6.0 g L^{-1} and also maximum growth was observed at 4.0 g L^{-1}. Thus the growth of the strain was increased with the increase of salt concentrations ranging from 4.0 to 6.0 g L^{-1}.

4   Conclusions

The present study found out the following salient observations for benzene degradation and consequently the growth of the indigenous Gram-positive bacteria *Aerococcus* sp. strain BPD-6. To our best knowledge, this is the first report of

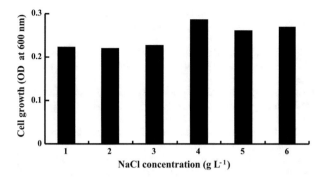

Fig. 8 Growth of *Aerococcus* sp. strain BPD-6 in MSM at 100 mg L^{-1} of benzene as sole source of carbon growth at different NaCl concentration (1.0–6.0 g L^{-1})

benzene utilization by *Aerococcus* sp. This indigenous bacterium *Aerococcus* sp. strain BPD-6 could be effectively used for biodegradation of benzene containing oily sludge. The lag phase of this bacterium was found to be around 6 h indicating that the strain is highly active and has the ability to degrade the contaminant in less time. The rate and inhibition constants were determined by correlating the experimental data with the kinetic models. High values of K_i indicated that substrate inhibited the growth of the strain at relatively high concentration. Percent toxicity removal was conducted by this strain with different initial benzene concentrations and maximum removal was observed at 100 mg L^{-1}. Thus the study revealed that the application of an indigenous microbial culture *Aerococcus* sp. could be treated of oily sludge containing highly recalcitrant compound such as benzene. Further studies are warranted to elucidate the mechanism of benzene degradation in terms of enzymatic relation which would facilitate in the development of enhanced remediation strategy for monoaromatic compounds from petroleum oil sludge.

Acknowledgements The authors are highly grateful to the DST FAST TRACK PROJECT for financial aid.

References

Abuhamed T, Bayraktar E, Mehmetoğlu T, Mehmetoğlu U (2004) Kinetics model for growth of *Pseudomonas putida* F1 during benzene, toluene and phenol biodegradation. Process Biochem 39(8):983–988

Aburto-Medina A, Ball AS (2015) Microorganisms involved in anaerobic benzene degradation. Ann Microbiol 65(3):1201–1213

Aiba S, Shoda M, Nagatani M (1968) Kinetics of product inhibition in alcohol fermentation. Biotechnol Bioeng 10(6):845–864

Arutchelvan V, Kanakasabai V, Elangovan R, Nagarajan S, Muralikrishnan V (2006) Kinetics of high strength phenol degradation using *Bacillus brevis*. J Hazard Mater 129(1–3):216–222

Ayed L, Chaieb K, Cheref A, Bakhrouf A (2009) Biodegradation of triphenylmethane dye Malachite Green by *Sphingomonas paucimobilis*. World J Microbiol Biotechnol. 25(4): 705–711

Cerqueira VS, Hollenbach EB, Maboni F, Vainstein MH, Camargo FAO, Peralbo MCR, Bento FM (2011) Biodegradation potential of oily sludge by pure and mixed bacterial cultures. Bioresour Technol 102:11003–11010

Chakraborty R, Coates JD (2005) Hydroxylation and carboxylation-two crucial steps of anaerobic benzene degradation by *Dechloromonas* strain RCB. Appl Environ Microbiol 71(9):5427–5432

Chen L, Liu Y, Liu F, Jin S (2014) Treatment of co-mingled benzene, toluene and TCE in groundwater. J Hazard Mater 275:116–120

Choudhary S, Sar P (2011) Uranium biomineralization by a metal resistant *Pseudomonas aeruginosa* strain isolated from contaminated mine waste. J Hazard Mater 186(1):336–343

Das R, Kazy SK (2014) Microbial diversity, community composition and metabolic potential in hydrocarbon contaminated oily sludge: prospects for in situ bioremediation. Environ Sci Pollut Res 21(12):7369–7389

Fahy A, Ball AS, Lethbridge G, McGenity TJ, Timmis KN (2008) High benzene concentrations can favour Gram-positive bacteria in groundwaters from a contaminated aquifer. FEMS Microbiol Ecol 65(3):526–533

Farhadian M, Duchez D, Vachelard C, Larroche C (2009) Accurate quantitative determination of monoaromatic compounds for the monitoring of bioremediation processes. Bioresour Technol 100(1):173–178

Gillespie IMM, Philp JC (2013) Bioremediation, an environmental remediation technology for the bioeconomy. Trends Biotechnol 31:329–332

Gomes R, Nogueira R, Oliveira J, Peixoto J, Brito A (2006) Kinetics of fluorene biodegradation by a mixed culture. In: Proceedings of the second IASTED international conference on advanced technology in the environmental field, Lanzarote, Canary Islands, Spain, 6–8 Feb 2006. Acta Press, Lanzarate, Canary Islands, Spain

Kim DJ, Choi JW, Choi NC, Mahendran B, Lee CE (2005) Modeling of growth kinetics for *Pseudomonas* spp. during benzene degradation. Appl Microbiol Biotechnol 69(4):456–462

Li J, de Toledob RA, Shim H (2017) Multivariate optimization for the simultaneous bioremoval of BTEX and chlorinated aliphatic hydrocarbons by *Pseudomonas plecoglossicida*. J Hazard Mater 321(1–3):238–246

Lin CW, Cheng YW (2007) Biodegradation kinetics of benzene, methyl tert-butyl ether, and toluene as a substrate under various substrate concentrations. J Chem Technol Biotechnol 82 (1):51–57

Loh KC, Wang SJ (1997) Enhancement of biodegradation of phenol and a nongrowth substrate 4-chlorophenol by medium augmentation with conventional carbon sources. Biodegradation 8 (5):329–338

Luo F, Devine CE, Edwards EA (2016) Cultivating microbial dark matter in benzene-degrading methanogenic consortia. Environ Microbiol 18(9):2923–2936

Maliyekkal SM, Rene ER, Philip L, Swaminathan T (2004) Performance of BTX degraders under substrate versatility conditions. J Hazard Mater 109(1–3):201–211

Mathur A, Majumder C (2010) Kinetics modelling of the biodegradation of benzene, toluene and phenol as single substrate and mixed substrate by using *Pseudomonas putida*. Chem Biochem Eng Q 24(1):101–109

Mayer DA, de Souza AAU, Fontana E, de Souza SMAU (2016) Kinetic study of biodegradation of BTX compounds in mono and multicomponent systems in reactor with immobilized biomass. Bioprocess Biosyst Eng. 39:1441–1454

Monteiro AA, Boaventura RA, Rodrigues AE (2000) Phenol biodegradation by *Pseudomonas putida* DSM 548 in a batch reactor. Biochem Eng J 6(1):45–49

Mukherjee AK, Bordoloi NV (2012) Biodegradation of benzene, toluene, and xylene (BTX) in liquid culture and in soil by *Bacillus subtilis* and *Pseudomonas aeruginosa* strains and a formulated bacterial consortium. Environ Sci Pollut Res 19(8):3380–3388

Munoz R, Diaz L, Bordel S, Villaverde S (2007) Inhibitory effects of catechol accumulation on benzene biodegradation in *Pseudomonas putida* F1 cultures. Chemosphere 68(2):244–252

Onysko KA, Budman HM, Robinson CW (2000) Effect of temperature on the inhibition kinetics of phenol biodegradation by *Pseudomonas putida* Q5. Biotechnol Bioeng 70(3):291–299

Raghuvanshi S, Babu BV (2010) Biodegradation kinetics of methyl iso-butyl ketone by acclimated mixed culture. J Biodegradation 21:31–42

Rahul, Mathur AK, Balomajumder C (2013) Performance evaluation and model analysis of BTEX contaminated air in corn-cob biofilter system. Bioresour Technol 133:166–174

Singh D, Fulekar MH (2010) Benzene bioremediation using cow dung microflora in two phase partitioning bioreactor. J Hazard Mater 175(1–3):336–343

Singh RK, Kumar S, Kumar S, Kumar A (2008) Biodegradation kinetic studies for the removal of cresol from wastewater using Gliomastix indicus MTCC 3869. Biochem Eng J 40(2):293–303

Skladany GJ, Baker KH (1994) Laboratory biotreatability studies. In: Baker KH, Herson DS (eds) Bioremediation. McGraw-Hill Inc, NY, pp 97–172

Yang R, Humphrey A (1975) Dynamic and steady state studies of phenol biodegradation in pure and mixed cultures. J Biotechnol Bioeng 17:1211–1220

Simultaneous Removal of High Nitrate and Phosphate from Synthetic Waste Water by Prolonged Anoxic Cycle in a Batch Reactor

Jyotsnarani Jena and Trupti Das

Abstract Phosphorus, as well as nitrate, is an essential nutrient for different agricultural activities, but its excess disposals and discharges into natural water sources result in eutrophication. This leads to water quality problems including increased purification costs, interference with the recreational and conservation value of water bodies. In the current study, attempts have been made to remove nutrients from synthetic wastewater using activated sludge under anoxic condition. Enhanced biological nutrient removal is an efficient and sustainable biological technology to remove overall COD from wastewater. It is based on the enrichment of activated sludge with phosphorous accumulating organisms (PAOs) and denitrifying phosphate accumulating organisms (DNPAOs) by introducing alternating anoxic and aerobic conditions, which favour specific bacterial growth over ordinary heterotrophic organisms. In a small reactor, anoxic condition prevails throughout the working time (24 h). In this reactor, the removal rate of phosphate and nitrate is ~ 75 and >90% respectively. Metagenomic analysis of the sludge confirmed the presence of DNPAOs in the anaerobic reactor. These metagenomic studies further confirmed the presence of DNPAOs. These DNPAOs removed high amount of nutrient in the presence of an optimum amount of carbon source in anaerobic condition using nitrate as electron acceptor.

Keywords DNPAOs · COD · Nitrate · Phosphate

1 Introduction

Use of biological processes for nutrient removal from wastewater is being studied due to an increasing demand to reduce the concentrations of nutrients such as nitrate and phosphate from various industrial as well as domestic effluents. Phosphorus, as well as nitrate, is an essential nutrient for different agricultural activities,

J. Jena (✉) · T. Das
CSIR-Institute of Minerals and Materials Technology, Bhubaneswar 751013, Odisha, India
e-mail: jyotsna97rta@gmail.com

© Springer International Publishing AG 2018
A. K. Sarma et al. (eds.), *Urban Ecology, Water Quality and Climate Change*,
Water Science and Technology Library 84,
https://doi.org/10.1007/978-3-319-74494-0_24

but its excess disposals and discharges into natural water sources result in eutrophication leading to several water quality problems (Metcalf & Eddy 1979). Extensive research is being pursued on the activated sludge treatment processes to obtain a high nutrient removal efficiency and economically feasible unit configuration.

Some researchers interpret that nitrate existing in the reactor could be used as an electron acceptor for the growth of non-poly P heterotrophs, this reduces the amount of substrate available for PAOs and hence cause the reduction of phosphorous removal (Hascoet and Florentz 1985). On other hand, reports showed that nitrate could be used as final electron acceptor for phosphorous removal instead of oxygen. The later theory has the potential to reduce the overall operational cost as well as resulting in less sludge production. Occurrence of Denitrifying Phosphate Accumulating Organisms (DNPAOs), capable of using nitrate instead of oxygen as an electron acceptor for phosphorous uptake (Bortone et al. 1999) in the reactor ensure the phosphate uptake as well as nitrate removal from the system. Several studies have proposed that DNPAOs have kind of similar physiological characteristics to those of PAOs based on the metabolic mechanism of polyphosphate, polyhydroxyalkanoates (PHA) and glycogen (Murnleitner et al. 1997). DNPOs use nitrate as the final electron acceptor and Volatile Fatty Acids (VFA) like acetate, as the electron donor (Ahn et al. 2002).

In the current study, an attempt has been made to study the effect of COD/NO_3^--N ratio on simultaneous removal of nitrate and phosphate from high strength synthetic wastewater, in a bioreactor, maintained under prolonged anaerobic conditions. Efforts have also been made to analyse the microbial community involved in anaerobic phosphate removal using molecular methods.

2 Methods

2.1 Effect of COD/N Ratio on Removal of Nitrate and Phosphate

2.1.1 Experimental Set-Up

A 2 L volume closed glass reactor was used for the experiments and was connected with mechanical stirrers to improve contact between the microorganism and the synthetic wastewater. The sequencing batch reactor (SBR) was operated continuously for 21 h under anoxic condition which was maintained constantly by purging nitrogen gas (XL grade ~99% purity) to the reactor. All Experiments were carried out at room temperature (25–30 °C) and at pH values between 7.5 and 8.

2.1.2   Synthetic Waste Water

Synthetic waste water (1.7 $\pm$ 0.1 g/L Ca(NO$_3$)$_2$4H$_2$O: as nitrate source, 0.043 g/L of KH$_2$PO$_4$: as PO$_4$ source, 6.86 g/L MgSO$_4$, 0.38 g/L peptone) along with 0.3 mL of nutrient solution (0.15 g/L FeCl$_3$·6H$_2$O, 0.15 g/L H$_3$BO$_3$, 0.03 g/L CuSO$_4$·5H$_2$O, 0.18 g/L KI, 0.12 g/L MnCl$_2$·4H$_2$O, 0.06 g/L Na$_2$MoO$_4$·2H$_2$O, 0.12 g/L ZnSO$_4$·7H$_2$O, 0.15 g/L CoCl$_2$·H$_2$O, 10 g/L EDTA) was used to feed the reactor, making the working volume of the SBR 1 L. Required amount of sodium acetate (0, 500, 800, 1000, 5000 ppm) was added to the synthetic waste water to achieve the COD/N ratio of 0.36, 0.98, 1.36, 1.62, 6.37 respectively.

2.1.3   Inoculation and Start-Up

Initially 1 L of synthetic wastewater consisting of definite COD/N ratio was inoculated with mixed bacterial sludge from a parent reactor which had a 2 L of working volume and was operational on a continuous mode for 180 days achieving stable and efficient phosphate removal under anaerobic conditions. The reactor was being operated with a 12 h cycle consisting of 6 h anaerobic-5 h aerobic and 1 h for settle/decant/refill.

2.1.4   Analysis

Performance of bioreactor was monitored by measuring the phosphate, nitrate-N, COD, TSS, VSS and pH. Samples were taken once every hour and immediately filtered and analyzed for N-nitrate, PO$_4$–P, COD concentration. In these studies experimental COD/N ratio was investigated. Nitrate and phosphate concentration was monitored spectrophotometrically at 420 and 880 nm respectively (APHA 1998). Analysis of TSS, VSS, COD were conducted as described in the standard methods (APHA 1998). TSS and VSS were maintained at 5 $\pm$ 0.4 gm and 3 $\pm$ 0.5 gm respectively.

2.2   Metagenomic Work

2.2.1   Total DNA Isolation

2 ml of sludge sample from the reactor were suspended in 1.5 ml of lysis buffer (NaCl, Na$_2$EDTA, Tris-HCL, pH adjusted to 8) containing 100 µg/ml of lysozyme and subjected to vigorous vortex. 2 ml of proteinase K (10 mg/ml) was added to the mixture and kept in water bath for 1.5 h. After the centrifugation of this mixture for 15 min at 12,000 rpm (PerkinElmer), the supernatant was recovered. The recovered aqueous supernatant was incubated with RNase (5 µg/ml of supernatant) for 1 h in

35 °C. 1 ml of Phenol-chloroform (1:1) solution was added to the incubated solution and centrifuged at 10,000 rpm for 15 min. The clear soup was transferred to a new tube after the centrifugation followed by the addition of 0.8 volume of isopropanol (0.8 ml/1 ml of supernatant). The collected pellets were subjected to 70% ethanol and centrifuged at 10,000 rpm for 5 min. The pellet obtained after centrifugation was dried and suspended in 100 of nuclease-free water for storage.

2.2.2 PCR Amplification

The region corresponding to positions 341 and 907 in the 16S rDNA of *Escherichia coli* was PCR-amplified using the forward primer 27F(5′-AGAGTTTGATCCT GGCTCAG-3′) and reverse primer 1492R(5GGTTACCTTGTTAC GACTT 3′AAGTCGTAACAAGGTAACC) PCR amplification was conducted in an auto-mated thermal cycler (BIO RAD PCR System). Amplified PCR products were electrophoresed on 2% (wt/vol) agarose gel in 1X TAE for 70 min for 50 V and then checked with ethidium bromide staining.

2.2.3 Cloning into a Vector

The amplified PCR product was cloned into TA cloning vector and then transferred into an appropriate host through transformation (Fermentas TA cloning kit). Blue-white screening was performed for the screening of transformed bacterial cells. The cloning was again conformed by the restriction enzyme treatment of few selected clones. DNA sequencing was done in Bhat biotech, New Delhi. Partial 16S rRNA gene sequences were analysed using BLA search facility.

3 Result and Discussion

3.1 Nitrate Removal Efficiency

Nitrate removal rate was negligible in absence of acetate (when the COD/N ratio was 0.36 as shown in Fig. 1, due to apparent inhibition of heterotrophic denitrifi-cation (Carrera et al. 2004). Maximum removal (98%) of nitrate was observed in presence of 1000 ppm of acetate (with COD/N ratio >1) whereas there was a marked decline in the rate (81%) with a five-time increase in the acetate concen-tration (COD/N ratio >6) in the medium specifying the ineffectiveness of organic carbon to enhance the overall reaction kinetics beyond a certain threshold limit. There was a marked rise in the denitrification rate as well as the pH, followed by the supply of external carbon source (Yuana et al. 2008). Due to denitrification, the pH of the medium was observed to go beyond 8.0, within 1 h of reaction.

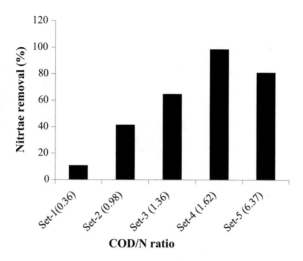

Fig. 1 Nitrate removal efficiency with respect to different COD/N ratio

3.2 *PO₄ Removal Efficiency*

In spite of a varying organic load and high nitrate concentration, stable PO_4 removal was achieved in all the sets (Fig. 2) and maximum (60–70%) uptake occurred within the first 3–4 h of reaction. Hardly any PO_4 release into the medium might be due to the availability of nitrate throughout the experimental period. PO_4 uptake in the absence of organic carbon can be justified by the utilization of the internally stored PHA (Poly HydroxyAlkanoate) in the DNPAOs (Ahn et al. 2002).

Fig. 2 Phosphate removal efficiency with respect to different C/N ratio

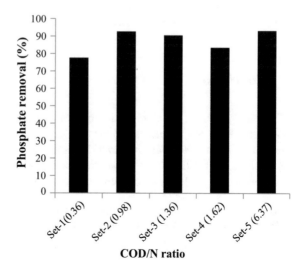

Fig. 3 COD removal efficiency with respect to different COD/N ratio

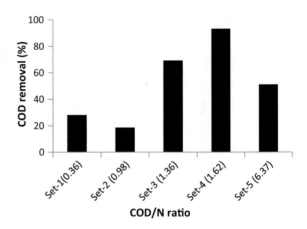

3.3 COD Removal Efficiency

With a gradual increase in organic carbon load (as well as COD/N ratio) there was a marked variation in COD removal efficiency. In presence of 1000 mg/L of acetate as carbon source (COD/N ratio 1.62), maximum COD ($\sim$98%) removal was achieved (Fig. 3). The trend of COD removal rate was synchronous to that of nitrate removal with respect to carbon load.

3.4 Exploration of Microbes in the Reactor

The white colonies from the plate were selected and tested for the presence of the desired sequence by restriction digestion method. Sequencing of these confirmed, the presence of DNPAOs like *Dokdonella* sp, *Rhodanobacter* etc. in the reactor.

4 Conclusion

DNPAOs present in the reactor (as screened in the metagenomic analysis) uptake phosphate by using nitrate as the electron acceptor. Successful removal of phosphate, nitrate and COD was achieved in an anoxic batch reactor with an optimum COD/N ratio of 1.62. Scale up studies needs to be carried out to establish the process as energy efficient, economic and eco-friendly technique.

Acknowledgements Authors are thankful to the Director, IMMT, Bhubaneswar for his encouragement and to DST, New-Delhi for financial support.

References

Ahn J, Daidou T, Tusenda S, Hirata A (2002) Transformation of phosphorous and relevant intracellular compounds by a phosphorous accumulating enrichment culture in the presence of both electron acceptor and electron donor. Biotechnol Bioeng 79:1

American Public Health Association (APHA) (1998) Standard methods for the examination of water and waste water, 17th edn. Washington, DC

Bortone G, Libelli SM, Tilche A, Wanner J (1999) Anoxic phosphate uptake in the DEPHANOX process. Water Sci Technol 40:177–185

Carrera J, Vicent T, Lafuente J (2004) Effect of influent COD/N ratio on biological nitrogen removal (BNR) from high-strength ammonium industrial waste water. Process Biochem 39:2035–2041

Hascoet MC, Florentz M (1985) Influence of nitrate on biological phosphorous removal from waste water. Water SA 11:1–8

Metcalf & Eddy (1979) Waste water Engineering: treatment disposal reuse, 2nd edn

Murnleitner E, Kuba T, van Loosdrecht MCM, Heijnen JJ (1997) An integrated metabolic model for the aerobic and denitrifying biological phosphorus removal. Biotechnol Bioeng 54:434–450

Yuana L, Zhangb C, Zhangb Y, Dingb Y, Xi D (2008) Biological nutrient removal using an alternating of anoxic and anaerobic membrane bioreactor (AAAM) process. Desalination 221:566–575

Effects of Heavy Metals on the Environment by Utilization of Urban Waste Compost for Land Application: A Review

Jiwan Singh and Ajay S. Kalamdhad

Abstract Composting is becoming a more acceptable and economical method for treating urban waste including sewage sludge, municipal solid waste, tannery waste etc., especially for cities with a high population density. However, the major disadvantage of composting these wastes is the high heavy metal contents in the end-product then it is harmful to the environment. Heavy metals are toxic to soil, plants, aquatic life, and human health if their concentration is high in the compost. Heavy metals exhibit toxic effects toward soil biota by affecting key microbial processes and decrease the number and activity of soil microorganisms. Even low concentration of heavy metals may inhibit the physiological metabolism of the plant. Uptake of heavy metals by plants and subsequent accumulation along the food chain is a potential threat to animal and human health. Contaminants in aquatic systems, including heavy metals, stimulate the production of reactive oxygen species (ROS) that can damage fishes and other aquatic organisms. Hence the compost has to be used for agriculture it should be free from heavy metals. Therefore, the present study evaluated the effects of heavy metal containing compost on soil, plants, human health, and aquatic life.

Keywords Composting · Heavy metals · Human health · Aquatic life

1 Introduction

Rapid industrialization and population explosion in India has led to the migration of people from villages to cities, which generate thousands of tons of MSW daily (Sharholy et al. 2008). Composting is becoming a more acceptable and economical method for treating urban waste including sewage sludge, municipal solid waste, tannery waste etc., especially for cities with a high population density.

J. Singh (✉) · A. S. Kalamdhad
Department of Civil Engineering, Indian Institute of Technology Guwahati (IITG), Guwahati, India
e-mail: s.jiwan@iitg.ernet.in

© Springer International Publishing AG 2018   329
A. K. Sarma et al. (eds.), *Urban Ecology, Water Quality and Climate Change*,
Water Science and Technology Library 84,
https://doi.org/10.1007/978-3-319-74494-0_25

Composting can be defined as the process in which of organic waste treatment by aerobic microorganisms; as such, it comprises three major phases: mesophilic and thermophilic stages and cooling (the compost stabilization stage) (Neklyudov et al. 2008). It can reduce the solid waste volume by 40–50% (Zorpas et al. 2002), pathogens are destroyed by the metabolic heat generated by the thermophilic phase, degrade a big number of hazardous organic pollutants, and make available a final product that can be used as a soil improvement or fertilizer (Cai et al. 2007). If the final product contains high heavy metals concentration it may be noxious to soil, plants, and human health. Heavy metals uptake by plants and successive accumulation in human tissues and biomagnifications through the food chain causes both human health and environment concerns (Wong and Selvam 2006).

Heavy metals are commonly defined as being those metallic elements with a density >5000 kg m^{-3} (Talbot 2006). Heavy metals are considered one of the major sources of soil pollution. Heavy metal pollution of the soil is caused by various metals, especially Cu, Ni, Cd, Zn, Cr, and Pb (Karaca et al. 2010). Some heavy metals (like Fe, Zn, Ca, and Mg) have been reported to be of bio-importance to man and their daily medicinal and dietary allowances had been recommended. However, some others (like As, Cd, Pb, and methylated forms of Hg) have been reported to have no known bio-importance in human biochemistry and physiology and consumption even at very low concentrations can be toxic (Duruibe et al. 2007).

Heavy metals exert poisonous effects on soil microorganism, therefore, results in the change of the diversity, population size, and overall activity of the soil microbial communities (Ashraf and Ali 2007). Prominent Pb in soils may decrease soil productivity and a very low Pb concentration may inhibit some vital plant processes i.e., photosynthesis, mitosis, and water absorption with toxic symptoms of dark green leaves, wilting of older leaves, stunted foliage and brown short leaves, and stunted foliage and brown short roots (Bhattacharyya et al. 2008). The metal plant uptake from soils at high concentrations may result in a great health risk taking into consideration food-chain implications (Jordao et al. 2006). Uptake of heavy metals by plants and succeeding accumulation along the food chain is a prospective threat to human health. The consumption of heavy metal contaminated food can seriously deplete some essential nutrients in the body that are further responsible for decreasing immunological defenses, intrauterine growth retardation, disabilities associated with malnutrition and high prevalence of upper gastrointestinal cancer rates (Khan et al. 2008).

Heavy metals containing agricultural runoff enter in aquatic environment it may toxic to aquatic plants and animals. If compostable waste such as sewage sludge, municipal solid waste, and pig manure contain heavy metals, it may change the composting process by inhibiting bacterial growth. In the vermicomposting process, heavy metals affect earthworm life cycle. Therefore, the aim of the present study was to assess the effects of heavy metal containing compost on soil, plants, human health, and aquatic life as well as effects of heavy metal containing compostable material on composting process.

2 Effects on Soil

Soil pollution by heavy metals is of most important anxiety all the way through the industrialized world (Hinojosa et al. 2004). Heavy metal pollution not only results in unfavorable effects on various parameters relating to plant quality and yield but also cause changes in the size, composition, and activity of the microbial population (Yao et al. 2003). Therefore, heavy metals are considered as one of the major sources of soil pollution. Heavy metal pollution of the soil is caused by various metals especially Cu, Ni, Cd, Zn, Cr, and Pb (Hinojosa et al. 2004). The adverse effects of heavy metals on soil biological and biochemical properties are well predictable. The soil properties i.e., organic matter, clay contents, and pH have major influences on the degree of the effects of metals on biological and bio-chemical properties (Speira et al. 1999).

Heavy metals indirectly affect soil enzymatic activities by shifting the microbial community which synthesizes enzymes (Shun et al. 2009). Heavy metals exhibit toxic effects toward soil biota by affecting key microbial processes and decrease the number and activity of soil microorganisms. Conversely, long-term heavy metal effects can increase bacterial community tolerance as well as the tolerance of fungi such as arbuscular mycorrhizal (AM) fungi, which can play an important role in the restoration of contaminated ecosystems (Mora et al. 2005). Chen et al. (2010) recommended that heavy metals caused a decrease in bacterial species richness and a relative increase in soil actinomycetes or even decreases in both the biomass and diversity of the bacterial communities in contaminated soils.

Karaca et al. (2010) reported that the enzyme activities are influenced in different ways by different metals due to the different chemical affinities of the enzymes in the soil system. Cadmium is more toxic to enzymes than Pb because of its greater mobility and lower affinity for soil colloids. Cu inhibits b-glucosidase activity more than cellulose activity. Pb decreases the activities of urease, catalase, invertase, and acid phosphatase significantly. Phosphatase and sulfatase are inhibited by As (V) but that urease was unaffected. Cd contamination has a negative effect on the activities of protease, urease, alkaline phosphatase, and arylsulfatase but no sig-nificant effect on that of invertase. Each soil enzyme exhibits a different sensitivity to heavy metals. The order of inhibition of urease activity generally decreased according to the sequence Cr > Cd > Zn > Mn > Pb.

Diversity and activity of soil microbes play significant roles in the recycling of plant nutrients, maintenance of soil structure, detoxification of noxious chemicals, and the control of plant pests and plant growth communities are important indices of soil quality. It is important to investigate the functioning of soil microorganisms in ecosystems exposed to long-term contamination by heavy metals (Wang et al. 2007).

Chromium is commonly present in soils as Cr(III) and Cr(VI), which are characterized by distinct chemical properties and toxicities. Cr(VI) is a strong oxidizing agent and is highly toxic, whereas Cr(III) is a micronutrient and a non-hazardous species 10–100 times less toxic than Cr(VI) (Garnier et al. 2006). Cr(VI) has been reported to cause shifts in the composition of soil microbial populations and known to cause detrimental effects on microbial cell metabolism at high concentrations (Shun et al. 2009). Ashraf and Ali (2007) also reported that the heavy metals exert toxic effects on soil microorganism hence results in the change of the diversity, population size, and overall activity of the soil microbial communities and observed that the heavy metal (Cr, Zn, and Cd) pollution influenced the metabolism of soil microbes in all cases. In general, an increase of metal concentration adversely affects soil microbial properties e.g., respiration rate, enzyme activity, which appears to be very useful indicators of soil pollutions. In case of soil contaminated with lead (Pb) slight change was observed in the soil microbial profile.

3   Effects on Plants

Some of these heavy metals i.e., As, Cd, Hg, Pb, or Se are not essential for plants growth since they do not perform any known physiological function in plants. Others i.e., Co, Cu, Fe, Mn, Mo, Ni, and Zn are essential elements required for normal growth and metabolism of plants, but these elements can easily lead to poisoning when their concentration is greater than optimal values (Garrido et al. 2002; Rascio and Izzo 2011). The use of compost to improve agricultural yield without caring with possible negative effects might be a problem since the waste composts are most applied to improve soils used to grow vegetables. Considering the edible part of the plant in most vegetable species, the risk of transference of heavy metals from soil to humans should be a matter of concern (Jordao et al. 2007).

Uptake of heavy metals by plants and subsequent accumulation along the food chain is a potential threat to animal and human health (Sprynskyy et al. 2007). The absorption by plant roots is one of the main routes of the entrance of heavy metals in the food chain (Jordao et al. 2007). Absorption and accumulation of heavy metals in plant tissue depend upon many factors which include temperature, moisture, organic matter, pH, and nutrient availability. The uptake and accumulation of Cd, Zn, Cr, and Mn in *Beta vulgaris* (Spinach) were higher during the summer season, whereas Cu, Ni, and Pb accumulated more during the winter season (Karaca et al. 2010). It may be expected that during the summer season the relatively high decomposition rate of organic matter is likely to release heavy metals in soil solution for possible uptake by plants. The higher uptake of heavy metals i.e., Cd, Zn, Cr, and Mn during the summer season may be due to high transpiration rates as compared to the winter season due to high ambient temperature and low humidity (Sharma et al. 2007).

Heavy metal accumulation in plants depends upon plant species and the efficiency of different plants in absorbing metals is evaluated by either plant uptake or soil to plant transfer factors of the metals (Khan et al. 2008). Elevated Pb in soils may decrease soil productivity, and a very low Pb concentration may inhibit some vital plant processes, such as photosynthesis, mitosis, and water absorption with toxic symptoms of dark green leaves, wilting of older leaves, stunted foliage, and brown short roots (Bhattacharyya et al. 2008). Heavy metals are potentially toxic and phytotoxicity for plants resulting in chlorosis, weak plant growth, yield depression, and may even be accompanied by reduced nutrient uptake, disorders in plant metabolism and reduced ability to fixate molecular nitrogen in leguminous plants (Guala et al. 2010). Seed germination was gradually delayed in the presence of increasing concentration of lead (Pb), it may be due to prolonged incubation of the seeds that must have resulted in the neutralization of the toxic effects of lead by some mechanisms e.g., leaching, chelation, metal binding or/and accumulation by microorganisms (Ashraf and Ali 2007).

4 Effects on Aquatic Environment

Heavy metals are highly persistent, toxic in trace amounts, and can potentially induce severe oxidative stress in aquatic organisms. Thus, these contaminants are highly significant in terms of ecotoxicology. Moreover, metals are not subject to bacterial degradation and hence remain permanently in the marine environment (Woo et al. 2009). Contamination of a river with heavy metals may cause devastating effects on the ecological balance of the aquatic environment, and the diversity of aquatic organisms becomes limited with the extent of contamination (Ayandiran et al. 2009). Heavy metals released into aquatic systems are generally bound to particulate matter, which eventually settle down and become incorporated into sediments. Surface sediment, therefore, is the most important reservoir or sink of metals and other pollutants in aquatic environments. Sediment-bound pollutants can be taken up by rooted aquatic macrophytes and other aquatic organisms (Peng et al. 2008).

Because a major fraction of the trace metals introduced into the aquatic environment eventually become associated with the bottom sediments, environmental degradation by metals can occur in areas where water quality criteria are not exceeded, yet organisms in or near the sediments are adversely affected (Gurrieri 1998). Diatom community structure can be affected by high levels of micropollutants, and in particular by metals, which are often found in rivers (Morin et al. 2007; Jongea et al. 2009). Once heavy metals are accumulated by an aquatic organism, they can be transferred through the upper classes of the food chain. Carnivores at the top of the food Chain including humans, obtain most of their heavy metal burden from the aquatic ecosystem by way of their food, especially where fish are present so there exists the potential for considerable biomagnifications (Ayandiran et al. 2009).

Contaminants in aquatic systems, including heavy metals, stimulate the production of reactive oxygen species (ROS) that can damage fishes and other aquatic organisms (Woo et al. 2009). Fish is a commodity of potential public health concern as it can be contaminated with a range of environmentally persistent chemicals, including heavy metals. The consumption of fish containing elevated levels of metals is a concern because chronic exposure to heavy metals can cause health problems (Soliman 2006). Mercury (Hg) is one of the most important pollutants both because of its effect on marine organisms and it is potentially hazardous to humans. Methyl-mercury, which is formed in aquatic sediments through the bacterial methylation of organic mercury, is toxic chemicals compound of mercury, in fact, nearly all of the mercury in fish muscles occurs as Methyl-mercury (Soliman 2006). Transport of metals in fish occurs through the blood where the ions are usually bound to proteins. The metals are brought into contact with the organs and tissues of the fish and consequently accumulated to a different extent in different organs or tissues of the fish. There are five potential routes for a pollutant to enter a fish. These routes are through the food, nonfood particles, gills, oral consumption of water, and the skin. Once the pollutants are absorbed, they are transported by the blood to either a storage point (that is, bone) or to the liver for transformation and storage. If the pollutants are transformed by the liver, they may be stored there or excreted in the bile or passed back into the blood for possible excretion by the gills or kidneys, or stored in fat, which is an extra hepatic tissue (Ayandiran et al. 2009).

Benthic macroinvertebrate assemblages contain species with various sensitivities to contaminants and have been widely used to evaluate the ecological impacts of metal contamination in streams. They play vital roles in lotic food webs by forming a major link between primary producers and higher trophic levels and in lotic ecosystems by regulating organic matter decomposition and nutrient cycling. However, the impact of heavy metals on macroinvertebrates has not been evaluated in terms of their food value for fish, even though invertebrates are an important food source for many moving-water fish species. It is of particular importance to evaluate the effects of heavy metal pollution on drift-prone macroinvertebrates, on which most commercially or recreationally important salmonid species depend (Iwasaki et al. 2002).

5   Effects on Human Health

The plant uptake of heavy metals from soils at high concentrations may result in a great health risk taking into consideration food-chain implications. Utilization of food crops contaminated with heavy metals is a major food chain route for human exposure. The food plants whose examination system is based on exhaustive and continuous cultivation have a great capacity of extracting elements from soils. The cultivation of such plants in contaminated soil represents a potential risk since the vegetal tissues can accumulate heavy metals (Jordao et al. 2006).

Heavy metals become toxic when they are not metabolized by the body and accumulate in the soft tissues (Sobha et al. 2007). Chronic level ingestion of toxic metals has undesirable impacts on humans and the associated harmful impacts become perceptible only after several years of exposure (Khan et al. 2008). Cadmium (Cd) is a well known heavy metal toxicant with a specific gravity 8.65

times greater than water. The target organs for Cd toxicity have been identified as liver, placenta, kidneys, lungs, brain, and bones (Sobha et al. 2007). Depending on the severity of exposure, the symptoms of effects include nausea, vomiting, abdominal cramps, dyspnea, and muscular weakness. Severe exposure may result in pulmonary odema and death. Pulmonary effects (emphysema, bronchiolitis, and alveolitis) and renal effects may occur following subchronic inhalation exposure to cadmium and its compounds (Duruibe et al. 2007). The Itai-itai disease in Japan brought the dangers of environmental Cd to world attention. Cd has been associated to a lesser or greater extent with many clinical conditions including anosmia, cardiac failure cancers, cerebrovascular infarction, emphysema, osteoporosis, proteinuria cataract formation in the eyes. Yet, it has been difficult to tie down obvious links of environmental exposures with morbidity and mortality (Lalor 2008).

Zinc is considered to be relatively nontoxic, especially if taken orally. However, the excess amount can cause system dysfunctions that result in impairment of growth and reproduction. The clinical signs of zinc toxicosis have been reported as vomiting, diarrhea, bloody urine, icterus (yellow mucus membrane), liver failure, kidney failure, and anemia (Duruibe et al. 2007). Copper (Cu) is an essential element in mammalian nutrition as a component of metalloenzymes in which it acts as an electron donor or acceptor. Conversely, exposure to high levels of Cu can result in a number of adverse health effects. Exposure of humans to Cu occurs primarily from the consumption of food and drinking water. Acute Cu toxicity is generally associated with accidental ingestion; however, some members of the population may be more susceptible to the adverse effects of high Cu intake due to genetic predisposition or disease (Stern et al. 2007). Excessive human intake of Cu may lead to severe mucosal irritation and corrosion, widespread capillary damage, hepatic and renal damage, and central nervous system irritation followed by depression. Severe gastrointestinal irritation and possible necrotic changes in the liver and kidney can also occur. The effects of Ni exposure vary from skin irritation to damage to the lungs, nervous system, and mucous membranes (Argun et al. 2007).

Lead (Pb) is physiological and neurological toxic to humans. Acute Pb poisoning may results in a dysfunction in the kidney, reproduction system, liver, and brain resulting in sickness and death (Odum 2000). Pb heads the threats even at extremely low concentrations (Kazemipour et al. 2008). A notably serious effect of lead toxicity is its teratogenic effect. Lead poisoning also causes inhibition of the synthesis of hemoglobin; cardiovascular system and acute and chronic damage to the central nervous system (CNS) and peripheral nervous system (PNS). Other chronic effects include anemia, fatigue, gastrointestinal problems, and anoxia. Lead can cause difficulties in pregnancy, high blood pressure, muscle and joint pain (Odum 2000).

Other effects include damage to the gastrointestinal tract (GIT) and urinary tract resulting in bloody urine, neurological disorder, and can cause severe and permanent brain damage. While inorganic forms of lead, typically affect the CNS, PNS, GIT, and other biosystems, organic forms predominantly affect the CNS. Lead affects children; particularly in the 2–3 years old range by leading to the poor development of the grey matter of the brain, thereby resulting in poor intelligence quotient (IQ). Its absorption in the body is enhanced by Ca and Zn deficiencies (Duruibe et al. 2007).

Chromium (Cr) is the 10th abundant element in the earth's mantle and persists in the environment as either Cr(III) or Cr(VI). Cr(VI) is toxic to plants and animals, being a strong oxidizing agent, corrosive, soluble in alkaline and mildly acidic water, toxic and potential carcinogens (Shaffer et al. 2001; Jeyasingh and Philip 2005; Huang et al. 2009). The toxicity of Cr(VI) derives from its ability to diffuse through cell membranes and oxidize biological molecules (Shaffer et al. 2001). Mercury is toxic and has no known function in human biochemistry and physiology. Inorganic forms of mercury cause spontaneous abortion, congenital malformation, and gastrointestinal disorders (like corrosive esophagitis and hematochezia). Poisoning by its organic forms, which include monomethyl and dimethylmercury presents with erethism (an abnormal irritation or sensitivity of an organ or body part to stimulation), acrodynia (Pink disease, which is characterized by rash and desquamation of the hands and feet), gingivitis, stomatitis, neurological disorders, total damage to the brain and CNS, and are also associated with congenital malformation (Duruibe et al. 2007). As with lead and mercury, arsenic toxicity symptoms depend on the chemical form ingested. Arsenic acts to coagulate protein, forms complexes with coenzymes, and inhibits the production of adenosine triphosphate (ATP) during respiration. It is possibly carcinogenic in compounds of all its oxidation states and high-level exposure can cause death. Arsenic toxicity also presents a disorder, which is similar to, and often confused with Guillain-Barre syndrome, an anti-immune disorder that occurs when the body's immune system mistakenly attacks part of the PNS, resulting in nerve inflammation that causes muscle weakness (Duruibe et al. 2007).

6   Effects on Composting Process

Heavy metal effects are not limiting up to soil, plants, and human health but also affect the composting process by changing microbial diversity. Microorganisms are helpful in degradation of organic matter, detoxify some organic and inorganic pollutants, change mobility and bioavailability of heavy metals to plants. Since heavy metals can affect the microbial reproduction and cause morphological and physiological changes. So the biodegradation processes might be influenced by toxic heavy metals in the environment. Microbial enzymes might be affected by heavy metals due to the potential inhibition of both enzymatic reactions and complex metabolic processes (Huang et al. 2010). Heavy metals decrease the phosphatase synthesis during the composting process (Garcfa et al. 1995). Microorganisms have to cope with toxic Pb during their growth in the Pb-contaminated substrates and the exposure of microorganism to metals always inhibits microbial growth and activity (Huang et al. 2010).

Malley et al. (2006) studied the effect of heavy metals Cu and Zn on dehydrogenase and protease activity of the substrate during vermicomposting by adding three different dosages of Cu and Zn. In this study, it was reported that the highest levels of dehydrogenase activity were found in the control substrate compared to substrate with added Cu or Zn dosages. Activity levels decreased with the addition of Cu dosage (547, 1094, and 1642 mg/kg) and Zn dosage (1068, 1780, and 2492 mg/kg). The dehydrogenase level was highest in the control with no heavy metals added and decreased with increasing heavy metal dosage down to dosage 3. Heavy metals may inactivate enzyme reactions by complexing the substrate, by reacting with protein-active groups of enzymes, or by reacting with the enzyme–substrate complex or indirectly by altering the microbial community which synthesizes enzymes.

Heavy metals, in general, are potent inhibitors of enzymatic reactions. Cu and Cd in addition to binding with aromatic amino acid residues in enzyme molecules can also cause oxidative damage of proteins by the induction of oxidative stress associated with the production of reactive oxygen species like hydroxyl or superoxide radicals (Baldrian 2003). The significant impacts of high Cu and Zn on a vermicomposting system including accumulation in worm tissue and reductions in the number of juveniles produced. While *Eisenia fetida* earthworm has the potential to be used in the vermicomposting of organic waste, the level of Cu and Zn needs to be monitored to ensure the earthworm population in the system (Malley et al. 2006).

7 Conclusion

Heavy metals containing compost may change the physical, chemical, and biological properties of soil. These metals uptake by plants from the soil, it reduces the crop productivity by inhibiting physiological metabolism. Heavy metals uptake by plants and successive accumulation in human tissues and biomagnifications through the food chain causes both human health and environment concerns. Heavy metals containing agricultural runoff enter in the aquatic environment, and harm to aquatic plants and animals. Therefore, if the compost has to be applied in agriculture it should be free from pathogens and heavy metals.

References

Argun ME, Dursun S, Ozdemir C, Karatas M (2007) Heavy metal adsorption by modified oak sawdust: Thermodynamics and kinetics. J Hazard Mater 141:77–85

Ashraf R, Ali TA (2007) Effect of heavy metals on soil microbial community and mung beans seed germination. Pakistan J Bot 39(2):629–636

Ayandiran TA, Fawole OO, Adewoye SO, Ogundiran MA (2009) Bioconcentration of metals in the body muscle and gut of Clarias gariepinus exposed to sublethal concentrations of soap and detergent effluent. J Cell Animal Biol 3(8):113–118

Baldrian P (2003) Interactions of heavy metals with white-rot fungi. Enzyme Microb Technol 32:78–91

Bhattacharyya P, Chakrabarti K, Chakraborty A, Tripathy S, Powell MA (2008) Fractionation and bioavailability of Pb in municipal solid waste compost and Pb uptake by rice straw and grain under submerged condition in amended soil. Geosci J 12(1):41–45

Cai QY, Mo CH, Wu QT, Zeng QY, Katsoyiannis A (2007) Concentration and speciation of heavy metals in six different sewage sludge-composts. J Hazard Mater 147:1063–1072

Chen GQ, Chen Y, Zeng GM, Zhang JC, Chen YN, Wang L, Zhang WJ (2010) Speciation of cadmium and changes in bacterial communities in red soil following application of cadmium-polluted compost. Environ Eng Sci 27(12):1019–1026

Duruibe JO, Ogwuegbu MOC, Egwurugwu JN (2007) Heavy metal pollution and human biotoxic effects. Int J Phys Sci 2(5):112–118

Garcfa C, Moreno JL, Hernfindez T, Costa F (1995) Effect of composting on sewage sludges contaminated with heavy metals. Biores Technol 53:13–19

Garnier J, Quantin C, Martins ES, Becquer T (2006) Solid speciation and availability of chromium in ultramafic soils from Niquelandia, Brazil. J Geochem Explor 88:206–209

Garrido S, Campo GMD, Esteller MV, Vaca R, Lugo J (2002) Heavy metals in soil treated with sewage sludge composting, their effect on yield and uptake of broad bean seeds (Vicia faba L.). Water Air Soil Pollut 166:303–319

Guala SD, Vega FA, Covelo EF (2010) The dynamics of heavy metals in plant–soil interactions. Ecol Model 221:1148–1152

Gurrieri JT (1998) Distribution of metals in water and sediment and effects on aquatic biota in the upper Stillwater River basin, Montana. J Geochem Explor 64:83–100

Hinojosa MB, Carreira JA, Ruiz RG, Dick RP (2004) Soil moisture pre-treatment effects on enzyme activities as indicators of heavy metal-contaminated and reclaimed soils. Soil Biol Biochem 36:1559–1568

Huang S, Peng B, Yang Z, Chai L, Zhou L (2009) Chromium accumulation, microorganism population and enzyme activities in soils around chromium-containing slag heap of steel alloy factory. Trans Nonferrous Met Soc China 19:241–248

Huang DL, Zeng GM, Feng CL, Hu S, Zhao MH, Lai C, Zhang Y, Jiang XY, Liu HL (2010) Mycelial growth and solid-state fermentation of lignocellulosic waste by white-rot fungus Phanerochaete chrysosporium under lead stress. Chemosphere 81:1091–1097

Iwasaki Y, Kagaya T, Miyamoto K, Matsuda H (2002) Environmental effects of heavy metals on riverine benthic macroinvertebrate assemblages with reference to potential food availability for drift-feeding fishes. Toxicol Chem 28(2):354–363

Jeyasingh J, Philip L (2005) Bioremediation of chromium contaminated soil: optimization of operating parameters under laboratory conditions. J Hazardous Mat B 118:113–120

Jongea MD, Vijverb BV, Blusta R, Bervoetsa L (2009) Responses of aquatic organisms to metal pollution in a lowland river in Flanders: a comparison of diatoms and macroinvertebrates. Sci Total Environ 407:615–629

Jordao CP, Nascentes CC, Cecon PR, Fontes RLF, Pereira JL (2006) Heavy metal availability in soil amended with composted urban solid wastes. Environ Monit Assess 112:309–326

Karaca A, Cetin SC, Turgay OC, Kizilkaya R (2010) Effects of heavy metals on soil enzyme activities. In: Sherameti I, Varma A (eds) Soil heavy metals. Soil biology, vol 19. Springer, Heidelberg, pp 237–265

Kazemipour M, Ansar iM, Tajrobehkar S, Majdzadeh M, Kermani HR (2008) Removal of lead, cadmium, zinc, and copper from industrial wastewater by carbon developed from walnut, hazelnut, almond, pistachio shell, and apricot stone. J Hazard Mater 150:322–327

Khan S, Cao Q, Zheng YM, Huang YZ, Zhu YG (2008) Health risks of heavy metals in contaminated soils and food crops irrigated with wastewater in Beijing, China. Environ Pollut 152:686–692

Lalor GC (2008) Review of cadmium transfers from soil to humans and its health effects in the Jamaican environment. Sci Total Environ 400:162–172

Malley C, Nair J, Ho G (2006) Impact of heavy metals on enzymatic activity of substrate and on composting worms Eisenia fetida. Biores Technol 97:1498–1502

Mora AP, Calvo JJO, Cabrera F, Madejon E (2005) Changes in enzyme activities and microbial biomass after "in situ" remediation of a heavy metal-contaminated soil. Appl Soil Ecol 28:125–137

Morin S, Vivas NM, Duong TT, Boudou A, Coste M, Delmas F (2007) Dynamics of benthic diatom colonization in a cadmium/zinc-polluted river (Riou-Mort, France). Fundam Appl Limnol 168(2):179–187

Neklyudov AD, Fedotov GN, Ivankin AN (2008) Intensification of composting processes by aerobic microorganisms: a review. Appl Biochem Microbiol 44(1):6–18

Odum HT (2000) Back ground of published studies on lead and wetland. In: Odum HT (ed) Heavy metals in the environment using wetlands for their removal. Lewis Publishers, New York, p 32

Peng K, Luo C, Luo L, Li X, Shena Z (2008) Bioaccumulation of heavy metals by the aquatic plants Potamogeton pectinatus L. and Potamogeton malaianus Miq. and their potential use for contamination indicators and in wastewater treatment. Sci Total Environ 392:22–29

Rascio N, Izzo FN (2011) Heavy metal hyperaccumulating plants: how and why do they do it? And what makes them so interesting? Plant Sci 180:169–181

Shaffer RE, Cross JO, Pehrsson SLR, Elam WT (2001) Speciation of chromium in simulated soil samples using X-ray absorption spectroscopy and multivariate calibration. Anal Chim Acta 442:295–304

Sharholy M, Ahmad K, Mahmood G, Trivedi RC (2008) Municipal solid waste management in Indian cities—a review. Waste Manag 28:459–467

Sharma RK, Agrawal M, Marshall F (2007) Heavy metal contamination of soil and vegetables in suburban areas of Varanasi, India. Ecotoxicol Environ Saf 66:258–266

Shun HH, Bing P, Zhi Y, Li C, Li Z (2009) Chromium accumulation, microorganism population and enzyme activities in soils around chromium-containing slag heap of steel alloy factory. Trans Nonferrous Met Soc China 19:241–248

Sobha K, Poornima A, Harini P, Veeraiah K (2007) A study on biochemical changes in the fresh water fish, Catla catla (hamilton) exposed to the heavy metal toxicant cadmium chloride. Kathmandu Univ J Sci Eng Technol 1(4):1–11

Soliman ZI (2006) A study of heavy metals pollution in some aquatic organisms in Suez Canal in Port-Said Harbour. J Appl Sci Ress 2(10):657–663

Speira TW, Kettlesb HA, Percivalc HJ, Parshotam A (1999) Is soil acidification the cause of biochemical responses when soils are amended with heavy metal salts? Soil Biol Biochem 31:1953–1961

Sprynskyy M, Kosobucki P, Kowalkowski T, Buszewsk B (2007) Influence of clinoptilolite rock on chemical speciation of selected heavy metals in sewage sludge. J Hazard Mater 149:310–316

Stern BR, Solioz M, Krewski D, Aggett P, Aw TC, Baker S, Crump K, Dourson M, Haber L, Hertzberg R, Keen C, Meek B, Rudenko L, Schoeny R, Slob W, Starr T (2007) Copper and human health: biochemistry, genetics, and strategies for modeling dose-response relationships. J Toxicol Environ Health Part B 10:157–222

Talbot VL (2006) The chemical forms and plant availability of copper in composting organic wastes. Ph.D. dissertation, University of Wolverhampton

Wang YP, Shi JY, Wang H, Li Q, Chen XC, Chen YX (2007) The influence of soil heavy metals pollution on soil microbial biomass, enzyme activity, and community composition near a copper smelters. Ecotoxicol Environ Saf 67:75–81

Wong JWC, Selvam A (2006) Speciation of heavy metals during co-composting of sewage sludge with lime. Chemosphere 63:980–986

Woo S, Yum S, Park HS, Lee TK, Ryu JC (2009) Effects of heavy metals on antioxidants and stress-responsive gene expression in Javanese medaka (*Oryzias javanicus*). Comp Biochem Physiol Part C 149:289–299

Yao H, Xu J, Huang C (2003) Substrate utilization pattern, biomass and activity of microbial communities in a sequence of heavy metal-polluted paddy soils. Geoderma 115:139–148

Zorpas AA, Vassilis I, Loizidou M, Grigoropoulou H (2002) Particle size effects on uptake of heavy metals from sewage sludge compost using natural zeolite clinoptilolite. J Colloid Interface Sci 250:1–4

Fabrication of Mesoporous Silica MCM-41 Via Sol-Gel and Hydrothermal Methods for Amine Grafting and CO_2 Capture Application

Sravanthi Loganathan, Kishant Kumar and Aloke Kumar Ghoshal

Abstract In the current scenario, it is well known that the world's energy consumption increases at a rapid pace. This is because of increment in population growth as well as advanced development in economic conditions of growing nations. All over the world, coal remains as the major source of energy for electricity generation. Combustion of coal leads to emission of carbon dioxide (CO_2) in the atmosphere which is considered to be the major greenhouse gas responsible for global warming issue. Even though government policies enforce laws to utilize sustainable resources and other alternatives, complete eradication of coal resources for electricity generation cannot be possible in practice. This is due to the fact that global energy demand cannot be met in the future years by utilization of other alternatives alone rather than coal. Therefore, it becomes clear that combustion of coal cannot be avoided in order to meet the global energy demand. This in turn indicates the fact that there will be continuous rise in CO_2 emissions in the atmosphere due to combustion of coal. It is considered that concentration of CO_2 released from coal-based power plants is 500 times greater as compared to atmosphere. Therefore, carbon capture from power plants will be an appropriate option to address the global warming issue at this point of time. Thus, there prevails urging need for promoted R&D activities in order to develop a suitable technology for carbon capture application. Adsorption by solid amine-grafted mesoporous silica adsorbents is regarded as an effective technique for CO_2 capture. Much emphasis on the development of suitable support for amine grafting has not been provided in the literature. Hence, the current work investigates on the suitable synthesis method to develop mesoporous silica MCM-41 with high surface area and enhanced pore volume. In the present work, two different synthesis methods (sol-gel and hydrothermal) are used for the preparation of MCM-41. The MCM-41 sample

S. Loganathan (✉) · K. Kumar · A. K. Ghoshal
Department of Chemical Engineering, Indian Institute of Technology Guwahati,
Guwahati, India
e-mail: l.sravanthi@iitg.ernet.in

© Springer International Publishing AG 2018
A. K. Sarma et al. (eds.), *Urban Ecology, Water Quality and Climate Change*,
Water Science and Technology Library 84,
https://doi.org/10.1007/978-3-319-74494-0_26

synthesized via hydrothermal method demonstrated increased surface area and pore volume as compared to sol-gel technique. Thereafter, structural, thermal, and morphological characteristics of MCM-41 synthesized by hydrothermal method are discussed.

Keywords Coal · Global warming · CO_2 · Adsorption · Mesoporous silica MCM-41 · Amine grafting · Sol-gel and hydrothermal

1 Introduction

It is well known that fossil fuel-based power plants remain as major contributors for anthropogenic CO_2 emissions. This in turn leads to global warming and subsequent climate change issues. Thus, reduction of CO_2 emissions in the atmosphere has now become a great challenge to the modern society and needs to be addressed. Amidst of all the separation processes available for CO_2 removal, adsorption is regarded as the most energy-efficient and cost-effective method (Siriwardane et al. 2001; Karra and Walton 2010; Calleja et al. 1998; Sayari and Hamoudi 2001). For efficient removal of CO_2 from flue gas, an ideal adsorbent should encompass certain important features which include (i) high selectivity for CO_2, (ii) high adsorption capacity in the temperature range of 45–75 °C at 0.15 bar, (iii) Faster rate of adsorption, (iv) Ease of regeneration under mild conditions, and (v) high stability along with tolerance toward moisture as well as other impurities in the feed (Harlick and Sayari 2006).

A wide variety of adsorbents like zeolite 13X, activated carbons, metal organic frameworks, and mesoporous silica are investigated so far for CO_2 capture application (Siriwardane et al. 2001; Karra and Walton 2010; Calleja et al. 1998; Sayari and Hamoudi 2001). Recently, amine-functionalized adsorbents have gained increasing attention for CO_2 removal due to their attractive properties such as high selectivity, high CO_2 adsorption capacity, good mechanical strength, stable adsorption capacity after cycles, and adequate adsorption/desorption kinetics. Among the most investigated materials for the amine grafting, MCM-41 exhibits high surface area, high pore volume, availability of high surface functional sites for grafting of amine moieties, and high tolerance to moisture (Harlick and Sayari 2006; Sravanthi et al. 2013). Previously, various amine-impregnated MCM-41 adsorbents are also been reported for CO_2 removal. It is observed that amine-impregnated adsorbents demonstrated high amine loading capacity along with increased CO_2 adsorption capability. However, the enhanced adsorption capacity is accompanied by decreased rate of adsorption. Further, the weaker interaction that exists between amine moieties and porous support led to instability issue for amine-impregnated porous sorbents. Therefore, it is mandatory to perform regeneration of such adsorbents under strict control over the temperature conditions. This is because higher temperature conditions lead to evaporation or degradation of amines and lower temperature conditions result in incomplete regeneration of adsorbed species (Harlick and Sayari 2006; Sravanthi et al. 2013). With the aim to avoid the demerits in connection with amine-impregnated porous silica adsorbents, researchers have put forth considerable efforts to introduce grafting process for amine functionalization such that covalent bonding will be established among surface silanol groups of porous silica support and amine moieties.

Since very few reports are available for investigation on the synthesis procedure for development of suitable porous silica supports for amine grafting, the information available so far is not adequate enough to make use of adsorption technology in industrial scale. Further investigation needs to be done in order to enable the use of adsorption in industrial scale and also to develop potential porous supports suitable for amine grafting for efficient capturing of CO_2 from flue gas. Therefore, the present work investigates the surface area and pore size distribution characteristics of synthesized via variation in synthesis procedure. Thereafter, thermal, structural, and morphological characteristics of MCM-41 sample synthesized via hydrothermal method are discussed.

2 Materials and Methods

2.1 Chemicals

Tetraethyl orthosilicate (TEOS) was purchased from MERCK and used as silica source. Cetyl trimethyl ammonium bromide (CTMABr) was purchased from Spectrochem Pvt. Ltd. and used as a template or structure-directing agent. Aqueous ammonia purchased from MERCK was used as a mineralizing agent.

2.2 Experimental

2.2.1 Synthesis of MCM-41

MCM-41 samples are prepared by two methods viz. sol-gel and hydrothermal. In sol-gel method, CTMABr was dissolved in water and stirred for 30 min to obtain homogeneous solution. Followed by this, pH adjustment was made using a pH meter. The pH of the homogeneous solution was adjusted to be in the range of 11–12 with the addition of ammonia. Thereafter, TEOS was added to the solution and again stirred for 8 h. Then, the solution was washed with Millipore water and filtered. White precipitate obtained after filtration was dried in an oven at a temperature of 120 °C overnight. The dried white powder was subjected to calcination at 550 °C for 5 h. The calcination temperature of 550 °C was attained at a heating rate of 1 °C/min. In case of hydrothermal method, molar composition for the gel

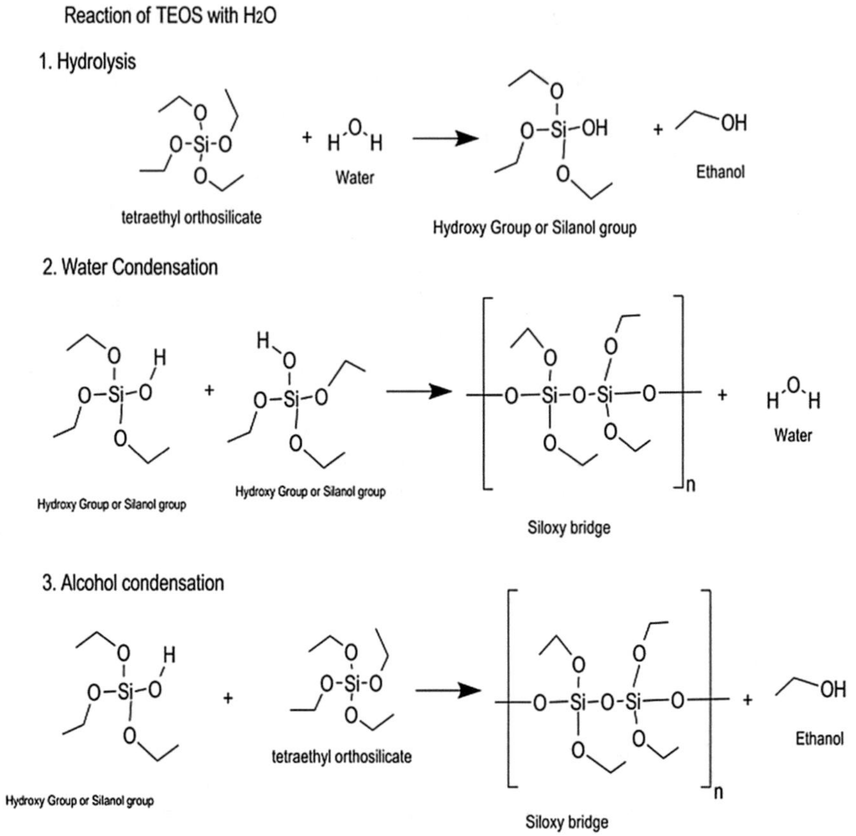

Reaction of TEOS with H$_2$O

1. Hydrolysis

2. Water Condensation

3. Alcohol condensation

Fig. 1 Hydrolysis and condensation reactions during synthesis of MCM-41

remained constant like sol-gel method. The white homogeneous suspension obtained after 8 h of stirring was subjected to hydrothermal treatment in an autoclave reactor at a temperature condition of 110 °C for 24 h. The hydrolysis and condensation reactions that occur during the synthesis of MCM-41 are shown in Fig. 1.

2.2.2 Characterization

Beckmann Coulter SA3100 surface area analyzer was used for measurement of surface area and pore size distribution characteristics. SHIMADZU Fourier transform infrared (FTIR) spectroscopy was used for identifying the functional groups present in MCM-41 samples. Thermal stability of the samples was determined using a METTLER TOLEDO thermogravimetric analyzer (TGA). Morphological as well as elemental analyses for the samples were performed using a JEOL field emission scanning electron microscopy (FE-SEM).

3   Results and Discussion

3.1   Surface Area Measurements

The specific surface area for the MCM-41 samples is calculated using Brunauer–Emmett–Teller (BET) method and pore size distribution is determined using Berrett–Joyner–Halenda (BJH) model. All the samples are subjected to degassing at 200 °C for 6 h prior to nitrogen adsorption–desorption measurements. The BET surface area for the MCM-41 sample synthesized by sol-gel method is found to be 1343 m^2/g and the pore volume is 0.8328 cc/g. The BET surface area for the MCM-41 sample synthesized by hydrothermal method is obtained to be 1,451 m^2/g and the pore volume is 0.9378 cc/g. Figure 2 shows the comparative analysis of pore size distribution for MCM-41 samples prepared by sol-gel and by hydrothermal methods. The hydrothermal method gave higher surface area and pore volume as compared to sol-gel method but the more narrow range pores are obtained by the latter method. Since there is no considerable difference in surface area as well as pore size distribution for both the samples, only MCM-41 sample prepared via hydrothermal method is subjected to rest of the characterization.

3.2   FTIR Analysis

FTIR spectrum for as-synthesized MCM-41 sample prepared by hydrothermal method is shown in Fig. 3a. It can be seen from the Figure that as-synthesized MCM-41 sample exhibits two intense peaks at 2,924 and 2,852 cm^{-1} which arise because of stretching of C–H groups present in hydrocarbon chain of surfactant. The respective bending vibration for C–H group can be seen at 1,477 cm^{-1}. The peak observed at 3,144 cm^{-1} corresponds to the presence of surface silanol

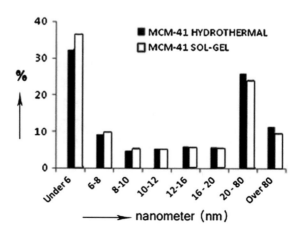

Fig. 2 Pore size distribution for MCM-41 samples synthesized by sol-gel and hydrothermal methods

moieties and physisorbed water molecules. The stretching vibrations associated with siloxane (Si–O–Si) bridges are characterized by the presence of peaks at 1,066 and 1,228 cm^{-1}. The bending vibration of siloxane bridges is confirmed by the presence of bands at 461 and 795 cm^{-1}. The FTIR spectrum for calcined MCM-41 prepared by hydrothermal method can be seen from Fig. 3b. It is evident from Fig. 3b that all the peaks corresponding to organic functionalities disappear in case of calcined MCM-41 sample. This is because surfactant used as structure-directing agent gets removed completely in calcination process and leads to the formation of porous structure in MCM-41 (Monash and Pugazhenthi 2010).

Fig. 3 FTIR spectrum for MCM-41 sample prepared by hydrothermal method **a** as-synthesized and **b** calcined at 550 °C

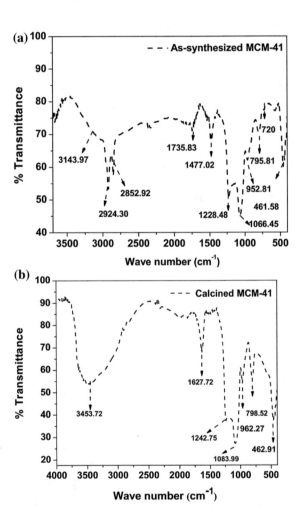

Fig. 4 Thermogravimetric analysis for as-synthesized and calcined MCM-41 prepared by hydrothermal method

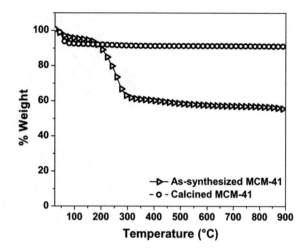

3.3 TGA Analysis

Thermogravimetric curve obtained for as-synthesized and calcined MCM-41 sample prepared by hydrothermal method is presented in Fig. 4. It can be seen that the TGA curve for as-synthesized MCM-41 exhibits three clear regions of weight loss. The weight loss that occurs below 100 °C corresponds to removal of physisorbed water molecules. Followed by this template, CTMABr gets decomposed till the temperature condition of 550 °C. In case of calcined MCM-41 sample, only two distinct regions of weight loss can be observed. The first region of weight loss is due to the water loss as mentioned earlier for as-synthesized MCM-41 sample. Thereafter, the weight loss is not so significant and calcined MCM-41 sample is found to be thermally stable throughout the temperature regime analyzed. The insignificant weight loss noticed after 100 °C in turn indicates that the template gets removed completely after subjecting the as-synthesized MCM-41 sample to calcination process (Monash and Pugazhenthi 2010; Kim et al. 2009).

3.4 Morphology Analysis

FE-SEM image obtained for calcined MCM-41 prepared by hydrothermal method is depicted in Fig. 5. It can be observed from Fig. 5 that MCM-41 demonstrates spherical shape.

Fig. 5 FE-SEM image for
calcined MCM-41 prepared
by hydrothermal method
obtained at a magnification of
36.29 KX

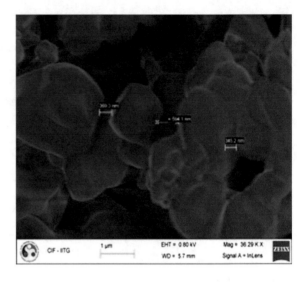

4 Conclusions

MCM-41 samples have been prepared by both sol-gel and hydrothermal methods.
BET surface area measurements show increased surface area and pore volume for
MCM-41 sample prepared by hydrothermal method. The MCM-41 sample prepared by hydrothermal method exhibits a surface area of 1,451 m^2/g and pore
volume of 0.94 cc/g. In case of MCM-41 sample prepared by sol-gel method,
surface area and pore volume are obtained to be $\sim 1,343$ m^2/g and 0.8328 cc/g,
respectively. This indicates that hydrothermal method adopted is effective for the
synthesis of MCM-41 with increased surface area and pore volume characteristics
which in turn are considered as significant parameters required for achieving
increased amine loading. FTIR analysis elucidates the presence of surface silanol
and siloxane bridges in the structure of MCM-41. TGA analysis indicates that the
calcined MCM-41 sample offers excellent thermal stability up to 900 °C. In
addition to this, FTIR and TGA analyses reveal the fact that complete removal of
surfactant CTMABr takes place after calcination. Future studies will be aimed at
grafting suitable amines onto MCM-41 synthesized by both the techniques in order
to compare the amine loading and CO_2 adsorption characteristics.

References

Calleja G, Pau J, Calles JA (1998) Pure and multicomponent adsorption equilibrium of carbon
    dioxide, ethylene, and propane on ZSM-5 zeolites with different Si/Al ratios. J Chem Eng Data
    43:994–1003
Harlick PJE, Sayari A (2006) Applications of pore-expanded mesoporous silicas. 3. triamine silane
    grafting for enhanced CO_2 adsorption. Ind Eng Chem Res 45:3248–3255

Karra JR, Walton KS (2010) Molecular simulations and experimental studies of CO_2, CO, and N_2 adsorption in metal-organic frameworks. J Phys Chem C 114:15735–15740

Kim JM, Chang SM, Kong SM, Kim KS, Kim J, Kim W (2009) Control of hydroxyl group content in silica particle synthesized by the sol-precipitation process. Ceram Int 35:1015–1019

Monash P, Pugazhenthi G (2010) Investigation of equilibrium and kinetic parameters of methylene blue adsorption onto MCM-41. Korean J Chem Eng 27:1184–1191

Sayari A, Hamoudi S (2001) Periodic mesoporous silica-based organic–inorganic nanocomposite materials. Chem Mater 13:3151–3168

Siriwardane RV, Shen MS, Fisher EP, Poston JA (2001) Adsorption of CO_2 on molecular sieves and activated carbon. Energy Fuels 15:279–284

Sravanthi L, Mayur T, Ghoshal AK (2013) Novel pore-expanded MCM-41 for CO_2 capture: synthesis and characterization. Langmuir 29:3491–3499

Water Hyacinth as a Potential Source of Biofuel for Sustainable Development

Deepmoni Deka, Saprativ P. Das, Rajeev Ravindran,
Mohammad Jawed and Arun Goyal

Abstract Water hyacinth (*Eichhornia crassipes*), a noxious weed and fast growing perennial aquatic plant found in lakes and ponds all over Guwahati is affecting the ecosystem in a deleterious manner. It tampers with aquatic life by deoxygenating the water and depleting nutrients for young fish in sheltered bays. It also blocks supply intakes for the hydroelectric plant, interrupting electrical power. Contrastingly, owing to high cellulose and hemicellulose content in its biomass and having a wide distribution in Assam, it was selected for our study. Therefore, using water hyacinth for bioethanol production can tackle the pollution problem owing to fossil fuel emissions and control its growing extent in water bodies. In the present research, water hyacinth was used for bioethanol production involving simultaneous hydrolysis and fermentation at shake flask and reactor level using different saccharifying enzymes and potential fermentative microbes. *Candida shehatae*, utilizing pentoses was used along with *Saccharomyces cerevisiae* in simultaneous saccharification and fermentation (SSF) experiments. The substrate was subjected to three pretreatments viz. wet oxidation, phosphoric acid–acetone treatment and ammonia fibre expansion (AFEX). Recombinant *E. coli* BL21 (DE3) and BL21 (plysS) cells harbouring expressing Glycoside hydrolase family 5 (GH5) and family 43 (GH43) genes from *Clostridium thermocellum* were employed for cellulase and hemicellulase production respectively. *Trichoderma reesei* cellulase and *Bacillus subtilis* AS3 producing thermostable cellulases were also engaged in saccharification process. The wet oxidation-treated water hyacinth conferred an ethanol titre of 0.90 g/L with *B. subtilis* cellulase and 1.26 g/L with *T. reesei* cellulase amid a fermentative microbial combination of *S. cerevisiae* and *C. shehatae*. A higher ethanol titre (1.69 g/L) was achieved with an enzymatic consortium of GH5

D. Deka · M. Jawed · A. Goyal
Centre for the Environment, Indian Institute of Technology Guwahati,
Guwahati 781039, Assam, India
e-mail: deepmoni@iitg.ernet.in

S. P. Das · R. Ravindran · A. Goyal (✉)
Department of Biotechnology, Indian Institute of Technology Guwahati,
Guwahati 781039, Assam, India
e-mail: arungoyl@iitg.ernet.in

© Springer International Publishing AG 2018
A. K. Sarma et al. (eds.), *Urban Ecology, Water Quality and Climate Change*,
Water Science and Technology Library 84,
https://doi.org/10.1007/978-3-319-74494-0_27

351

cellulase and GH43 hemicellulase along with *S. cerevisiae* and *C. shehatae.*
Contrastingly, AFEX pretreatment yielded a maximum ethanol titre of 1.98 g/L
with same enzymatic consortium and bioethanol producers. Consequently, on
scaling up the SSF experiments with 5% (w/v) AFEX-pretreated substrate in shake
flask and bioreactor along with *S. cerevisiae* and *C. shehatae,* an ethanol titre of
9.78 and 17.97 g/L were obtained respectively.

Keywords Simultaneous saccharification and fermentation · *Bacillus subtilis*
Trichoderma reesei · *Saccharomyces cerevisiae Candida shehatae*
Wet oxidation · Phosphoric acid · Acetone · AFEX

1 Introduction

Lignocellulosic ethanol has gained tremendous significance in recent years due to
its projection as a feasible alternative to petroleum-based fuels. The choice of
feedstock for bioethanol production is a major concern since the biomass either
directly or indirectly competes with the food crops. Use of locally available
agroresidues as feedstock will not only be cost effective but also will bring economic as well as social benefits for the rural community. Water hyacinth
(*Eichhornia crassipes*), a noxious weed and fast growing perennial aquatic plant
found in lakes and ponds all over Guwahati can be a suitable candidate for bioethanol production. Under the climatic conditions of Assam, (Singh et al. 1984)
reported a daily average water hyacinth biomass productivity of 0.26 ton of dry
biomass per hectare in all seasons. This plant is a typical menace infesting large
areas of water bodies causing ecological and socio-economic problems which
include diminution of biodiversity, blockage of rivers and drainage system,
depletion of dissolved oxygen with contribution in environmental pollution. Owing
to its high hemicellulose and less lignin content and advantage of a non-competitor
to food crops, water hyacinth can be a potential source for bioethanol production.
The major rate-limiting step in economically efficient conversion of lignocellulosic
substrates to liquid biofuels is efficacy of hydrolytic enzymes and degree of crystallinity of the lignocellulosic biomass. It is therefore important to employ more
effective hydrolytic enzymes with maximum release of utilizable sugars for bioethanol production. Pretreatments like wet oxidation, phosphoric acid (H_3PO_4)-
acetone treatment and ammonia fibre expansion (AFEX) down the lignocellulosic
complex composed of cellulose and lignin moieties bound by hemicellulose chains
and make the substrate accessible for enzymatic hydrolysis. The basic purpose of
pretreatment is to remove lignin and hemicellulose, reduce cellulose crystallinity
and increase the porosity and enzymatic hydrolysis (Mosier et al. 2005).

The thermophilic bacterium *Clostridium thermocellum* comprises of genes
coding for exocellular multienzyme complexes called cellulosomes exhibiting
endoglucanase and exoglucanase activity. Glycoside hydrolases (GH) are a set of
enzymes with varying substrate specificity including cellulase and hemicellulase

activity (Adam 2011). Industrially, these enzyme systems cloned and expressed in a suitable vector can be utilized for deriving monomeric sugars out of lignocellulosic biomass. *Bacillus* species lack the cellulase system to degrade crystalline cellulose but can act upon carboxymethyl cellulose (Crispen et al. 2000). *Bacillus subtilis* (AS3) produces a cellulase system which is fairly thermostable and can remain active in varying pH conditions. Deka et al. (2011) reported the optimum temperature for *Bacillus subtilis* activity to be fairly close to that of fungal cellulases. Consumable sugars released after hydrolysis of agricultural residues can be taken up by ethanol-producing yeasts like organisms *Saccharomyces cerevisiae* and *Candida shehatae*. Simultaneous saccharification and fermentation (SSF) appear as a promising alternative. In this process, the enzymatic hydrolysis of cellulose and the fermentation of monomeric sugars are performed in one single step, which means a reduction in the end product inhibition of the enzymatic complex (Saha et al. 2011). SSF process first described by (Takagi et al. 1977), scores over separate hydrolysis and fermentation in offering higher saccharification rate and ethanol yield by eliminating end product inhibition and decreased risk of contamination. A major challenge in improving the SSF process is matching the temperature conditions required for optimum performance of the enzyme and the fermenting microorganism. *Saccharomyces cerevisiae* provides a better choice for the production of ethanol in large-scale cultivation, due to their inherent ability to utilize various substrates, high ethanol tolerance and its robustness to withstand a range of metabolic inhibitions (Casey and Ingledew 1986). Although *S. cerevisiae* is reported to be highly efficient bioethanol producer but are unable to utilize pentose sugars forming major components of the biomass. Yeasts like *C. shehatae* having key enzymes xylose reductase and xylitol dehydrogenase can assimilate pentose sugars to ethanol by the pentose phosphate pathway (Kadam and Schmidt 1997).

The present study investigates the effective and comparative performance of fungal, naturally isolated bacterial and recombinant hydrolytic enzymes along with fermentative microbes like *S. cerevisiae* and *C. shehatae* on bioethanol production from variously pretreated water hyacinth in an efficient SSF process at shake flask and bioreactor level.

2   Materials and Methods

2.1   Chemicals, Reagents and Substrates

Ampicillin and Kanamycin, Potassium dichromate ($K_2Cr_2O_7$), LB and GYE medium and other reagents of analytical grade were procured from Merck and Himedia laboratories (India). Carboxymethylcellulose (low viscosity, 50–200 cP) was purchased from Sigma Aldrich (St. Louis, USA). Oat spelt arabinoxylan was purchased from Sigma Aldrich. Water hyacinth (*Eichhornia crassipes*) was collected from Guwahati, Assam, India (Fig. 1). The substrates are cut into small

Fig. 1 Water hyacinth selected for simultaneous saccharification and fermentation study. **a** Water hyacinth plant. **b** Leaves and stem cut into small fraction

(a)   (b)

pieces of length 4–5 cm. They were washed with water to remove unwanted dust particles and then dried at 80 °C in hot air oven. The substrates were further grinded in a mixer and passed through a 1 mm mesh size sieve prior to pretreatment.

2.2 Microorganisms and Culturing Conditions

Bacillus subtilis AS3 (Genbank accession No. EU754025) was a gift from Prof. D. Goyal, Thapar University, Patiala, India. *B. subtilis* inoculum was prepared by taking a loop full of culture from the nutrient agar slant in a test tube containing 5 mL of nutrient broth with 2% (w/v) glucose and incubated at 37 °C and 180 rpm in an incubator shaker for 16–18 h (to reach optical density (OD) at 600 nm = 0.6–0.8). For enzyme production, 2% (v/v) of the fresh inoculum culture was transferred to two 250 mL Erlenmeyer flasks each containing 50 mL optimized media (Deka et al. 2011). The pH was adjusted to 7.2 using 1 N NaOH before autoclaving. The flasks were incubated at 39 °C with shaking at 121 rpm. After 48 h, 1.0 mL of samples were collected, centrifuged at 10,000 g for 10 min at 4 °C and the cell-free supernatant was used as the crude enzyme for SSF experiment. *Trichoderma reesei* (MTCC 164) was procured from IMTECH, Chandigarh. *T. reesei* was maintained in potato dextrose medium at 4 °C. 1 mL of spore suspension (5×10^7 spores/mL) was inoculated to 100 mL of Potato dextrose broth and incubated at 28 °C with shaking at 120 rpm for 48 h. 1.0 mL of culture broth was centrifuged at 10,000 rpm for 15 min and the cell-free supernatant obtained was filtered twice and 1 mL of the filtrate was used as the crude enzyme for SSF experiment. The gene encoding family 5 glycoside hydrolase (GH5) was cloned and expressed as reported earlier (Taylor et al. 2005) and is now also commercially available with NZYTech, Lda, Lisbon, Portugal. *E. coli* BL21 (plysS) cells harbouring GH43 recombinant enzymes were cloned, cultured and maintained in our laboratory at IIT Guwahati. For production of recombinant cellulase (GH5), 1% of the *E. coli* Bl-21 cells from glycerol stock were inoculated into 5 ml of LB medium containing 100 µg/mL ampicillin and incubated at 37 °C for 16 h at 180 rpm. 1% of the culture inoculum was transferred to 250 ml of LB medium in 500 ml flask containing 100 µg/mL ampicillin and was incubated at 37 °C, 180 rpm till the culture reached mid-exponential phase (A_{600} 0.6). Isopropyl-β-D-thiogalactopyranoside (IPTG) was added to this mid-exponential phase culture to a final concentration of 1 mM followed by further 8 h incubation for protein induction. For production of recombinant hemicellulase (GH43), 5 mL of LB medium containing 50 μg/mL kanamycin was used for inoculation of 50 μL of the *E. coli* culture from glycerol stock and was incubated at 37 °C for 16 h at 180 rpm. Rest of the production process is same as for GH5. Both the *E. coli* cells were collected by centrifugation (9,000 g, 4 °C, 15 min) and were resuspended in 50 mM sodium acetate buffer adjusted to pH 5.2. The recombinant enzymes were expressed as soluble protein. The cell extract containing soluble enzyme was sonicated in an ice bath for 15 min followed by centrifugation (13,000 g, 4 °C, 30 min). The supernatant was used as the enzyme source for SSF experiment.

S. cerevisiae (NCIM 3215) and *C. shehatae* (NCIM 3500) were procured from NCL, Pune. Both *S. cerevisiae* and *C. shehatae* were maintained in MGYP agar slants containing (g/100 mL) Malt extract 0.3, Glucose 1, Yeast extract 0.3, Peptone 0.5 (Wickerman 1951) and stored at 4 °C. A loop full of these slant culture was transferred to GYE medium containing (g/100 mL) Glucose 1, Yeast extract 0.1 supplemented with KH_2PO_4 0.1, $(NH_4)_2SO_4$ 0.5 and $MgSO_4.7H_2O$ 0.05 and incubated at 30 °C with shaking at 120 rpm for 48 h. 0.5 mL each of the actively growing aerobic culture (3.6 × 10^8 cells/mL) was transferred to the fermentation medium for SSF experiments.

2.3 Pretreatment of Substrates

2.3.1 Wet Oxidation

One gram of powdered water hyacinth was mixed with 17 mL of water and 0.03 mL of concentrated sulphuric acid (36.5%, w/w). The mixture was kept at 120 °C for 1 h. Subsequently, cooling was done to room temperature. Finally, vacuum filtration was carried out and the leftover residue was subjected to SSF (Carlos and Anne 2007).

2.3.2 Phosphoric Acid (H_3PO_4)–Acetone

One gram of the powdered water hyacinth was taken in 250 mL Erlenmeyer flask with addition of 8 mL of concentrated phosphoric acid (of not less than 85%). The mixture was then incubated at 50 °C at 120 rpm for 1 h. The slurry was then poured into 24 mL chilled acetone and the mixture was centrifuged at 8000 rpm for 10 min. The pellet was collected and washed three times. During the third washing, pH was adjusted to 5–6 using NaOH (Li et al. 2009).

2.3.3 Ammonia Fibre Explosion (AFEX)

One gram of the powdered water hyacinth was taken in 250 mL Erlenmeyer flasks. 2.5 mL of ammonia solution was added and the mixture was then incubated at 170 °C for 1 h (Garlock et al. 2011).

2.4 Simultaneous Saccharification and Fermentation (SSF) of Water Hyacinth at Shake Flask and Bioreactor Level

In the first step, batch SSF experiments were performed at shake flask level. One gram each of the pretreated samples was autoclaved in a 250 mL flasks separately containing 100 mL of acetate buffer (pH 5.0, 20 mM) along with yeast extract (0.1%, w/v) and peptone (0.1%, w/v) and were used as fermentation media. Each batch of substrate underwent three different pretreatments and subjected to three different enzymatic consortiums. In total, nine SSF trials were carried out. The first SSF medium containing 1% (w/v) wet oxidized water hyacinth comprised of 1 mL of *B. subtilis* cellulase (3.3 U/mg, 0.5 mg/mL) along with 0.5 mL of *S. cerevisiae* inoculum (3.6×10^8 cells/mL) and 0.5 mL of *C. shehatae* (2.1×10^8 cells/mL) as the fermentative microbes. Similar SSF enzymatic and fermentative conglomerate was employed for acid–acetone and AFEX-treated substrate. The fourth SSF medium containing 1% (w/v) wet oxidized water hyacinth comprised of 1 mL of crude *T. reesei* cellulase (12.9 U/mg, 0.82 mg/mL) as the hydrolytic enzyme along with 0.5 mL of *S. cerevisiae* inoculum (3.6×10^8 cells/mL) and 0.5 mL of *C. shehatae* (2.1×10^8 cells/mL) as the bioethanol producers. Likewise, SSF experiments of acid–acetone and AFEX-treated substrate were carried out using same combination of hydrolytic enzyme and fermentative microbes. The seventh SSF medium containing 1% (w/v) wet oxidized water hyacinth comprised of 0.5 mL of crude recombinant cellulase (GH5) (5.6 U/mg, 0.44 mg/mL) and 0.5 mL hemicellulase (GH43) (0.30 mg/mL, 3.9 U/mg) for hydrolysis and 0.5 mL of *S. cerevisiae* (3.6×10^8 cells/mL) and 0.5 mL of *C. shehatae* (2.1×10^8 cells/mL) for bioethanol production. Similarly, SSF experiments were carried out for acid–acetone and AFEX-treated substrate employing alike enzymatic and fermentative combination. Each of the nine flasks was kept at 120 rpm and 30 °C. The samples were collected at every 6 h till 72 h with the estimation of various parameters like cell OD at 600 nm, ethanol concentration (g/L), reducing sugar (g/L) and specific activity (U/mg).

In the next step, the SSF experiment using AFEX-pretreated water hyacinth 5% (w/v) substrate along with mixed enzymes (GH5 and GH43) and mixed cultures (*S. cerevisiae*, *C. shehatae*) was scaled up in 250 mL shake flask with a working volume of 100 ml as well as in a 2L laboratory-scale fermenter (Applikon, model Bio Console ADI 1025, Holland) with a working volume of 1L. The fermentation conditions were also scaled up accordingly. In case of batch SSF in bioreactor, an aeration rate of 1 vvm was controlled by a mass flow controller. The batch was run till 72 h with the sample collection at every 6 h interval. There was a constant monitoring of parameters like cell OD at 600 nm, ethanol concentration (g/L), reducing sugar (g/L) and specific activity (U/mg).

2.5 Analytical Methods

2.5.1 Cellulose, Hemicellulose and Lignin Estimation

Cellulose, hemicellulose and lignin were determined by standardized methods of NREL, USA (Sluiter et al. 2008). 0.3 g of cellulosic substrate (leafy biomass) was mixed with 3 mL of 72% H_2SO_4 and incubated at 30 °C for 1 h. Then, there was an addition of 84 mL distilled water to bring down concentration of H_2SO_4 to 4% (v/v) with further autoclaving at 121 °C and 15 psi pressure for 1 h followed by vacuum filtration. The residue collected after filtration was weighed which is acid-insoluble lignin. The pH of the collected filtrate was neutralized by addition of $CaCO_3$. Finally, the filtrate was assayed for reducing sugar which is glucose from where cellulose is calculated (1 g cellulose = 1.1 g of glucose). The remaining content is hemicellulose.

2.5.2 Enzyme Assay and Protein Content

The enzyme assay of crude recombinant (GH5), (GH43) and *T. reesei* were carried out by incubating the enzyme with CMC for 10 min at 50 °C. The reaction mixture (100 µL) contained 10 µL of enzyme and 1.0% final concentration of CMC in 20 mM acetate buffer (pH 5.0). In case of *B. subtilis* AS3, the assay of cellulase was carried out in 100 µL of reaction mixture containing 1.3% final concentration of CMC (65 µL of 2% CMC) in 50 mM glycine NaOH buffer (pH 9.2) and 35 µL of cell-free supernatant and incubated at 45 °C for 10 min. The cellulase activity was measured by estimating the liberated reducing sugar following Nelson–Somogyi procedure (Nelson 1944; Somogyi 1945). The absorbance was measured at 500 nm using a UV-visible spectrophotometer (Perkin Elmer, Model lambda-45) against a blank with D-glucose as standard. One unit (U) of cellulase activity is defined as the amount of enzyme that liberates 1 µmole of reducing sugar (glucose) per min at 50 °C. The protein concentration was determined by the Bradford method using bovine serum albumin (BSA) as standard (Bradford 1976).

2.5.3 Ethanol Estimation

For ethanol content estimation, dichromate method was used where ethanol produced was converted to acid by reaction with dichromate (Seo et al. 2009).

The cell-free culture was diluted 10 times (reaction volume 10 mL) to which 2 mL of potassium dichromate ($K_2Cr_2O_7$) (3.37 g/100 mL) was added and absorbance was measured on spectrophotometer (Perkin Elmer, Model lambda-45) at 600 nm.

3   Results and Discussion

3.1   Cellulose, Hemicellulose and Lignin Determination

The structural polysaccharides showed that water hyacinth contains cellulose 20.91%, hemicellulose 44.88% and lignin 30.04%. The results suggested that water hyacinth can be used as the potential substrate for the production of bioethanol and was used for all SSF experiments.

3.2   SSF with Different Pretreatments and 1% (W/V) Substrate at Shake Flask Level

The comparisons of pretreatment of wet oxidation, acid acetone and AFEX on water hyacinth were investigated by subjecting it to simultaneous saccharification by *B. subtilis* cellulase, *T. reesei* cellulase recombinant cellulase (GH5) and recombinant hemicellulase (GH43) with consequent fermentation by *S. cerevisiae* and *C. shehatae*. The wet oxidized water hyacinth conferred an ethanol titre of 0.9 g/L with *B. subtilis* cellulase and 1.26 g/L with *T. reesei* cellulase amid a fermentative microbial combination of *S. cerevisiae* and *C. shehatae*. A higher ethanol titre (1.69 g/L) was achieved with an enzymatic consortium of recombinant cellulase (GH5) and hemicellulase (GH43) along with *S. cerevisiae* and *C. shehatae* (Table 1). Acid–acetone-pretreated water hyacinth gave an ethanol titre of 1.22 g/L (Table 1) with *B. subtilis* cellulase and 1.43 g/L with *T. reesei* cellulase and 1.70 g/L using mixed enzyme GH5 cellulase and GH43 hemicellulase with similar combination of bioethanol producers. On performing SSF using AFEX-pretreated water hyacinth, a higher ethanol concentration of 1.38 g/L with *B. subtilis* cellulase and 1.55 g/L with *T. reesei* cellulase were obtained. Interestingly, a maximum ethanol titre of 1.98 g/L (Table 1) was achieved with enzymatic consortium GH5 cellulase and GH43 hemicellulase with similar combination of fermentative microbes.

Thus, on the basis of ethanol titre (g/L) obtained from SSF experiments involving pretreated water hyacinth, AFEX pretreatment was found to be the best. Also, enzymatic consortium of GH5 cellulase and GH43 hemicellulase proved superior over *B. subtilis* cellulase and *T. reesei* cellulase owing to its ability for high saccharification rate producing more amounts of sugars with high ethanol titre. The SSF combination of mixed recombinant hydrolytic enzymes and mixed fermentative microbes was selected for further studies with 5% (w/v) water hyacinth.

Table 1 Comparative chart of different pretreatments on SSF combinations from water hyacinth (*Eichhornia crassipes*) using mixed recombinant enzymes (GH5 cellulase and GH43 hemicellulase) and mixed fermentative microbes (*S. cerevisiae* and *C. shehatae*)

S. No.	SSF combination	Pretreatment	Substrate concentration (%, w/v) and mode of SSF	Reducing sugar* (g/L)	Ethanol yield (g/g)	Ethanol titre* (g/L)
1	*B. subtilis* (AS3) cellulase + *S. cerevisiae* + *C. shehatae*	Wet oxidation	1% Shake flask	1.1 ± 0.02	0.09	0.90 ± 0.01
		Acid–acetone		1.3 ± 0.03	0.122	1.22 ± 0.05
		AFEX		1.4 ± 0.08	0.138	1.38 ± 0.03
2	*T. reesei* cellulase + *S. cerevisiae* + *C. shehatae*	Wet oxidation	1% Shake flask	1.28 ± 0.03	0.126	1.26 ± 0.02
		Acid–acetone		1.42 ± 0.01	0.143	1.43 ± 0.01
		AFEX		1.43 ± 0.02	0.155	1.55 ± 0.03
3	rec. cellulase (GH5) + rec. hemicellulase (GH43) + *S. cerevisiae* + *C. shehatae*	Wet oxidation	1% Shake flask	1.77 ± 0.06	0.169	1.69 ± 0.07
		Acid–acetone		1.78 ± 0.03	0.170	1.70 ± 0.01
		AFEX		1.90 ± 0.04	0.198	1.98 ± 0.10
4	rec. cellulase (GH5) + rec. hemicellulase (GH43) + *S. cerevisiae* + *C. shehatae*	AFEX	5% Shake flask	10.22 ± 0.02	0.196	9.78 ± 0.02
5	rec. cellulase (GH5) + rec. hemicellulase (GH43) + *S. cerevisiae* + *C. shehatae*	AFEX	5% Bioreactor	18.02 ± 0.03	0.359	17.97 ± 0.06

*The values correspond to the maximum reducing sugar and maximum ethanol titre at a particular time values are mean $\pm$ SE ($n = 3$)

3.3 SSF with AFEX Pretreatment and 5% (W/V) Substrate at Shake Flask Level

On performing the SSF process in shake flask level using 5% AFEX-pretreated water hyacinth, mixed recombinant enzymes GH5 cellulase and GH43 hemicellulase and mixed microbial combination *S. cerevisiae* and *C. shehatae* gave an ethanol titre of 9.78 g/L. The maximum reducing sugar amounted to 10.22 g/L (Table 1, Fig. 2) with a yield coefficient of 0.196 (g/g).

3.4 SSF with 5% (W/V) Substrate at Bioreactor Level

Owing to the controlled conditions of pH and aeration, the AFEX-pretreated water hyacinth (*Eichhornia crassipes*) was used for SSF experiment in laboratory-scale reactor along with *S. cerevisiae* and *C. shehatae* as the fermentative microbe. A maximum reducing sugar concentration peaked at 18.02 g/L. The maximum ethanol titre obtained was 17.97 g/L (Table 1, Fig. 3) with a yield coefficient of 0.359 (g/g).

A number of different pretreatment methods have been investigated such as wet oxidation, phosphoric–acetone and ammonia fibre explosion (AFEX) method. Table 1 shows the comparison of the different pretreatment methods used during SSF process. AFEX pretreatment proved to be better than wet oxidation and acid acetone technique on analysis of SSF parameters. The other two pretreatments are

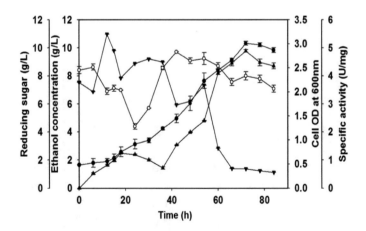

Fig. 2 SSF profile of 5% (w/v) Water hyacinth (*Eichhornia crassipes*) after AFEX pretreatment using recombinant cellulase (GH5), recombinant hemicellulase (GH43) along with *S. cerevisiae* and *C. shehatae* at shake flask level showing variation of (●) cell OD measured at 600 nm, (▲) ethanol concentration (g/L), (▼) reducing sugar (g/L) and (o) specific activity (U/mg) with time (h) respectively

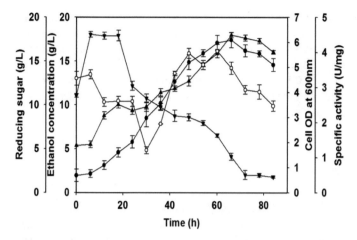

Fig. 3 Simultaneous saccharification and fermentation (SSF) profile of 5% (w/v) water hyacinth (*Eichhornia crassipes*) after AFEX pretreatment using recombinant cellulase (GH5), recombinant hemicellulase (GH43) along with *S. cerevisiae* and *C. shehatae* at bioreactor level showing variation of (●) cell OD measured at 600 nm, (▲) ethanol concentration (g/L), (▼) reducing sugar (g/L) and (o) specific activity (U/mg) with time (h) respectively

found to generate cellulase inhibitors interfering with the enzymatic hydrolysis, cost intensive during recovery of solvents and low sugar and ethanol yield. AFEX pretreatment has the advantage of not producing inhibitors for the downstream biological processes and high temperature (>90 °C) and pH (<12.0) minimize formation of sugar degradation products resulting in higher yields. AFEX-pretreated water hyacinth releases maximum amount of sugars in least pretreatment time with high ethanol concentrations and yield. It was reported to be more effective for herbaceous crops and grasses than for hardwood and softwood (McMillan 1994). The ethanol titre obtained in all SSF experiments involving different pretreatments was maximum for recombinant hydrolytic enzymes as compared to the natural ones. These findings evidently prove the efficiency of thermostable recombinant *C. thermocellum* cellulases and hemicellulases higher over *B. subtilis* and *T. reesei* cellulases, due to having highest rates of cellulose utilization (Fontes and Gilbert 2010). SSF studies, using recombinant cellulase (GH5) and hemicellulase (GH43) along with hexose and pentose sugars utilizing *S. cerevisiae* and *C. shehatae* gave a better choice of bioethanol production. The concentration of reducing sugar fell rapidly by 30 h of fermentation, thereafter the decline was gradual. With the fall in reducing sugar, there was simultaneous production of ethanol with maximum ethanol production after 60 h of fermentation and then a decrease gradually. Maintaining reaction parameters such as temperature, pH, aeration and agitation in the bioreactor accounted to better ethanol yield.

The values of ethanol concentration obtained in our study are comparable with the other reported literatures. Reddy et al. (2010) obtained an ethanol concentration of 2.2 g/L from 1% of banana waste using a coculture of *Clostridium thermosaccharolyticum* HG8 and *Thermoanaerobacter ethanolicus* ATCC 31937. An ethanol titre of 1.4 g/L using recombinant cellulase isolated from *E. coli* BL21 cells transformed with full-length gene *Ct*Lic26A-GH5-CBM11 from *Clostridium thermocellum* on 1% Jamun (*Syzygium cumini*) leafy biomass as the substrate (Mutreja et al. 2011). A SSF experiment involving 30% solid content with commercial cellulase enzyme and traditional fermentative organism gave an ethanol concentration of 60 g/L (Santos et al. 2010).

4   Conclusion

The SSF profile of different combinations describes the relationship between the rate of saccharification and rate of sugar utilization along with the rate of ethanol formation. Mixed cultures of *S. cerevisiae* and *C. shehatae* have shown better ethanol yield due to its potential to use both hexose and pentose sugars present in water hyacinth. AFEX pretreatment proved to be better with the release of maximum amount of reducing sugars in least pretreatment time with high ethanol yield on analysis of SSF parameters. Recombinant cellulase (GH5) in combination with recombinant hemicellulase (GH43) showed promising saccharification rates with maximum release of utilizable sugars in the SSF process using water hyacinth. Scale up in laboratory-scale bioreactor employing mixed enzyme and mixed culture have the potential to maximize the volumetric productivity while minimizing the production costs in the bioethanol industry. In the present study, a 4.9-fold enhancement in bioethanol production was demonstrated using recombinant GH5 cellulase and GH43 hemicellulase (9.78 g/L) in shake flask and 9.08-fold increases (17.97 g/L) in bioreactor on increasing the substrate concentration from 1 to 5% (w/v). The controlled conditions of various parameters in the bioreactor had an added advantage on the growth and ethanol titre.

Thus, an increase in the substrate concentration along with enzyme loadings and inoculum size has resulted in relative acceleration in ethanol titre and yield.

References

Adam EG (2011) Cellulase, types and action, mechanisms and uses. Nova Science Publishers, New York

Bradford MM (1976) A rapid and sensitive method for quantitation of microgram quantities of protein utilizing the principle of protein-dye binding. Anal Biochem 72:248–254

Carlos M, Anne BT (2007) Wet oxidation pretreatment of lignocellulosic residues of sugarcane, rice, cassava and peanuts for ethanol production. J Chem Technol Biotechnol 82:174–181

Casey GP, Ingledew WM (1986) Ethanol tolerance in yeasts. Crit Rev Microbiol 13:219–280

Crispen M, Rajni H, Remigio Z, Bo M (2000) Purification and characterization of cellulases produced by two Bacillus strains. J Biotechnol 83:177–187

Deka D, Bhargavi P, Sharma A, Goyal D, Jawed M, Goyal A (2011) Enhancement of cellulase activity from a new strain of *Bacillus subtilis* by medium optimization and analysis with various cellulosic substrates. Enzyme Res. https://doi.org/10.4061/2011/151656

Fontes CMGA, Gilbert HJ (2010) Cellulosomes: highly efficient nanomachines designed to deconstruct plant cell wall complex carbohydrates. Ann Rev Biochem 79:655–681

Garlock RJ, Wong YS, Balan V, Dale BE (2011) AFEX pretreatment and enzymatic conversion of black locust (*Robinia pseudoacacia* L.) to soluble sugars. Bioenergy Res. https://doi.org/10.1007/s12155-011-9134-6

Kadam KL, Schmidt SL (1997) Evaluation of *Candida acidothermophilum* in ethanol production from lignocellulosic biomass. Appl Microbiol Biotechnol 48:709–713

Li H, Kim NJ, Jiang M, Kang JW, Chang HN (2009) Simultaneous saccharification and fermentation of lignocellulosic residues pretreated with phosphoric acid-acetone for bioethanol production. Biores Technol 100:3245–3251

McMillan JD (1994) Pretreatment of lignocellulosic biomass. In: Himmel ME, Baker JO, Overend RP (eds) Enzymatic conversion of biomass for fuels production. ACS symposium Series, Washington, vol 566, pp 292–324

Mosier NS, Wyman C, Dale B, Elander R, Lee YY, Holtzapple M, Ladisch MR (2005) Features of promising technologies for pretreatment of lignocellulosic biomass. Biores Technol 96:673–686

Mutreja R, Das D, Goyal D, Goyal A (2011) Bioconversion of agricultural waste to ethanol by SSF using recombinant cellulase from *Clostridium thermocellum*. Enzyme Res. https://doi.org/10.4061/2011/340279

Nelson N (1944) A photometric adaptation of the Somogyi method for the determination of glucose. J Biol Chem 153:375–380

Reddy HK, Srijana M, Reddy MD, Reddy G (2010) Coculture fermentation of banana agro-waste to ethanol by cellulolytic thermophilic *Clostridium thermocellum* CT2. Afr J Biotechnol 9:1926–1934

Saha BC, Nichols NN, Qureshi N, Cotta MA (2011) Comparison of separate hydrolysis and fermentation and simultaneous saccharification and fermentation processes for ethanol production from wheat straw by recombinant *Escherichia coli* strain FBR5. Appl Microbiol Biotechnol 92:865–874

Santos DS, Camelo AC, Rodrigues KCP, Carlos LC, Pereira N Jr (2010) Ethanol production from sugarcane bagasse by *Zymomonas mobilis* using simultaneous saccharification and fermentation (SSF) process. Appl Biochem Biotechnol 161:93–105

Seo HB, Kim HJ, Jung HK (2009) Measurement of ethanol concentration using solvent extraction and dichromate oxidation and its application bioethanol production process. J Ind Microbiol Biotechnol 36:285–292

Singh HD, Nag B, Sharma AK, Baruah JN (1984) Nutrient control of water hyacinth growth and productivity. In: Thyagarajan G (ed) Water hyacinth. UNEP report and proceedings series 7, UNEP, Nairobi, pp 243–263

Sluiter B, Hames R, Ruiz C, Scarlata J, Sluiter D, Templeton D (2008) Determination of structural carbohydrates and lignin in biomass. Laboratory Analytical Procedures (LAPs). Technical Report NREL/TP-510, 42618

Somogyi M (1945) A new reagent for the determination of sugars. J Biol Chem 160:61–68

Tagaki M, Abe S, Suzuki S, Emert GH, Yata N (1977) A method for production of alcohol directly from cellulose using cellulase and yeast. In: Proceedings of bioconversion of cellulosic substances into energy, chemicals and microbial protein. pp 551–571

Taylor EJ, Goyal A, Guerreiro CIPD, Prates JAM, Money VA, Ferry N, et al (2005) How family 26 glycoside hydrolases orchestrate catalysis on different polysaccharides: structure and activity of a *Clostridium thermocellum* lichenase, Ctlic26A. J Biol Chem 280:32761–32767

Wickerman LJ (1951) Taxonomy of yeasts. In: US Department of Agriculture Technical Bulletin No. 1029, Washington, pp 1–56

Part V
Ground Water Contamination

Saline Water Intrusion into the Coastal Aquifers of the Periyar River Basin, Central Kerala, India

K. T. Damodaran and P. Balakrishnan

Abstract Successful management of coastal groundwater resources depends largely on the accurate assessment and prediction of aquifer behaviour and the saline water–freshwater interface both under the natural and the anthropogenic conditions. The prime objective of this paper is to assess the extent of saline water intrusion into an urban groundwater aquifer and to know the impact of this phenomenon on the quality of groundwater. For this purpose, the coastal belt of Kochi near where the Periyar River debouches into the Arabian Sea is selected, considering the gravity of the problem prevailing in this region. The Periyar River is the longest river and the river with the largest discharge potential in the state. It is one of the few perennial rivers in the region and provides drinking water for several major towns of Central Kerala. The coastal zone of the Periyar River Basin is one of the most densely populated areas in the country. Geologically, the area has three distinct formations, viz., the crystalline rocks of the Precambrian age overlain by the Tertiaries and the Recent alluvium. The analysis of apparent resistivity variation with change in electrode spacing helped to deduce the depth and resistivity distribution of various subsurface units which in turn are interpreted in terms of various geological formations. The resistivity profiling coupled with resistivity sounding was found to be a highly effective method for determining the freshwater areas and the saline water contaminated zones at different depth levels. The electrical resistivity curves obtained expressed a dominant trend of decreasing resistivity with depth (thus increasing salinity). The rock matrix, salinity and water saturation are the major factors controlling the resistivity of the formation. Moreover, the fresh and saline groundwater inter-phase has been investigated, and it is observed that the saline water intrusion into the aquifers can be accurately

K. T. Damodaran (✉)
School of Marine Sciences, Cochin University of Science and Technology, Kochi 682016
Kerala, India
e-mail: ktdamodaran@gmail.com

P. Balakrishnan
State Groundwater Department, Regional Data Centre, Ernakulam, Kochi 682030
Kerala, India
e-mail: krishnagiri55@yahoo.com

© Springer International Publishing AG 2018
A. K. Sarma et al. (eds.), *Urban Ecology, Water Quality and Climate Change*,
Water Science and Technology Library 84,
https://doi.org/10.1007/978-3-319-74494-0_28

mapped using surface DC resistivity method. The iso-apparent resistivity map shows that contour patterns are almost parallel to the coast (north–south trend), with the increasing values towards east. The apparent resistivity along the eastern side is comparatively high when compared with the western margin of the area, which is an indication that the salinity is decreasing towards the eastern side. It is also observed that the resistivity values decrease with increase in depth of investigation. The very low resistivity is an indication that the aquifers are of poor chemical quality. Based on the analysis of the above observations, the freshwater saline water interface of the study area is demarcated.

Keywords Groundwater · Periyar river basin · Sand mining · Resistivity Saline water intrusion

1 Introduction

Groundwater plays a vital role in the development of agriculture, industry and also for drinking purposes in the Periyar River Basin, as the surface water resource of the area is highly inadequate to fulfil the overall water demand in the area. Over-exploitation of the scarce potable groundwater resource in the coastal areas of Kochi and nearby areas which form a part of the Periyar River Basin caused detrimental environmental problems such as lowering of water table, quality deterioration, saline water intrusion, etc. Therefore, there is a need to monitor the water-level declines in the area in a comprehensive and accurate manner and to reduce the rates and effects of groundwater depletion and the saline water intrusion. The paper closely examines the problems as to how much and where all the freshwater depletion takes place due to saline water intrusion and illegal sand mining on the basis of a detailed survey of the area from the geological and geophysical perspectives. The coastal zone of the Periyar River Basin is one of the most densely populated areas in the country. Major public drinking water projects of the district are depended on this river and increasing reliance on these sources raises significant concern over water safety and sustainability. The natural balance between the freshwater and the saline water is disturbed by the rampant illegal sand mining and also the excess groundwater withdrawals. An analysis of the hydrographs from the study area indicated that sand mining lowers the groundwater levels, reduces freshwater flow into the coastal aquifers and ultimately causes saline water intrusion. Hydrographs of 13 stations from the study area showed a decrease in water levels since the year 2000 or so. Due to over-pumping, one of the wells indicated a reverse trend since 1990. The paper also discusses the deterioration of water quality due to saline water intrusion, as a part of the investigation. The analysis of apparent resistivity variation with change in electrode spacing helped to deduce the depth and resistivity distribution of various subsurface units which in turn are interpreted in terms of various geological formations. The resistivity profiling coupled with resistivity sounding was found to be a highly effective method for determining the freshwater areas and the saline water contaminated zones at different depth levels. The electrical resistivity curves obtained expressed a dominant trend of decreasing resistivity with depth (thus increasing salinity). In general, the presence of four distinct resistivity zones was delineated, viz., (i) the unconsolidated clayey sand having resistivity values ranging between 72 and 262 Ωm with a layer thickness ranging from 1.2 to 3.2 m representing the first layer (surface layer). This is the only freshwater aquifer in the area. (ii) The surface layer is underlain by the sandy clay having resistivity values ranging from 14 to 75 Ωm with a layer thickness varying from 0.66 to 5.4 m. This is a transitional or mixing zone as the resistivity decreases with depth. (iii) The resistivity of the third layer varies from 24 to 75 Ωm with a layer thickness varying from 2.9 to 34 m. (iv) The fourth subsurface layer is characterised by a resistivity from 2 to 45 Ωm with a layer thickness of 2.9–34 m. The subsurface formations with resistivity values generally below 4 Ωm are an indication of the aquifer possibly containing brine. Resistivity profiling delineates the lateral changes in resistivity that can be correlated with steeply dipping interfaces between two subsurface geological formations of the area. DC resistivity sounding determines the thickness and resistivity of different horizontal or low dipping subsurface layers including the aquifer zone. Based on the analysis of the above observations, the freshwater saline water interface of the study area is demarcated.

2 Study Area

The study area falls within the Central Kerala forming the part of the Periyar River Basin and lies between Lat. 9° 55′ 00″N–10° 21′ 00″N and Long. 76° 08′ 38″–76° 55′ 00″E (Fig. 1).

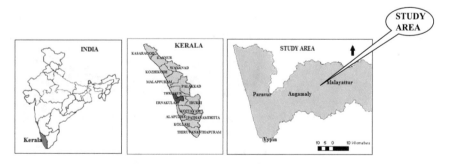

Fig. 1 Location map of the study area

2.1   Climate, Physiography and Geology of the Area

The study area enjoys a typical tropical climate with an average rainfall of about 300 cm. Physiographically, the area can be divided into three, namely, the coastal plains in the west, the midland region at the centre and the hill ranges in the east. The coastal plains have an elevation of less than 6 m above MSL. The midland region has an elevation up to 80 m above MSL, and the highland region has an elevation above 80 m above MSL. The general slope of the area is towards west.

Geologically, the study area has three distinct formations; the crystalline rocks of Precambrian age, the Tertiary formations and the Recent alluvium (Soman 1997; Paulose and Narayanaswami 1968; GSI 1995) as given in Table 1 and Fig. 2.

Table 1 Generalised geological succession of the study area (modified after Paulose and Narayanaswami 1968; Soman 1997)

Age		Formation	Lithology	
Quaternary	Recent	Alluvium	Sands, clay	
	Subrecent	Laterite	Laterite derived from the Tertiary sediments and the Precambrian crystalline	
Tertiary	Upper Miocene	Warkali beds	Fine- to medium-grained sand, variegated clay and lignite	
	Lower Miocene	Quilon beds	Limestone, calcareous clay and marl	
	Eocene to Oligocene	Vaikom beds	Gravel, coarse to fine sand, carbonaceous clay and lignite	
Precambrian		Migmatite group	Hornblende-biotite gneiss, biotite gneiss, etc.	Intruded by felsic and mafic intrusive of age varying from Proterozoic to Tertiary
		Charnockite group	Charnockites	

Fig. 2 Geology of the area

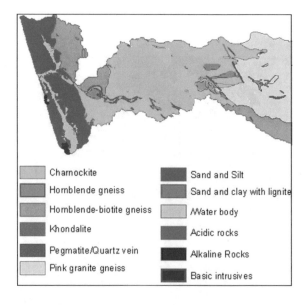

However, due to weathering and erosion, such a complete sequence may not be present uniformly throughout the area.

2.2   Types of Aquifers of the Study Area

Groundwater occurs under unconfined, semi-confined and confined conditions, in pores, rock fractures, cavities and other openings in the geologic formations. The most important water-bearing formations of the study area are unconsolidated sands and gravels, consolidated sands, laterites, weathered and the fractured crystalline rocks. Groundwater occurs under phreatic conditions in laterites, unconsolidated coastal sediments weathered and fractured crystalline rocks. It occurs under semi-confined to confined conditions in the deep-seated fractured aquifers in the crystalline and Tertiary formations.

2.2.1   Behaviour of Groundwater in the Study Area

As part of the present study, groundwater-level observations pertaining to 47 National Hydrograph Stations (NHS) (Fig. 3) maintained by the Kerala State Ground Water Department were collected for the period from 1985 to 2007, and this data is also used for the present study. Based on the data analysed, the water table fluctuation graphs, water table contour maps and hydrographs of the study area are prepared (Figs. 4, 5, 6 and 7). The hydrographs of the selected network stations in general show corresponding peaks and troughs with rainfall distribution. The best fit line for the hydrographs will bring out the trend of the water table with time. The peaks represent the low groundwater storage whereas the troughs correspond to the increased storage of groundwater. The overall trend of the water table during the period from 1985 to 2006 is given in Table 2. The normal,

Fig. 3 Location of the National Hydrograph Stations (NHS) of the study area

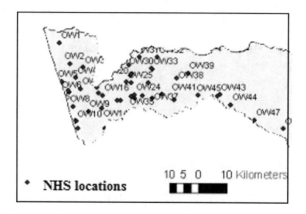

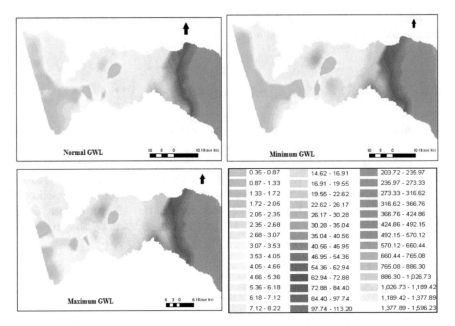

Fig. 4 Normal, minimum and maximum groundwater level of the area above MSL (1985–2006)

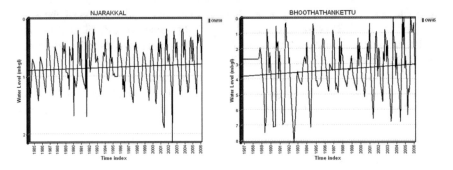

Fig. 5 Hydrographs showing increasing trend

minimum and maximum water levels are given in Fig. 4. The maximum water level drawdown in the lowland area is observed as 3.56 m at Kottappuram, and the minimum is 0.12 m at Varapuzha. The average water level of the low land region of the basin is 1.64 m. In the midland region, the maximum drawdown of 9.4 m is observed in the well at Angamali, the minimum is 0.27 m at Pindimana and the

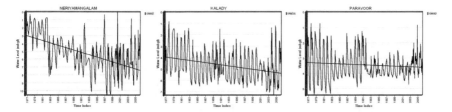

Fig. 6 Hydrographs showing decreasing trend

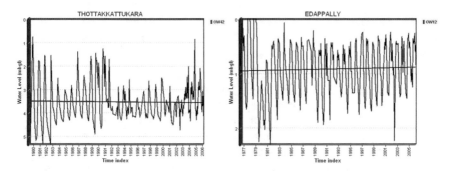

Fig. 7 Hydrographs showing steady groundwater level

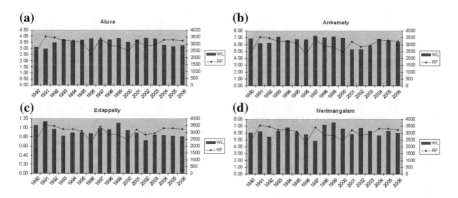

Fig. 8 a–d Comparison of groundwater level with rainfall

average drawdown of the region is 4.84 m. Taking into consideration of the entire area, the maximum fluctuation of 14.45 m has been observed at Poyya and the minimum of 0.01 m at Munambam during the period. The fluctuation pattern consequent upon the rampant and illegal sand mining will be discussed subsequently.

Table 2 Long-term groundwater-level trend of the observation wells in the study area from 1985 to 2006

Well No.	Location	District	Taluk	Rise (m/year)	Fall (m/year)	Trend (m/year)
OW01	Pappinivattom	Thrissur	Kodumgallur	0.018	0.000	0.018
OW02	Lokamalleswaram	Thrissur	Kodumgallur	0.024	0.000	0.024
OW03	Poyya	Thrissur	Kodumgallur	0.187	0.000	0.187
OW04	North Parur	Ernakulam	Paravur	0.009	0.000	0.009
OW05	Munambam	Ernakulam	Paravur	0.047	0.000	0.047
OW06	Pallipuram	Ernakulam	Kochi	0.013	0.000	0.013
OW07	Paravur Jn.	Ernakulam	Paravur	0.000	0.008	−0.008
OW08	Edavanakad	Ernakulam	Kochi	0.001	0.000	0.001
OW09	Varapuzha	Ernakulam	Paravur	0.000	0.026	−0.026
OW10	Njarakkal	Ernakulam	Kochi	0.007	0.000	0.007
OW11	Malipuram	Ernakulam	Kochi	0.000	0.013	−0.013
OW12	Edappally	Ernakulam	Kanayannur	0.009	0.000	0.009
OW14	Elur	Ernakulam	Paravur	0.014	0.000	0.014
OW15	Paravur	Ernakulam	Paravur	0.015	0.000	0.402
OW16	Kottappuram	Ernakulam	Paravur	0.013	0.000	0.013
OW17	Chalakkal	Ernakulam	Aluva	0.000	0.012	−0.012
OW18	Chengamanad	Ernakulam	Aluva	0.010	0.000	0.010
OW19	Alwaye town	Ernakulam	Aluva	0.000	0.010	−0.110
OW20	Karukutty	Ernakulam	Aluva	0.000	0.032	−0.032
OW21	Chalakkal	Ernakulam	Aluva	0.012	0.000	0.012
OW23	Chowara	Ernakulam	Aluva	0.000	0.015	−0.115
OW24	Vazhakkulam	Ernakulam	Aluva	0.000	0.016	−0.016
OW25	Nedumbassery	Ernakulam	Aluva	0.026	0.000	0.026
OW26	Kaladi	Ernakulam	Aluva	0.048	0.000	0.048
OW28	Angamali	Ernakulam	Aluva	0.022	0.000	0.022
OW29	Angamaly RH	Ernakulam	Aluva	0.009	0.000	0.009
OW30	Manjapara Town	Ernakulam	Aluva	0.000	0.011	−0.011
OW31	Attara	Ernakulam	Aluva	0.006	0.000	0.006
OW32	Sulli	Ernakulam	Aluva	0.000	0.025	−0.025
OW33	Manjapra In.	Ernakulam	Aluva	0.023	0.000	0.023
OW34	Kalady Jn.	Ernakulam	Aluva	0.000	0.027	−0.027
OW35	Kanjur	Ernakulam	Aluva	0.000	0.045	−0.045
OW37	Vallom	Ernakulam	Aluva	0.000	0.077	−0.077
OW38	Malayattur	Ernakulam	Kunnathunad	0.000	0.027	−0.027
OW39	Panamkuzhi	Ernakulam	Kunnathunad	0.012	0.000	0.012
OW40	Kuruppampady	Ernakulam	Kunnathunad	0.000	0.015	−0.015
OW41	Kottapadi	Ernakulam	Kothamangalam	0.008	0.000	0.008
OW42	Aluva	Ernakulam	Aluva	0.002	0.000	0.002

(continued)

Table 2 (continued)

Well No.	Location	District	Taluk	Rise (m/year)	Fall (m/year)	Trend (m/year)
OW43	Thattekad	Ernakulam	Kothamangalam	0.074	0.000	0.074
OW44	Kuttamangalam	Ernakulam	Kothamangalam	0.002	0.000	0.002
OW45	Pindimana	Ernakulam	Kothamangalam	0.038	0.000	0.038
OW46	Keerampara	Ernakulam	Kothamangalam	0.000	0.050	−0.050

2.2.2 Classification of Groundwater-Level Fluctuations

Davis and De Wiest (1967) classified groundwater-level fluctuation in general into four types, namely, (i) fluctuation owing to change in groundwater storage, (ii) fluctuation brought about by atmospheric pressure in contact with water surface in wells, (iii) fluctuation resulting due to deformation of aquifer and (iv) fluctuation owing to disturbances within the well. Minor fluctuations are attributed to chemical or thermal changes in and around the wells. The magnitude of the water fluctuations depends on the usage, climate, drainage, topography and the geological conditions. Long-term fluctuations in groundwater-level studies will help us to understand the depletion and recharging conditions of an aquifer. The stress and strain in water level due to groundwater recharge, discharge and intensity of rainfall are reflected in water-level fluctuations with time. The rate and magnitude of fluctuation during any period portray the net effects of recharge or discharge during that period, in relation to inherent characteristics of the aquifer media (Rangarajan and Athavale 2000).

2.2.3 Comparison of Groundwater Table with Rainfall

The groundwater level of some selected NHS of the study area for the period from 1990 to 2006 was compared with the corresponding rainfall. The fluctuation in groundwater level of these wells was almost concordant with the corresponding rainfall, Fig. 8a–d.

2.2.4 Analysis of Hydrographs of the Study Area

The water levels fluctuate with annual recharge and discharge conditions. A high water table occurs during the wet season and low water table during dry seasons, indicating a gain or loss of water in the aquifer storage. Water table data is a direct indicator of the status of groundwater of the area. The hydrographs are showing declining trend where the groundwater exploitation is quite high and showing an

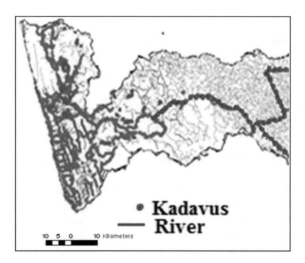

Fig. 9 Sand mining kadavus of the study area

increasing trend where the exploitation is low. Average water level is used as a tool that conveys the status of the groundwater of the area. From the analysis of the water-level data, it is observed that 22 stations are showing a declining trend and another 23 stations are showing an increasing trend while 4 stations are maintaining almost a steady level during the period (Table 2 and Figs. 4, 5, 6 and 7). The average increase in trend at the above area is 0.02 m/year, and the average decline trend of the wells is 0.02 m/year. The analysis of the hydrographs reflects the volume of groundwater stored or discharged from an aquifer. The analysis shows that the groundwater trend is not uniform throughout the basin from 1985 to 2007. A close analysis of the pre-monsoon hydrographs shows that 18.75% of the NHS exhibit 0–2 m of water-level fluctuation, 43.75% show the 2–5 m fluctuation, 31.25% show variation between 5 and 10 m and 6.25% show variation above 10 m. The maximum percentage of well frequency is in the range of 2–5 m. The natural balance between the freshwater and the saline water is further disturbed by the illegal sand mining in the area. An analysis of the hydrographs from the study area revealed that excessive sand mining lowers the groundwater levels, reduces freshwater flow into the coastal aquifers and ultimately causes saline water intrusion. Hydrographs of 22 stations showed a decrease in groundwater levels. Due to over-exploitation, one of the wells indicated a reverse trend since 1900. The water quality as well has considerably deteriorated (Table 3). The stations where the declining tread of groundwater table observed are the main area of sand mining.

Table 3 Chemical parameters of groundwater of the low lands of the study area during the pre-monsoon period

Well No.	Location	Na	CO₃	Alk	SO₄	Fe	TH	EC	pH	TDS	HCO3	N03	Ca	Mg	K	Cl
OW01	Pappinivattom	16.50	0.00	40.00	11.90	0.39	85.00	220.00	8.20	132.00	47.04	2.80	25.00	4.90	1.50	23.00
OW02	Lokamalleswaram	39.40	6.00	110.00	8.30	0.10	118.00	635.00	8.70	381.00	132.00	4.70	29.60	10.80	5.20	57.00
OW03	Poyya	6.40	0.00	10.80	2.30	0.28	21.00	65.00	7.10	39.00	10.80	2.32	9.00	2.00	1.30	10.00
OW04	North Parur	22.20	4.70	96.00	22.59	2.80	92.10	350.00	8.50	210.00	102.80	1.90	29.90	2.40	8.20	31.40
OW05	Munambam	39.00	30.00	51.00	15.00	3.45	189.00	312.00	8.90	199.00	138.00	4.80	60.00	4.90	6.80	73.00
OW06	Pallipuram	24.20	27.90	267.90	15.50	2.70	293.00	580.00	8.90	348.00	202.00	2.48	76.90	16.10	9.10	45.00
OW07	Ezhikkara	22.20	4.70	98.00	22.59	2.10	103.00	350.00	8.49	210.00	102.80	2.50	32.00	3.30	8.20	31.40
OW08	Edavanakad	136.30	26.40	234.00	44.12	2.44	367.00	1180.00	8.90	708.00	231.80	4.26	95.20	29.20	16.00	185.10
OW09	Varapuzha	75.50	32.50	152.00	27.50	2.09	148.00	710.00	8.90	426.00	175.30	13.53	99.00	14.80	16.00	115.90
OW10	Njarakkal	19.00	6.98	194.00	13.60	2.80	182.00	410.00	8.70	246.00	222.40	2.50	52.50	15.30	4.30	28.80
OW11	Malipuram	27.80	24.00	180.00	38.00	2.02	227.00	500.00	8.80	300.00	170.80	4.80	78.90	7.20	7.20	40.00
OW12	Edappally	27.40	21.28	156.20	18.30	0.50	201.00	490.00	8.60	294.00	126.88	2.78	71.00	4.80	8.20	39.50
OW13	Elur North	932.00	30.20	272.80	23.40	3.00	596.00	5550.00	8.90	3330.00	342.40	5.34	128.50	115.80	65.00	1523.00
OW14	Pathalam	31.70	2.10	4.50	25.00	1.30	186.00	690.00	8.50	414.00	20.00	3.80	28.00	7.73	9.50	42.30
OW15	Paravur	18.00	24.00	224.20	19.10	1.17	280.00	490.00	8.80	294.00	224.80	2.60	78.50	8.10	8.19	27.50
OW17	Chalakkal	37.00	0.00	112.00	28.00	0.70	94.00	407.00	8.30	282.00	127.00	3.70	21.50	12.00	4.90	55.00
OW18	Chengamanad	24.00	8.00	3.60	2.00	0.50	43.00	220.00	8.00	132.00	14.00	24.00	7.90	5.40	4.50	37.00
OW19	Alwaye	5.50	0.00	87.80	6.00	0.72	98.00	220.00	8.40	132.00	68.32	2.96	93.00	1.20	3.60	10.30
OW20	Karukutty	10.00	4.65	106.70	3.40	0.62	86.00	230.00	8.50	138.00	120.70	2.10	36.30	4.90	2.50	16.60
OW21	Aluva East	19.00	4.80	66.00	11.90	1.21	109.00	250.00	8.43	150.00	70.80	3.84	22.80	11.00	3.60	27.30
OW22	10th Mile	11.40	0.00	57.70	2.40	0.80	55.80	159.00	8.10	95.40	60.48	2.57	12.00	6.20	2.70	15.70
OW23	Chowara	19.60	0.00	9.50	3.90	0.36	37.00	190.00	7.30	114.00	21.30	8.82	16.10	3.70	2.00	29.00
OW24	Vazhakkulam	4.40	0.00	10.20	3.70	0.35	26.00	70.00	7.40	42.00	15.20	2.10	8.00	2.00	1.70	8.20

(continued)

Table 3 (continued)

Well No.	Location	Na	CO₃	Alk	SO₄	Fe	TH	EC	pH	TDS	HCO3	N03	Ca	Mg	K	Cl
OW25	Nedumbassery	9.60	0.00	29.40	2.51	1.80	34.80	133.00	7.79	79.80	35.90	3.48	10.00	2.40	1.80	13.20
OW26	Kaladi	19.20	0.00	52.00	22.70	0.80	75.00	245.00	8.20	147.00	42.60	3.68	16.10	5.50	2.50	27.30
OWZ7	Angamali	14.80	0.00	29.10	2.90	0.15	52.00	140.00	7.90	106.00	35.20	11.10	10.20	3.70	2.60	21.00
OW29	Vapalassery	8.20	0.00	13.70	5.22	0.75	32.50	102.00	7.90	61.20	16.70	3.40	11.00	2.50	1.70	13.70
OW30	Manjapara	9.80	0.00	16.00	5.63	0.35	27.00	123.00	8.20	74.00	23.50	2.19	4.20	2.10	3.10	17.00

2.3 Impacts of Sand Mining on the Groundwater Resource of the Area

The demand for sand has considerably increased since 1990, as there is an abrupt increase in construction works in Cochin and the adjacent areas. Sand and gravel are the main source of all civil constructions for thousands of years, and today, the demand for sand and gravel continues to increase. Now the Government of Kerala imposed restrictions on mining the sand, and permit system has been introduced. In the study area, at present, permit was issued to mine sand from 44 sand mining kadavus (locations) (Fig. 9) with a permissible quantity of 273 loads (1636 m^3 approximately) per day. Although people should confine their mining operations to the area specified in their permit, these conditions are seldom respected. As a result, the sand removed is four to six times the permitted quantum. In some cases, mining continues even after the expiry of the permit period. The mining operations, both legal and illegal, are being carried out in a number of places in the area, and the norm regarding the depth of the mine is often flouted. Thus, over-exploitation of river sand results in the depletion of the groundwater resource and thereby the destruction of agricultural practices, across the districts.

An aquifer can be compared to a bank account (USGS 2003). Just as a bank account must be balanced, withdrawals from an aquifer must be balanced by increased recharge or decreased discharge of storage. Indiscriminate, unregulated and unscientific development of the groundwater resources is the direct cause of the alarming decline in groundwater level. The sand mining could cause severe environmental, social and economic issues that are not reversible. The indiscriminate mining of sand from the study area has become a serious environmental issue now, such as the depletion of groundwater resource, lesser availability of water for industrial, agricultural and drinking purposes, salinization of the Aluva drinking water source many times due to the ingression of seawater into the river, destruction of agricultural land, loss of employment to farm workers, threat to livelihoods, human rights violations, damage to roads and bridges, etc. The river sand mining directly affects the natural equilibrium. The sand mining operations, both legal and illegal, have been noticed in a number of places, and the norm regarding the depth of the mine is often flouted.

2.4 The Geo-electric Survey

Generally in hard rock area, groundwater is controlled by secondary porosity, i.e. vertical and horizontal joints, fractures and cavities. Identification of these zones in the subsurface will be helpful to augment the drinking water supply. Vertical electrical soundings were carried out in 49 predetermined locations of the study area using Schlumberger electrode arrangements with an objective of analysing geophysical parameters quantitatively and qualitatively (Fig. 10 and Table 4). During

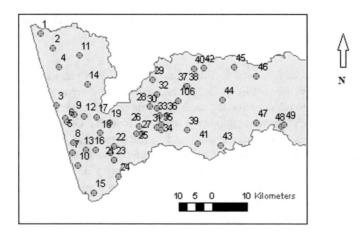

Fig. 10 VES locations of the study area

the field survey, the apparent resistivities obtained for different electrode spacing are plotted against current electrode spacing on a double-log graph sheets to get the field curves which in turn are interpreted by curve matching techniques and are given in Table 4. From the data analysed, it is observed that the resistivity of the first layer ($\rho 1$) of this basin ranges from 45 Ωm at Munambam to as high as 210 Ωm at Chengamanad, and the thickness varies from 1.1 to 3.2 m. The apparent resistivity and the corresponding thickness show that thickness of the first layer varies based on the topography and the soil types. The unconsolidated loose soils are more susceptible to erosion. In majority of the stations in the highland and midland regions, the topmost layer is hard laterite soil as capping and in the coastal plains, it is unconsolidated sandy alluvium where the chemical quality of groundwater is poor. The resistivity of the second layer ($\rho 2$) ranges from 14 Ωm at Ezhikkakara to 4350 Ωm at Chengamanad (Table 4) and generally consists of hard laterite in the midland and parts of the highland regions and weathered hard rocks as lenses in the upland regions and sandy soil in the coastal plains, and the thickness varies from 0.45 to 12.4 m. The resistivity of the third layer varies from 2 Ωm at Ezhikkakara to 1860 Ωm at Pindimana, and the thickness varies from 0.54 to 63.5 m. The clayey sand is the fourth layer aquifer in the coastal planes, and the crystalline formations are encountered as the fourth layer in all other parts of the study area. The fourth layer is deeper and acts as good aquifer in places like Neriamangalam of the highland region. The fifth layer acts as a good aquifer in the midland region, except at Illithodu, Malayattoor, Thattekkad and Edakuna areas, where the aquifers are poor yielding. The fifth layer is present only at places like Neriamangalam, but with high resistance values exhibiting the massive nature of the subsurface layer. The resistivity of the fifth layer varies from 1 Ωm at Vypin, Narakkal and Ezhikkakara to 1820 Ωm at Thattekkad with depths varying from 6.2 to 81 m. The very high subsurface resistivity of various layers (>1000 Ωm) is observed at locations in the highland and midland regions of the basin. This is an

Table 4 Analysed vertical electrical sounding (VES) data of the study area h1–h5 are layer thickness and ρ1–ρ5 are apparent resistivities)

VES No.	Location	Lat	Long	h1 (m)	h2 (m)	h3 (m)	h4 (m)	h5 (m)	ρ1	ρ2	ρ3	ρ4	ρ5	ρ6	Curve type
1	Matilakam	76.14	10.32	1.50	0.90	13.20	9.50	0.00	234	45	7	2	1	0	Q
2	Pappinivattom	76.165	10.293	1.10	0.66	0.54	2.90	6.20	262	46	22	8	4	3	Q
3	Munambam	76.172	10.172	1.80	2.20	18.80	22.10	0.00	45	24	8	2	1	0	Q
4	Eriyad	76.18	10.25	1.60	1.20	13.50	10.30	0.00	195	39	12	8	4	0	Q
5	Cherai	76.192	10.144	2.20	3.10	21.60	24.20	0.00	85	25	9	6	3	0	Q
6	Pallipuram	76.195	10.137	1.70	4.80	17.50	24.00	0.00	65	28	7	6	3	0	Q
7	Nayarambalam	76.207	10.072	2.10	3.20	19.80	24.20	0.00	52	34	9	2	2	0	Q
8	Edavanakad	76.208	10.092	1.80	2.40	22.50	25.60	0.00	96	31	6	4	2	0	Q
9	Paravur	76.211	10.152	1.20	3.60	6.90	24.00	0.00	68	36	19	11	3	0	Q
10	Njarakkal	76.218	10.045	1.90	2.70	20.40	23.60	0.00	59	32	8	3	2	0	Q
11	Vellangallur	76.22	10.28	1.90	3.40	18.20	11.50	34.50	620	1355	275	450	624	870	AH
12	Ezhikkakara	76.231	10.148	1.80	5.40	22.40	16.60	0.00	65	14	2	1	1	0	Q
13	Kongorpilly	76.236	10.077	2.40	4.30	19.20	19.40	0.00	138	75	32	21	8	0	Q
14	Poyya	76.239	10.216	1.80	3.60	16.20	10.20	31.60	532	1324	221	421	288	568	AH
15	Vypin	76.25	9.99	1.70	2.60	21.50	25.30	0.00	72	33	13	5	1	0	Q
16	Varapuzha	76.256	10.077	1.70	0.85	13.50	17.00	0.00	120	35	6	21	8	0	Q
17	Mannam	76.258	10.146	3.20	5.80	6.80	34.00	0.00	650	425	325	185	45	0	Q
18	Alangad	76.266	10.114	2.40	5.80	2.10	33.20	0.00	1370	975	425	226	625	0	K
19	Chengamanad	76.287	10.136	1.50	7.50	2.40	26.40	0.00	2110	4350	552	242	780	0	AH
20	N.Kalamassery	76.290	10.072	2.30	6.80	3.50	35.50	0.00	876	1154	642	225	354	0	KH
21	Kalamassery	76.297	10.056	1.50	4.80	3.20	7.20	18.00	505	1235	224	875	14	86	KQ
22	Elur North	76.297	10.086	1.10	2.20	3.10	22.40	0.00	192	66	10	8	15	0	Q
23	Pathalam	76.297	10.086	2.50	10.00	19.50	30.00	15.00	740	570	340	220	185	435	H
24	Edappally	76.307	10.022	1.50	1.35	0.50	28.40	0.00	245	145	105	45	7	0	Q

(continued)

Table 4 (continued)

VES No.	Location	Lat	Long	h1 (m)	h2 (m)	h3 (m)	h4 (m)	h5 (m)	ρ1	ρ2	ρ3	ρ4	ρ5	ρ6	Curve type
25	Aluva	76.346	10.112	1.80	6.10	9.60	23.50	0.00	620	1110	1450	265	675	0	KH
26	Desom	76.352	10.127	1.40	2.80	6.90	15.20	62.10	1225	675	298	145	495	890	H
27	Alwaye	76.356	10.112	1.85	2.85	13.44	47.60	65.79	1113	1652	565	175	295	450	AH
28	Nedumbassery	76.374	10.170	1.30	2.30	5.44	5.00	0.00	843	252	67	434	654	0	HK
29	Karukutty	76.381	10.225	1.50	3.10	7.20	16.20	81.00	1210	722	275	144	225	725	H
30	Kaladi	76.389	10.167	2.40	4.20	2.10	21.40	32.40	1154	865	325	225	653	910	H
31	Chowara	76.389	10.126	1.50	0.45	3.91	30.20	0.00	1810	1082	152	288	292	0	HA
32	Chowara	76.389	10.126	1.60	6.40	21.30	0.00	0.00	382	152	342	1110	0	0	HA
33	Angamaly	76.389	10.195	1.40	0.70	15.60	36.00	0.00	1030	3475	375	150	285	0	H
34	Sreemoolanagaram	76.395	10.150	1.50	9.60	11.20	21.20	54.2	1450	954	325	165	725	1050	H
35	Aluva East	76.399	10.119	2.10	1.20	57.60	60.60	0.00	702	1050	210	442	650	0	AH
36	Vazhakkulam	76.400	10.131	1.90	9.60	10.50	48.00	0.00	802	696	272	1225	544	0	HA
37	Kanjur	76.410	10.150	2.10	6.80	1.50	14.60	45.30	976	1454	675	275	554	990	KH
38	Manjapra	76.453	10.213	1.60	6.34	2.10	5.50	33.00	170	340	105	423	54	238	AH
39	Manjapara	76.455	10.212	1.80	8.60	23.70	0.00	0.00	832	1050	543	365	679	0	K
40	Vallom	76.456	10.120	3.20	6.90	4.70	42.20	0.00	689	998	430	175	375	0	KH
41	Sulli	76.471	10.249	2.10	7.20	12.40	19.60	0.00	880	975	1354	245	775	0	KH
42	Edakuna	76.491	10.251	1.90	5.20	13.20	0.00	0.00	660	886	1230	1395	0	0	A
43	Odakkali	76.529	10.087	1.95	2.96	19.20	10.30	0.00	453	674	164	155	165	0	AH
44	Malayattur	76.533	10.183	1.90	1.52	18.30	0.00	0.00	537	925	168	720	0	0	AH
45	Illithodu	76.557	10.253	1.90	1.90	55.80	59.60	0.00	2100	210	1400	1800	1910	0	A
46	Panamkuzhi	76.605	10.233	2.30	4.80	12.30	15.80	0.00	1654	1155	875	454	1	0	H
47	Pindimana	76.660	10.128	1.50	0.45	3.20	20.30	18.20	510	1253	1860	625	1820	1980	A
48	Thattekad	76.666	10.132	1.30	1.69	3.20	6.80	26.00	402	805	688	174	420	985	AH
49	Thattekad	76.666	10.132	3.10	12.40	63.50	0.00	0.00	2600	1300	350	1050	0	0	H

Table 5 Comparison of the yield of the borewells with the geology of the study area

No.	Location	Taluk	Total depth drilled (m)	Thickness (m)				Rock type	Yield in lph.
				Soil	Laterite	Lithomargic clay	Weathered rock		
1	Kanjukuzhy	Udumbanchola	100.00	1.00	2.00	0.00	4.00	Charnockite	750.00
2	Udumbanchola	Udumbanchola	110.00	0.00	0.00	0.00	3.00	Migmatite	Dry
3	Karikkode Jn.	Udumbanchola	50.00	0.50	2.00	0.00	18.00	Charnockite	5000.00
4	Adimali	Devikolam	130.00	0.50	2.00	3.00	13.50	Charnockite	1000.00
5	Chithirapuram	Devikolam	45.00	0.00	0.00	0.00	12.30	Migmatite	Dry
6	Munnar	Devikolam	45.00	1.00	2.00	2.00	5.00	Migmatite	1000.00
7	Nedumkandom	Udumbanchola	30.00	1.00	0.00	0.00	8.50	Charnockite	1500.00
8	Kanjoor	Aluva	100.00	2.00	8.00	4.00	1.00	Charnockite	1500.00
9	Mambra	Aluva	66.00	1.00	4.00	0.00	1.00	Charnockite	8000.00
10	Kuruppampady	Kunnathunad	80.00	1.00	6.00	1.00	1.00	Gabbro	750.00
11	Mookkannur	Aluva	77.00	3.00	10.00	0.00	3.00	Charnockite	4000.00
12	Manjapra	Aluva	70.00	1.00	9.00	0.50	2.50	Charnockite	3000.00
13	Nedumbassery	Aluva	54.00	3.00	10.00	2.00	2.00	Charnockite	10000.00
14	Koovappady	Kunnathunad	80.00	0.00	0.00	0.00	3.50	Charnockite	5000.00
15	Odakkali	Kothamangalam	85.00	1.00	0.00	0.00	2.00	Charnockite	250.00

Source State Groundwater Department

indication of the massive nature of the subsurface formations. The resistivity of the subsurface layers is high where the surface and the subsurface runoff are also high and the infiltration is very low. It is also observed that the charnockites are good aquifers when compared to the other aquifers of the study area (Table 5).

2.4.1 Correlation of the Resistivity Results with Borewell Lithology

Lithologic information obtained from the borewell log of the area is used to substantiate the VES field curve. The Kerala State Groundwater Department drilled some borewells adjacent to some of the VES locations, and the lithology of the selected borewells was collected and analysed. All these borewells are located in the midland region of the basin. The details of the analysed layer resistivity and actual layer thickness in each location are given in Figs. 11a, 12, 13 and 14b. In general, the electrical resistivity field data shows a satisfactory match with each layer earth section.

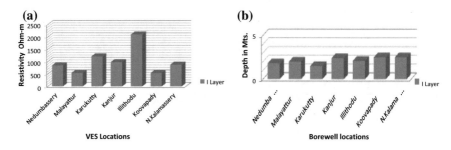

Fig. 11 **a** and **b** Comparison of **a** resistivity with the **b** borewell lithology of the first layer

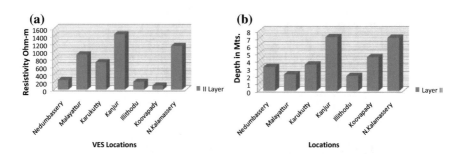

Fig. 12 **a** and **b** Comparison of **a** resistivity with the **b** borewell lithology of the second layer

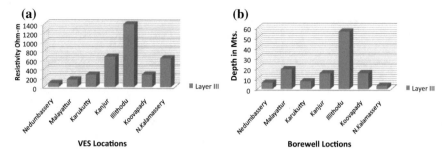

Fig. 13 **a** and **b** Comparison of **a** resistivity with the **b** borewell lithology of the third layer

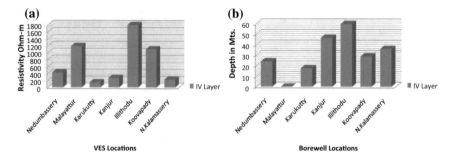

Fig. 14 **a** and **b** Comparison of **a** resistivity with the **b** borewell lithology of the fourth layer

2.5 Delineation of Coastal Aquifers of the Area

The subsurface formations with resistivity values generally below 4 Ωm are reflecting an aquifer possibly containing brine. The rock matrix, salinity and water saturation are the major factors controlling the resistivity of the formation. Moreover, the inter-phase of the freshwater–saline water qualities has been investigated, and it is seen that the saline water intrusion into the aquifers can be accurately mapped using surface DC resistivity method. The iso-apparent resistivity map shows that the contour patterns are almost parallel to the coast (north–south trend), with the increasing values towards east. The apparent resistivity along the eastern side is comparatively high when compared with the western margin of the area. It is also observed that the resistivity values decrease with increase in depth of investigation at the coastal area. The very low resistivity is an indication that the aquifers are of poor chemical quality. Groundwater samples were also collected at few stations of the study area during the pre-monsoon period and found that the quality of groundwater of the coastal plains is poor during the peak summer season due to the saline water intrusion into the coastal aquifers (Table 3). Based on the above observations, the freshwater saline water interface is demarcated (Fig. 15).

Fig. 15 Fresh–saline
groundwater interface of the
study area

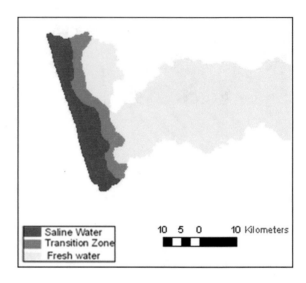

3   Results and Discussion

The present study deals with the investigation to assess the aquifer conditions and to
infer adequate knowledge of the groundwater resource to protect groundwater
supplies as a source of water for the study area and also the extent of saline water
intrusion. The vertical electrical soundings were carried out in the selected 49
locations, distributed approximately in a grid pattern to study the subsurface
hydrogeological conditions of the study area with current electrode spacing ranging
from 100 to 400 m. The apparent resistivities thus obtained for different electrode
spacings were recorded and plotted against current electrode spacing in double-log
graph sheets to get the field curves. The field curves are interpreted from curve
matching technique. From the interpretation of the resistivity curves, the different
layer resistivities and corresponding thickness and depth of the different subsurface
layers are identified, and dimension of the aquifer and type of bed rock are also
indicated. The boundary of the aquifers has been identified as zones with high,
moderate and poor yield potential, which were estimated for future development of
groundwater of the basin. Based on the analysis and interpretation of the VES data,
it is inferred that groundwater occurs under phreatic conditions in the laterite,
lithomargic clay, unconsolidated coastal sediments, weathered crystalline and under
semi-confined conditions in the fractured crystalline rocks in the area. The movement of groundwater in the area is mainly controlled by joints and other planes of
structural weakness, openings in rocks, interconnection of fractures and their extent.
From the analysis and Interpretation of the VES data, it is further inferred that the
distribution of the various subsurface layers is not uniform within the basin. The
sandy soil is the dominant phreatic aquifers of the coastal plains, and laterite and
lithomargic clay are the dominant phreatic aquifers of other lowland and the
midland regions.

4 Conclusion

The coastal zone in the Periyar River Basin is one of the most densely populated areas in the country. The natural balance between the freshwater and the saline water in this area is disturbed by groundwater withdrawals and other human activities that lower the groundwater level, reduce freshwater flow to coastal waters and ultimately cause saline water intrusion. The deterioration of water quality in the coastal zones of the area is due to saline water infiltration into the freshwater aquifer and has become a major concern. With the aim of providing valuable information on the hydro-geologic conditions of the aquifers, the subsurface lithology and delineating the groundwater salinity zones, vertical electrical resistivity soundings were carried out with electrode spacing varying between 1 and 400 m. The DC resistivity surveys revealed significant variations in subsurface resistivity. Along the low land regions, the electrical resistivity curves showed a dominant trend of decreasing resistivity with depth (thus increasing salinity). In general, the presence of four distinct resistivity zones was delineated (Table 4).

The river sand mining has caused lowering of groundwater in many parts of the Periyar River Basin, which in turn results in many severe environmental problems. These problems need integrated approaches to mitigate and manage to achieve sustainable development. The resistivity data adds valuable information about the presence of the conducting subsurface fracture and groundwater movement of the area. Based on the geophysical survey, the freshwater–saline water interface of the coastal aquifers has been demarcated around 10 km eastwards almost parallel to the coast (Fig. 15). The concomitant groundwater quality variations are evident as in Table 3.

Acknowledgements The authors are thankful to the Cochin University of Science and Technology for providing necessary research facilities to carry out the work, to the State Groundwater Department for providing the groundwater levels of the NHS and the State Mining and Geology Department for providing the details of the sand mining kadavus of the study area.

References

Davis SN, De Wiest RJM (1967) Hydrogeology. Wiley, New York, USA 463 p

GSI (1995) Geological and mineral map of Kerala. Geological Survey of India Publications, Calcutta

Paulose KV, Narayanaswami S (1968) The tertiaries of Kerala coast. Mem Geol Soc India 2:300–308

Rangarajan R, Athavale RN (2000) Annual replenishable ground water potential of India—an estimate based on injected tritium studies. J Hydrol 234(1–2):38–53

Soman K (1997) Geology of Kerala. Geological Society of India, Bangalore 280 p

USGS (2003) Ground-water depletion across the nation, US Geological Survey Fact Sheet 103-03

A Geo-Environmental Study on Groundwater Recharge Zones and Groundwater Management in the Guwahati Municipal Area

Neelkamal Das and Dulal C. Goswami

Abstract The city of Guwahati, in spite of being located on the bank of the mighty river Brahmaputra, has been facing the problem of scarcity of water, especially during the lean season. At present, only about 27% of the population has access to piped water supply primarily based on the Brahmaputra River, while the rest depends on sources such as dug wells, shallow tube wells and deep tube wells for their domestic water requirements. However, during the last few years, a rapid increase in the population of the city has resulted in unprecedented exploitation of the groundwater resource. This, coupled with a decrease in infiltration rate due to increased concretisation and other landuse changes, has considerably depleted the groundwater table in many parts of the city. The groundwater resource of an area basically has two components, viz., dynamic and static. The dynamic groundwater resource of the city is around 11 mcm, while the static groundwater resource down to the depth of 200 m is about 625 mcm. Although this indicates the presence of ample groundwater resource, yet, its utilisation should be done in a scientific and systematic manner with due emphasis on the prevailing hydrogeological conditions of the area. It is imperative that designing of wells should be based on the aquifer characteristics, hydrogeological setup and water requirement of the area. Moreover, it is observed that the city of Guwahati experiences an average annual rainfall of around 162 cm with about 110–115 rainy days per year. The city thus has enough potential for harvesting the rainwater it receives, instead of allowing it to flow untapped. Rainwater outlets can be connected to storage tanks or allowed to pass into gravel-filled trenches, pits, existing open wells and borewells. Initiatives can also be taken to reclaim and revive the various wetlands and ponds in and around the city, as these water bodies act as natural groundwater recharge zones.

Keywords Water table · Aquifers · Groundwater resource · Groundwater recharge

N. Das (✉) · D. C. Goswami
Department of Environmental Science, Gauhati University, Guwahati, India
e-mail: neelkamaldas@yahoo.com

D. C. Goswami
e-mail: dulalg@yahoo.com

© Springer International Publishing AG 2018
A. K. Sarma et al. (eds.), *Urban Ecology, Water Quality and Climate Change*,
Water Science and Technology Library 84,
https://doi.org/10.1007/978-3-319-74494-0_29

389

1   Introduction

The importance of groundwater for the existence of human society is unparallel. It is a major source of drinking water in both urban and rural areas, besides being a significant source of water for the agricultural and industrial sectors. The city of Guwahati, in spite of being located on the bank of the mighty Brahmaputra, depends heavily on the groundwater resource for its water requirements. About 69.90% of the households in the city use groundwater, while 27% depend on piped water supply primarily from the Brahmaputra River and about 1.30% on surface water obtained mainly from streams (Goswami et al. 2005). But since the last few years due to excessive growth of population and the subsequent over-exploitation of groundwater, the water table in many parts of the city has been showing a declining trend. Another root cause of domestic water crisis in this part of the world is the system of water rights under the common law in India which gives ownership of groundwater to the landowner, despite the fact that groundwater is a shared resource from a common pool of aquifers.

1.1   Study Area

The study area covers part of the city of Guwahati—mainly the Guwahati Municipal Corporation (GMC) area. Guwahati is the premier city and gateway of

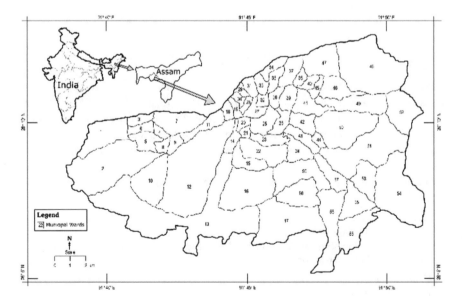

Fig. 1 Location map of Guwahati city showing its municipal ward boundaries

Northeast India. It is bounded by 26° 5′N to 26° 12′N latitudes and 91° 34′E to 91° 51′E longitudes. Situated on the southern bank of the river Brahmaputra, the southern and the eastern sides of the city are bounded by rows of hills which are extensions of the Khasi hills of Meghalaya (Pathak 2001). Geologically, Guwahati rests upon the typical Precambrian rock units which are overlain by young and recent alluvium. The river Bharalu dissects the main city for a length of about 9 km (Barman 1993). Small rivulets like *Panchadhara*, *Kalmoni* and *Khonanadi* flow through the fringe zone of the city in the south and south–west. The city is located at an elevation of about 54 m above mean sea level. The location map of Guwahati city with its municipal ward boundaries is shown in Fig. 1.

2 Methodology

The methodology applied in this study involves both empirical analysis and field surveys. Data for the study were collected from both primary and secondary sources. Primary data on various parameters pertaining to the study area based on field surveys using a questionnaire specially designed for the purpose and making field measurements on selected geohydrological parameters like groundwater level in wells and volume of surface water bodies have been collected from the field, satellite imagery (geocoded False Colour Composite (FCC) and black and white paper prints). The secondary data source comprises maps, Survey of India topographical sheets, statistics, published research papers and journals. These secondary data were mainly collected from various organisations and departments such as Central Groundwater Board (CGWB), Assam Remote Sensing Application Centre (ARSAC), Regional Meteorological Centre, National Institute of Hydrology (NIH) and Directorate of Geology & Mining (DGM).

The Survey of India (SOI) topographical sheets no 78 N/12, 78 N/16 on 1: 50,000 scale were used to prepare the base map and to acquire various information about the area. Though the SOI topographical sheet is in 1:50,000 scale, the required portion is enlarged up to 1:25,000 scale. All the thematic details were transferred to the base map from SOI topographical sheet. Preliminary interpretation of the satellite imagery was done, and landuse/landcover map and geomorphological were drawn from the imagery. The IRS imagery of 1997 on 1:50,000 scale was also used. In addition to these the hydrogeomorphological map, groundwater potential map and panel diagram of the study area were procured from various relevant organisations. The maps were then analysed and integrated using GIS techniques. To acquire the groundwater level in different parts of the study area, water levels of dug wells were taken, and location of the wells was collected using GPS.

3 Analysis and Results

From the analyses of the data and information, it has been observed that almost the entire western part of the study area is occupied by the alluvial plain (both older alluvial and younger alluvial plains). Lithologically, these formations comprise intercalated beds of sand, silt and clay in varying proportions (Goswami and Goswami 1996; Goswami et al. 1994). However, the thickness of the upper clay layer is only a few metres, while that of the sandy layer varies from 40 to 50 m. Hence, the infiltration of surface water is higher in this zone compared to others. This zone, which includes the Deepor Beel, Azra, Kahikuchi and Dharapur areas surrounding the LGBI airport, therefore forms a high recharge zone.

In the eastern part of the study area, there is a fault trending NNE-SSW (Fig. 2) which gives rise to the formation of Silsako and Hahchora Beels. Another prominent fault is seen in the central part of the study area trending along NNE-SSW direction. This lies in the corridor between the Fatasil hills and the Narakasur hills. These faults are overlain by weathered rocks or alluvial plain. This zone, which includes areas such as Garchuk, Betkuchi, Fatasil-Ambari, Birkuchi, the Silsako and Hahchora Beels, has a negligible thickness of clay layer in their soil profile, while the thickness of the sandy layer varies from 50 to 60 m, thus resulting in easy and convenient infiltration of surface water through them. This zone therefore serves as a high recharge zone.

The denudational hills, which include the Narengi hill, Japorigog hill, Narakasur hill, Sonaigul hill, Fatasil hill, Silapahar hill and the Kamakhya hill, consist mainly of highly metamorphosed quartzofeldspathic gneisses. The rocks are highly

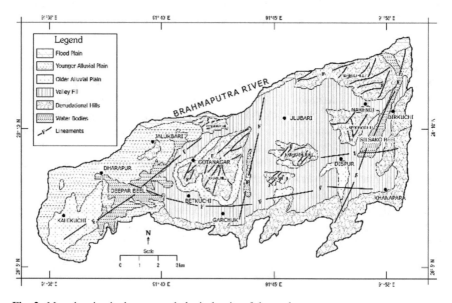

Fig. 2 Map showing hydrogeomorphological units of the study area

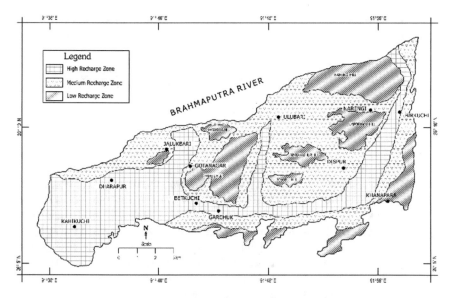

Fig. 3 Map showing potential groundwater recharge zones of the study area

dissected and deeply weathered (Goswami and Goswami 1996). These areas act mainly as run-off zone. As the primary porosity and permeability are low in this unit because of the presence of hard rocks, the amount of recharge is also very low.

Apart from the above-mentioned areas, there are few other valley-fill areas such as Chandmari, Dispur, Ulubari, Rupnagar, Mackhowa, Bhutnath and Maligaon that comprise soils with the layer varying between 40 and 60 m, yet the surface water cannot penetrate through them into the underground aquifer system. This is because these sandy layers are overlain by a thick layer of clay having thickness of the order of 10–20 m, thus preventing the percolation of water through such impermeable layer. Hence, these areas act as medium to low recharge zones. The potential groundwater recharge zones of the study area are shown in Fig. 3.

3.1 Development and Management of Groundwater Resource in the Study Area

As per hydrogeological studies conducted by the Central Ground Water Board during 2004–2006 (CGWB 2007), the net annual dynamic groundwater availability in Guwahati has been estimated to be in the tune of 11,045.31 ha-m or 11 mcm with a static groundwater resource of 625,152 ha-m or 625 mcm, up to the depth of 200 m. The current annual gross groundwater draft for all uses has been estimated to be around 2806 ha-m.

Based on the long-term groundwater trend for a minimum period of 10 years, the stage of groundwater development (in %) can be expressed as (CGWB 2007).

$$\text{Stage}(\%) = \frac{\text{Gross Groundwater Draft}}{\text{Net Annual Groundwater Availability}} \times 100$$
$$\text{Stage}(\%) = \frac{2806}{11045.31} \times 100$$
$$\text{Stage}(\%) = 25.40\%$$

Although this indicates the presence of ample groundwater resource and the stage of development is also within manageable proportions, it should however be taken into consideration that the utilisation of groundwater resource should be done in a more scientific and systematic manner, with due emphasis on the prevailing hydrogeological conditions of the area. Groundwater development and management is the key for sustainable upliftment of a particular area. Proper management practices will ensure effective utilisation of the groundwater resource. It is imperative that designing of wells should be based primarily on the aquifer characteristics, hydrogeological setup and water requirement.

3.2 Rainfall Regime of the Study Area

Rainfall is highly seasonal in the study area. It comes mainly under the regime of the south–west monsoon. The city of Guwahati experiences an average annual rainfall of about 162 cm, which is considerably less than the average annual rainfall of about 220 cm for the state of Assam as a whole. The average number of rainy days per year in the city is about 110–115 days. The average mean rainfall in the study area is shown in the form of bar diagram in Fig. 4.

The ratio of groundwater recharge to rainfall in hard rock terrain is customarily between 9 and 13% (Limaye 2003). In the study area, the denudational hills, which include Narengi hills, Japorigog hills, Fatasil hills, Silapahar hills, Kamakhya hill,

Fig. 4 Mean monthly rainfall in Guwahati city

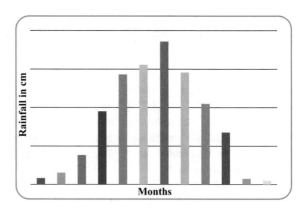

Narakasur hill and Sonaigul hill, are composed of hard rugged topography, due to which most of the run-off flows over the surface. The basements of these denudational hills are composed of hard rocks like granite and gneisses (Goswami and Goswami 1996). In hand specimen, these types of rocks may not have any connected porosity but on macroscopic scale, the storage coefficient is about 0.03–0.3% (Limaye 2003). Only the areas with thick forest cover and lineaments, which include the foothills along the flanks of the Hengrabari–Japorigog hills and the Fatasil hill, have a relatively good rate of recharge into underground aquifers.

In the alluvial plain, the ratio of recharge to rainfall is between 15 and 25% depending upon the nature of the topsoil (Limaye 2003). In the study area, alluvial plain includes places like Gotanagar, Betkuchi, Garchuk, Birkuchi and Garigaon. These areas consist mainly of unconsolidated materials like sand, gravel, pebble, silt and clay (Goswami and Goswami 1996). Hence, the rainwater can easily percolate through these materials and infiltrate into the underground aquifers. Thus, these areas can be considered as most favourable sites for recharge.

Moreover, it has been observed that the post-monsoon high water levels in the wells gradually deplete to a lower position towards the end of the summer. In the study area, the maximum rainfall occurs during the monsoon period and the minimum during the post-monsoon period, as evident from Fig. 5. Some of the dug wells dry up during the lean period (post-monsoon and pre-monsoon) of the year as the groundwater draft exceeds recharge of the underground aquifers. This is mainly because there are no structures to intercept and retain the excess rainfall that occurs during the monsoon period, and huge amount of rainwater is lost as run-off. It therefore becomes imperative that long-term measures be taken to entrap and retain rainwater during the monsoon period to utilise them in the lean periods.

The present study also reveals that areas having good permeability with high water holding capacity such as the alluviums are excellent zones for groundwater recharge. In the study area, a significant portion is occupied by the denudational hills, where the primary porosity and permeability are very low due to the presence

Fig. 5 Seasonal distribution of rainfall in Guwahati city

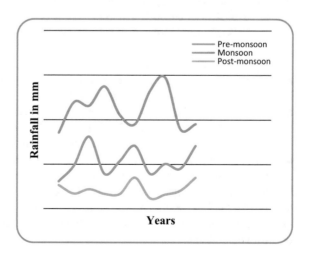

of hard rocks; hence, surface water cannot penetrate through these hard rocks and replenish the underground aquifers. This unit mainly acts as a high run-off zone. Hence, artificial structures can be constructed in these areas to entrap surface run-off water and facilitate groundwater recharge in the foothills and valley-fill areas. In general, the western part of the study area and some isolated valley-fill areas scattered in the eastern and southern part are highly favourable for groundwater recharge.

However, it has been observed that a significant portion of the potential recharge areas have already been utilised for settlement and infrastructure development purposes. Construction of paved and concrete roads and other developmental structures have covered most of the naturally occurring recharge zones and have converted them into run-off zones. Hence, awareness should be created among the general public about the importance of these potential recharge zones. Appropriate plans and programmes shall have to be formulated to reclaim and revive these areas that have already been deteriorated during the course of the time. The areas in the vicinity of the Deepor Beel, Silsako Beel and Hahchora Beel should be protected from unauthorised encroachment so that the water retention capacity of these water bodies is enhanced.

4   Observation and Discussion

The city of Guwahati is underlain by geological formation of Precambrian age followed by Quaternary formation. The Precambrians are usually represented by inselbergs, denudational hills, granite, schists, amphibolites, pegmatites and basic/acidic intrusions. The Quaternary formation is represented by sands of various grades, pebbles, cobbles, gravels, clay and silts. In these formations, groundwater is stored in various conditions. In the foothills of Fatasil hill, Narengi hill and Kamakhya hills containing Quaternary deposits, water is stored in porous fractures, and in case of hard consolidated foothills, it is stored in the fissures, joints and structurally weak formations. The hydrogeomorphological map of the study area is shown in Fig. 2.

It has been observed that the groundwater level in the study area varies according to local topographic conditions, i.e. in areas close to the undulating inselbergs/residual hills, the water level is deeper, viz., Basistha, Borbari, Panjabari, Mathgharia and Birubari areas, and in areas situated in relatively flat alluvial plains and valley-fill areas, the water level is generally shallower, viz., Rukminigaon, Wireless, Kacharibasti, Gotanagar, and Jalukbari areas. In valley-fill areas, depth to water level is variable depending on the thickness of the residuum. Overall depth to water level gradually reduces from elevated eastern and south-eastern areas towards the flat-lying alluvial plains in the west. The depth to water level at Nepali Mandir area is found to be deep though it is situated on valley-fill deposits. This may be due to the high amount of groundwater withdrawal in the area for both commercial and non-commercial purpose.

Hydrogeological studies revealed the presence of groundwater just under water table conditions in case of shallow aquifers; however, in case of deeper aquifers, it is available within the semi-confined to confined conditions. In the loose unconsolidated formations, depth to water in the open dug wells ranges from 2 to 4 m below ground level during pre-monsoon period. Dug wells located in the foothills zone, however, show deeper groundwater level ranging between 5 and 10 m below ground level during pre-monsoon period. Shallow tube wells constructed in the loose formation down to 30 m by Public Health Engineering Department yield around 2000–3000 litres per hour (CGWB 2007), and the well yield shows consistent behaviour throughout the year. Normal dug wells constructed in the pediment formation covering the valley parts of the city down to maximum depth of 15 m store a good quantity of water irrespective of seasonal change and can be pumped at the rate of 10 m^3/day (CGWB 2007). However, normal dug wells constructed in the weathered formation of the hill areas down to maximum depth of 25 m having water level around 5–7 m during monsoon period and more than 10 m during lean period show the erratic behaviour of storage depending upon the structural pattern of the country rock and seasonal rainfall. Deep tube wells constructed in the valley portion down to maximum depth of 200 m in the western parts of the city show a very good discharge of about 70–100 m^3/h for nominal drawdown. In the central part of the city, deep tube wells down to maximum depth of 100 m give yield up to 80 m^3/h. In the eastern and southern parts of the city near to hillocks, the discharge of deep tube wells down to maximum depth of 80 m give yields up to 30 m^3/h for considerable drawdown. The hard rocks found in the hillocks also are potential sites for construction of borewells. However, it depends totally on degree of structural weakness formed due to tectonic events that would have occurred in the ancient periods. Borewells constructed down to maximum depth of 200 m in the hard rocks have been found to be effective for groundwater development. Fractures, fissures and joints developed during tectonic events act as good water repository in these hard rocks. Maximum yield of such wells at particularly in the Beltola and Odalbakra areas reveals that water can be drawn at the rate of 80 m^3/h for 6 to 8 hours daily (Devi 1998; Konwar 2004).

The natural drainage system of the city is not uniform and is very much influenced by the landscape pattern of the city. Major streams and rivers originate from the southern and north-eastern highlands and flow through natural slope gradient. These include *Amchang, Bahini, Basistha Bahini, Bharalu, Khanajan* and *Mora Bharalu*. Wetlands are located in parts of the central, south-eastern and western regions of the city. These are primarily depressed valley areas with remnants of palaeochannels of Brahmaputra being traced through various studies and surveys. From Fig. 6, it is observed that natural drainage exists in at least 30 wards (there were 60 wards under GMC during the year 2012) of the city with the highest density of about 26 km^2 in ward number 46.

Moreover, it is evident from Fig. 7 that at least eight wards of the city, viz., 1, 2, 13, 37, 46, 47, 51 and 52, have more than 5 km^2 of its area under wetlands, and these wetlands facilitate in maintaining the moisture content of soil in its vicinity and also in elevating the water table of the surrounding area.

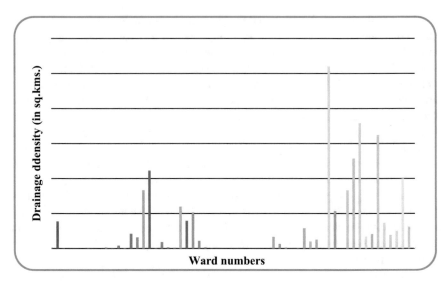

Fig. 6 The natural drainage density in different wards of Guwahati city

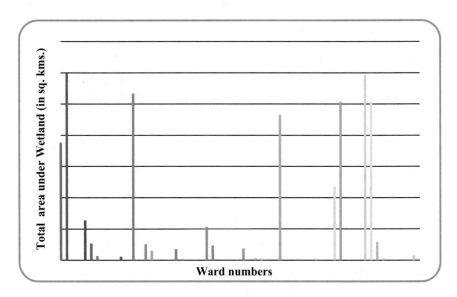

Fig. 7 Total area under wetland in different wards of Guwahati city

5 Conclusion

The ever-increasing rate of population growth coupled with an expanding urban sprawl has resulted in overdependence on groundwater for meeting the daily water requirements of a major chunk of the city population. This has led to a rapid decline of water table due to over-exploitation of groundwater in recent years. The irony of the situation is that in spite of being located on the bank of one of the world's largest rivers—the Brahmaputra, most of the inhabitants of Guwahati city face the problem of acute shortage of water. The rolling topography of the city marked by several hills and hillocks provides a perfect setting for locating large-size storage reservoirs and tanks which can serve different parts of the city through a well-organised network of water distribution system.

It is also worth mentioning that, in a place like Guwahati, which experiences high intensity but temporally variable rainfall, rainwater can be harvested through rooftop harvesting and stored in underground tanks specifically constructed for this purpose, which can then be used during lean periods. This practice is almost non-existent in the city, with the exception of a limited number of households, who tap the rainwater in vessels and containers primarily for domestic use. A survey conducted as a part of this study revealed that only 7% of the households surveyed harvest rainwater and of these nearly 97% use the water for washing and gardening purpose and the rest 3% use the water for cooking and drinking. Normally in Guwahati, rainwater is left to flow into the drains. As such, rainwater outlets could be connected through a pipe to a storage tank and allowed to pass into gravel-filled trenches, pits or existing open wells or borewells. This method is simple, less expensive and if followed at micro levels by all households, will help in the improvement of the groundwater capacity of the area very effectively. In addition, community rainwater collection schemes can be adopted in various apartment complexes of the city, and rainwater from the terraces of the buildings may be collected and led into large underground tanks or to nearby ponds. The water can also be diverted to ponds located at hydrologically favourable zones, and production well can be constructed on downstream side of the pond to tap the water for drinking and other domestic purposes after necessary filtration. Deep pits can also be dug away from the building foundation where the soil is more pervious. These pits can be filled with pebbles in the first layer, followed by gravel in the next and then sand for better percolation of the surface water into the underground aquifers. The depleted aquifers can also be utilised to store the rainwater. Moreover, initiatives can also be taken to activate and revive the various wetlands and ponds in and around the city, as these water bodies serve as one of the most significant natural groundwater recharge zones and also facilitate in maintaining the microclimate of the surrounding area.

References

Barman DK (1993) Remote sensing contribution for augmentation of urban water supply in Greater Guwahati area. In: National symposium on remote sensing applications for resource management with special emphasis on NE Region, pp 490–496

Central Ground Water Board (CGWB) (2007) Hydrogeological framework and impact of urbanization on groundwater regime in Greater Guwahati Area, Assam. Central Groundwater Board (CGWB), NER, Guwahati, India

Devi S (1998) Hydrogeological condition and ground water development potentialities in Greater Guwahati with special emphasis on drinking water requirement; AAP 1997–98. Unpublished CGWB Report; NER

Goswami DC, Goswami ID (1996) Estimating groundwater potentiality in and around Guwahati using remote sensing technique. In: National symposium on remote sensing for natural resource management with special emphasis on water management, pp 176–178

Goswami DC, Goswami ID, Duarah BP, Deka PP (1994) Geomorphological mapping of Assam using remote sensing techniques. In: 16th Indian Geographers Congress

Goswami DC, Kalita NR, Kalita S (2005) Pattern of availability and use of domestic water in Guwahati city. In: Symposium on 150 years of Guwahati under Public Administration—a critical assessment of its development, pp 71–80

Konwar M (2004) Integrated ground water development and management studies in Kamrup district, Assam; AAP 2002–2003. Unpublished CGWB Report; NER; Technical Series D

Limaye SD (2003) Some aspects of sustainable development of groundwater in India, IGC's 8th Prof. Jhingran Memorial Lecture, Indian Science Congress, Bangalore

Pathak B (2001) Study of some geophysical properties of the basement and its overlying sediments of the Greater Guwahati area, district Kamrup, Assam. Unpublished Doctoral Thesis, Gauhati University

Scope and Challenges of Sustainable Drinking Water Management in Assam

Runti Choudhury and C. Mahanta

Abstract Results of the Arsenic Screening and Surveillance Program in Assam revealed 29% of a total 56,180 public sources analyzed to be above the WHO safety guideline of 10 µg/L, exposing an unguarded population of 8,47,064 to the risk of arsenic contamination and resultant health hazard. Water security and water safety have thus emerged as a major concern in Assam. With surface water resource often contaminated with bacteriological, inorganic, and organic contaminants, groundwater continues to be the major source of drinking water especially in rural Assam. Arsenic contamination of groundwater sources and unreliability of surface water sources have led to an increased concern regarding safe drinking water supply in affected areas. Results of the Arsenic Screening and Surveillance Program demonstrated that the Titabor area lying along the southern bank of the Brahmaputra River in Assam is one of the arsenic hot spots in the region. This led to the formulation of the largest piped water supply scheme in the state to provide arsenic-free safe water to a design population of 1,90,000. Surface water from two tributaries of the Brahmaputra River, viz., Dhansiri and Dayong, is used to supply arsenic-free treated safe water. While adequate surface water availability in the vicinity of such high arsenic-contaminated aquifers has provided a viable alternative, this is infeasible in areas with discreet contaminated sources with scarce surface water sources nearby. Thus, an integrated approach based on groundwater and surface water can be a strategy; the choice often depends on surface water availability and sustainability of deeper safe aquifers subject to economic viability. Composite solutions like integration of rainwater harvesting structures, water recharging, delineation and safeguarding safe aquifers, reuse and recycling, water saving technologies, and smart systems are part of a composite option necessitated for addressing the gaps and inadequacies in ensuring safe drinking water in the region.

Keywords Water security · Contamination · Groundwater · Surface water

R. Choudhury (✉) · C. Mahanta
Department of Civil Engineering, IIT Guwahati, Guwahati, India
e-mail: runti.choudhury@gmail.com

C. Mahanta
e-mail: mahantaiit@gmail.com

© Springer International Publishing AG 2018
A. K. Sarma et al. (eds.), *Urban Ecology, Water Quality and Climate Change*,
Water Science and Technology Library 84,
https://doi.org/10.1007/978-3-319-74494-0_30

1 Introduction

Assam, the gateway to the northeastern region of India, lies between graticules 89° 42' to 95° 16'E longitude and 24° 08' to 28° 09'N latitudes. With an approximate population of 26.64 million population (Census of India 2001), the state covers a total geographic area of 78,438 km^2 which is 2.4% of total geographical area of India. The state consists of a total of 23 districts and 219 blocks, each block being further divided into gaon panchayats (GP), villages, and habitations. The greater part of Assam is within the Brahmaputra Valley, while the southernmost part lies in the Barak Valley, separated from the Brahmaputra Valley by the Central Assam range. The inselbergs (isolated hill, knob, ridge, or small mountain that rises abruptly from a plain or gently sloping surrounding) situated in the central districts south of the Brahmaputra are distinct features of the area. The Brahmaputra Valley is about 800 km long and 130 km wide. The Brahmaputra River along with its numerous tributaries is the prime sources of surface water resource in the state. Numerous other surface water resources in the form of ponds, wetlands, etc., also contribute to the surface water repository of the state.

About 80% of the rural population depend on groundwater resources for meeting their water needs. However, with the recent discovery of fluoride, arsenic, and other heavy metal contaminants along with bacteriological contamination in groundwater of the region, there has been an increased concern regarding the supply of safe and sustainable drinking water in the region. The first reported case of fluoride contamination in Assam was in May 1999 in the Tekelangjun area of Karbi Anglong, where fluoride concentrations were found to be as high as 5–23 mg/L (SOS 2000) which is much beyond the BIS permissible limit of 1.5 mg/L. Gradually with the advent and initiatives of more rigorous water quality programs, the highly fluoride-affected areas of Assam, viz., Kamrup, Nagaon, Karimganj, Golaghat, and Karbi Anglong, were explored, and the occurrence of fluorosis in these areas has been reported in different studies (Kotoky et al. 2010; Sushella 2007; Chakravarti et al. 2000) (Tables 1 and 2).

Similarly, a relatively recent Arsenic Screening and Surveillance Program in Assam conducted at IIT Guwahati for the state of Assam revealed 29% of a total 56,180 sources analyzed to have concentrations above the WHO safety guidelines of 10 µg/L, affecting an unguarded population of 8,47,064 to the risk of arsenic contamination (Table 3, Fig. 1). 794 arsenic-affected sources are located in different schools in the affected areas causing a mass threat.

Table 1 Concentration ranges of fluoride in different parts of the state of Assam

Author	Region	Concentration range of arsenic in mg/L
Paul (2000)	Karbi Anglong	5–23
Das et al. (2007)	Guwahati	0.18–6.88
Borah et al. (2010)	Darrang	0.01–0.98

Table 2 District-wise fluoride-affected blocks in Assam

S. No.	District	Total no. of blocks	No. of affected blocks	Name of affected block (fluoride conc. above 1.5 mg/L)
1	Karimganj	7	1	Loairpowa
2	Golaghat	8	1	Golaghat East
3	Nagaon	19	5	Jogijan, Binnakandi, Paschim Kaliabor, Kaliabor, Kathiatoli
4	Karbi Anglong	11	5	Howraghat, Samelangso, Longsompi, Lumbazong, Rongkhong
5	Kamrup	18	2	Bejera, Chandrapur (GMC Area)

Source Singh et al. (2011)

Table 3 Concentration ranges of arsenic in different parts of the state of Assam

Author	Region		Concentration range of arsenic in µg/L
Singh (2004)	Nagaon	Southern part of the Brahmaputra River in Assam, India	481–112
	Jorhat		194–657
	Golaghat		100–200
	Lakhimpur	Northern part of the Brahmaputra River in Assam, India	50–550
	Nalbari		100–422
	Dhubri		100–200
	Darrang		0–200
	Barpeta		100–200
	Dhemaji		100–200
Enmark and Nordborg (2007)	Darrang and Bongaigaon	Northern part of the Brahmaputra River in Assam, India	5–606
Chetia et al. (2008)	Golaghat	Southern part of the Brahmaputra River in Assam, India	1–73
Mahanta et al. (2009)	All over Assam		0–above 300

Bacterial contamination of water is another major cause of concern regarding safe drinking water in the region. The major pathogenic organisms responsible for waterborne diseases in the region are bacteria (*E. coli*, *Shigella*, and *V. cholera*), viruses (*Hepatitis A*, *Polio Virus*, and *Rota Virus*), and parasites (*E. histolytica*, *Giardia*, and *Hookworm*) (Wateraid 2001). Improper and inadequate sanitary habits are the primary cause of bacterial contamination in any region.

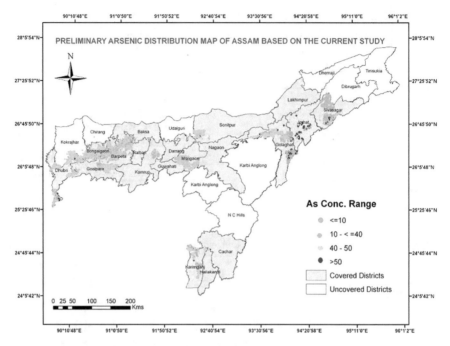

Fig. 1 Arsenic contamination map of Assam

2   Initiatives Undertaken for Arsenic Mitigation in the Brahmaputra Floodplains in Assam

With the discovery of fluoride and arsenic in groundwater of the region, mitigation initiatives to overcome the health implications of consumption of contaminated groundwater have been undertaken. Supply of surface water through piped networks to the communities in the affected areas has emerged as major initiative.

In the arsenic-affected hot spot areas delineated through the Arsenic Screening and Surveillance Program, major projects for using surface water and distribution through piped networks are being initiated. The Greater Titabor Piped Water Supply Scheme is one such initiative where surface water from two major rivers—Dhansiri and Dayong Rivers is used to supply arsenic-free safe water to a design population of 1,90,000 in the affected areas in Titabor in the Jorhat District in Assam, one of the arsenic hot spot areas in the region. Due to huge availability of surface water resources, such initiative for arsenic mitigation could prove to be a viable option in the Brahmaputra floodplain.

For fluoride mitigation, spring water in some of the affected areas is being used to supply safe waters to the affected communities in the region. Mitigation measures, notwithstanding cost factors, are being undertaken in the affected areas of arsenic and fluoride contaminated hot spot areas in the region. However, with issues of climate change, unpredictability of surface water sources, poor operation and maintenance (O&M) of piped schemes, lack of ownership, and acceptability, the feasibility of such alternative options with huge cost implications is greatly jeopardized.

3 Challenges to Provision of Safe Water

3.1 Unpredictability of Water Sources

For the state of Assam, rich in water resources, provision of safe waters through alternative surface water sources has been a viable option in areas with elevated groundwater contamination due to fluoride, arsenic, and other heavy metal contamination. While adequate surface water availability in the vicinity of such high arsenic-contaminated aquifers has provided a viable alternative, this has become infeasible in areas with discreet contaminated sources with limited surface water sources nearby.

Further, there seems to be an ambiguity regarding the availability of these surface water sources. Unpredictability and flashiness of some of the tributaries of Brahmaputra coupled with contamination of the surface sources due to industrial and agricultural effluents and discharge of sewage are some of the serious issues concerning the surface water sources. Cost implications of treating such contaminated surface waters sources and ambitious piped water supply schemes are high.

3.2 Poor O&M of Some Existing Water Sources

Mitigation strategy in areas of high incidence of arsenic, fluoride, and other such contaminants can entail removal but because of the costs and operational complexities of the technologies required, the alternative safe water source is always a better option. However, several factors are to be considered prior to the provision of safe water sources. Piped water supply schemes with huge capital investments are at many times found to be de nonfunctional primarily due to the issues of O&M of such schemes. Poor O&M of piped water supply schemes is one of the primary causes of concern regarding the sustainability of alternate water sources. Although initiatives for community participation in the O&M of water supply schemes are underway, failure of schemes due to poor O&M cannot be overruled. Even where supply quantity is reasonable, inadequate maintenance creates waste and contamination through leakage and unreliable flow. Apathy or unwillingness on the part of the communities to handle operation and maintenance is thus a major cause hindering supply of safe drinking water. The systems are either poorly planned and designed or operated without adequate maintenance, which means that the existing services are often of poor quality (Khatri and Vairavamoorthy 2007a, b). The high rate of water losses from distribution systems is also hindering the successful implementation of piped water supply. Many studies revealed that water losses in cities of developing countries are at levels of between 40 and 60% of water supplied (Arlosoroff 1999). Further, gaps and leaks in some sections of the water distribution system and intermittent water supply service allow the system to depressurize. The depressurization and gaps in water service lines may in turn again lead to intrusion of contaminants into the system (Hubbard et al. 2011), thereby minimizing the service life of the piped water system.

3.3   Community Ownership and Acceptability

Lack of community ownership and acceptability of piped water supply schemes are other issues inhibiting the success of such alternate schemes. Unless the local communities in the affected areas are made aware of the importance of water security and source sustainability, any initiatives on the part of the government would be meaningless. Lack of encouraging consultation with communities is perhaps one of the key factors hindering community ownership and acceptability of alternate safe water sources.

4   Water Security and Alternate Water Sources Composite Solutions for Water Security

Water Security means that every rural person has adequate safe water for drinking, cooking, and other domestic basic needs on a sustainable basis. Safe water should be readily and conveniently accessible at all times and in all situations (WSP 2000). However, most often initiatives for water security and safety are tailored with one or the other issues as mentioned above. Therefore, composite solutions like integration of rainwater harvesting structures, water recharging, delineation and safeguarding safe aquifers, reuse and recycling, water saving technologies, and smart systems are part of a composite option to minimize the gap arising and inadequacies in ensuring safe drinking water in the region.

4.1   Water Harvesting

With approximately 75–90% of rainfall concentrated in the summer months, June to September (Mooley and Parthasarathy 1984), India has a long history of rainwater harvesting (RWH) (Sakthivadivel 2007; Shah 2001). However, given the current threat of groundwater depletion and the potential of increasing recharge, the implementation and planning of RWH continues to grow in India (Agarwal and Narain 1997).

The northeastern region consists some of the world's highest rainfall areas where annual rainfall varies from 2000 to 4000 mm (Sarmah 2011). This holds unique potential for recharge and harvesting in the region. Rainwater harvesting is thus a concept which needs immediate development for minimizing the water scarcity in the region. Rooftop harvesting at households, schools, apartment buildings, etc., can be made for capturing rainfall and subsequently be stored for future use with proper treatment facilities. There is also a need to establish water harvesting projects at various places based on certain criteria that take into consideration the socioeconomic and the physical characteristics of the targeted area.

In terms of water harvesting, Mizoram tops the list in northeast India, with the rigorous promotion of rooftop rainwater harvesting structures. With most buildings in the state having sloping roofs made of corrugated galvanized iron (CGI) sheets conducive to rainwater harvesting, the state has made tremendous achievement in terms of harvesting the ample amount of rainfall that favors the state. Traditional water harvesting systems of the region also need to explored and promoted in the region (Table 3).

Restoration and development of water storage tanks and reservoirs in the region could be another viable option for harvesting water. With further treatment of the harvested water, provisions for supply can be made to facilitate the water demand at its vicinity.

4.2 Water Recycling

Water recycling is another important aspect that can be explored and developed in the region. Recycling and reuse of water is an increasingly important element of sustainable water management strategies in both water-stressed and water-surplus regions (Messalem 2006). Although at its nascent stage, water recycling and reuse is a vital option which can minimize the demand-supply gap. With very little scopes for water recycling in the present state, the concept needs to be promoted. Recycled wastewater can be used for purposes like gardening, car washing, etc. However, the key to successful implementation of reuse is user acceptance (Jeffrey and Jefferson 2001) and the assessment of risk. Therefore, with the greater issue of social acceptability of the concept in the region, people should realize the importance of the depleting water resource and the need to conserve the said resource in all possible forms.

4.3 Bioremediation of Wetlands

With wetlands characterizing most parts of the state, bioremediation of these wetlands could provide a possible solution for meeting the water demand in the region. Wetlands can reclaim water and provide a natural, free service in water purification, combating waterborne disease. Use of constructed wetlands for wastewater treatment is already in use in several countries around the world. Wetlands play an important role in influencing the flow and quality of water. Bioremediation is based on the natural biological and physical principles, which have been responsible for the deposition of ore bodies. Microbes and microorganisms have favored the accumulation and deposition of metals and minerals by fixing them in the soils from their dissolved states or by incorporating them into cell structures (Internet 2010). Thus, bioremediation could be utilized to effectively correct existing disparities and eliminate the need for chemical treatment (Davidson 1993).

4.4 Groundwater Recharge

Artificial recharge of groundwater basins is becoming important in groundwater management and primarily where conjunctive use of surface water and groundwater resources is considered in the context of integrated water resources management (Asano and Cotruvo 2004). The marshy lands including the static water bodies in the region can be used as groundwater recharge zones with a proper blend of technological interventions. This would also include restoration of the wetlands to an extent such that water can be made available from the restored wetlands at least to the communities residing near the wetland areas.

4.5 River Bed Filtration Technique

Bank filtration is the process where the groundwater table is lowered below that of an adjoining surface water table, whereby the difference in water levels causes the surface water to infiltrate through the permeable riverbed and bank or lakebed into the aquifer (Sandhu et al. 2010). In case the surface water body is a river, bank filtration is specifically called river bed filtration (RBF). The aquifer serves as a natural mechanical filter and with the natural attenuation processes of filtration, sorption, acid–base reaction, oxidation, reduction, hydrolysis, biochemical reactions, etc., in play, the potential contaminants are filtered (Dash et al. 2010). The process of bank filtration is different from direct surface water abstraction in that it eliminates bacteria, biodegradable compounds, parasites, particles, suspended solids, and viruses; partly eliminates adsorbable compounds; and equilibrates temperature changes and concentrations of dissolved constituents in the bank filtrate (Hiscock and Grischek 2002).

River bed filtration techniques are already being used in several cities in India as a means for providing safe drinking water. Some of the cities where RBF technologies are being used are Dehradun (Bandal River), Srinagar (Alaknanda River), Nainital (Nainital Lake), Rishikesh (Ganga River), Haridwar (Ganga River), Delhi (Yamuna River), Mathura (Yamuna River), Patna (Ganga River), Ahmedabad (Sabarmati River), Vadodara (Mahi River), and Medinipur (Kangsabati River) (Sandhu et al. 2010). Most if not all of these rivers from which water is being used for filtration are highly polluted rivers with higher concentrations of organic contaminants. Removal of high organic contaminants requires high doses of chlorination prior to flocculation, thus increasing the risk of generation of carcinogenic compounds. As such, bank filtration proves to be advantageous as a pretreatment prior to the use of reduced doses of chlorination followed by flocculation (Sandhu et al. 2010). The adoption of river bed filtration technique in the above listed cities has proved to be effective in providing drinking water in all these cities. Further, in areas with limited resource availability, river bank filtration technique can be used as a feasible pretreatment step in safe drinking water production and in some cases serve as a final step in providing potable water (Dash et al. 2010).

4.6 Soil Aquifer Treatment (SAT)

Soil aquifer treatment is another process whereby appropriately pretreated wastewater is allowed to infiltrate through the aerated unsaturated soil zone where it undergoes purification through unit operations and processes, viz., filtration, adsorption, chemical processes, and biodegradation. This treated water after reaching the water table the soil treated wastewater further moves laterally for some distance through the underlying saturated zone (aquifer) where it receives additional purification by dispersion and dilution. With the involvement of both soil and aquifer in the renovation process, the treatment process is largely known as the soil aquifer treatment (SAT) system (Nema et al. 2001). The technology still at its nascent stage in India as a whole and Assam in particular can perhaps be explored, and feasibility be ascertained prior to taking up the technology on a larger scale (Table 4).

4.7 Smart Technologies

Use of ICT (Information and Communication Technology) tools for efficient data management and information can help in proper monitoring of efficient water use. A recent study by TERI and NASSCOM reported that big cities save up to Rs. 27 lakh a day and at least 15% of water that is lost owing to leakages and pilferage using smart meters and sensors (NASSCOM 2011). For example, there

Table 4 Some of the traditional water harvesting techniques being used in different parts of NE India

Region	Indigenous water harvesting technology	Method
Nagaland	Zabo	Rainwater is allowed to fall on a patch of protected forest on the hilltop. As the water runs off along the slope, it passes through various terraces. The water is collected in pond-like structures in the middle terraces. Below these terraces are cattle yards, and toward the foot of the hill are paddy fields, where the runoff ultimately meanders into
Nagaland	Ruza	Another form of Zabo system which tailors water conservation with forestry, agriculture, livelihood and animal care
	Cheo-ozihi	River water is brought down by a long channel. From this channel, many branch channels are taken off, and water is often diverted to the terraces through bamboo pipes
Assam	Dongs	Water harvesting ponds constructed by the Boro community of Assam to harvest water for irrigation
Meghalaya	Bamboo drip irrigation	Tapping of stream and spring water by using bamboo pipes to irrigate plantations

Source WaterHarvesting.org

are two kinds of smart meters available in the country—the automated meter reading (AMR) and the advanced meter infrastructure (AMI). The AMR is a one-way communication to the utility, which conveys usage data, while the AMI is a two-way communication system between the device and the utility and can monitor real-time usage, can send alarms for excessive use, identify leakages, thefts, etc. Application of ICT tools would be useful not only in generation of scientific data and information but also its real-time dissemination for effective and timely decision-making which would further facilitate improving performance benchmarks of the water utilities and thus promote accountability and transparency in the water sector (NASSCOM 2011).

Use of such metered systems may perhaps in a way ensure effective water use in safe water-stressed areas in the region.

Further, the global mobile communications systems are also being used in different sectors to address water security challenges in different parts of the globe. Such systems for monitoring of assets, etc., are already in use in some of the Indian states like Bihar, Maharashtra, etc. Similar use of such systems in Assam to monitor the use of piped water systems, duration of water supply, and amount of water supplied can go a long way in creating a sustainable water supply network in the region. ICT tools can also help pare water leakage and pilferage. Establishment of a framework for efficient and dynamic management of database for water through the use of latest technologies including ICT tools has already been proved to be effective in some of the other states in India.

4.8 Management of Assets

Concurrent monitoring of the piped water supply networks and its different components would be essential for successful implementation of water supply schemes meant to serve as alternate water sources in areas with groundwater contamination. Community consultations can be made and they can be encouraged to take the responsibility of O&M of the assets. Theft, leakages and pilferage, and power supply issues are some of the common factors leading to failure of schemes. Community ownership can perhaps to a large extent help overcome these issues and facilitate the smooth functioning of schemes. Transferring powers to communities is envisioned to give them better control so that they can do better job managing resources than their state counterparts (O'Reilly and Dhanju 2012). One of the most important mechanisms linked to high performance of drinking water sources is the result of the dynamic interaction of a set of working rules enforced by the local communities and properly defined local accountability, as well as the capacity of local leaders to generate appropriate incentives to involve the community in sustainable solutions for collective action problems (Madrigal and Alpizar 2011). The success of such community managed schemes and Village Water and Sanitation Committees (VWSC) has already reported the effectiveness of community involvement and asset management in a number of states in India (WSP 2000).

5 Conclusion

Sustainability of rural water supply schemes is still an elusive goal. Most initiatives for ensuring water safety and security in the region are often beset with inherent issues of unsustainability, community unacceptability, inadequate supply, and poor O&M. A multitude of linkages between water supply inefficiency and distributional inequity are apparent. Therefore to eliminate the inherent gaps attributed to water security and safety in the region, composite solutions through use of rainwater harvesting, water recharging, delineation and safeguarding safe aquifers, reuse and recycling, water saving technologies, and smart systems needs to be explored, adopted, and augmented for the region based on the suitability and feasibility of each of these techniques.

Future efforts to assess sustainability should involve a holistic approach for quality assessment of water sources and alternative sources from supply sources to consumer. Aspects of possible climate change impacts on water availability and security also need to be integrated into the broader framework of alternative water supply schemes that are being provided.

Reference

Agarwal A, Narain S (eds) (1997) Dying wisdom. Rise, fall and potential of India's traditional water harvesting systems. Centre for Science and Environment, New Delhi

Arlosoroff S (1999) Water demand management. Paper presented at IECT–WHO international symposium on efficient water use in urban areas: innovative ways of finding water for cities, Kobe, Japan, 6–10 June 1999

Asano T, Cotruvo JA (2004) Groundwater recharge with reclaimed municipal wastewater, health and regulatory considerations. Water Res 38:1941–1951

Borah KK, Bhuyan B, Sarma HP (2010) Lead, arsenic, fluoride, and iron contamination of drinking water in the tea garden belt of Darrang district, Assam, India. Environ Monit Assess 169(1–4):347–352

Census (2001) Retrieved from the website http://censusindia.gov.in/

Chakravarti D, Chanda CR, Samanta G, Chowdhury UK, Mukherjee SC, Pal AB et al (2000) Fluorosis in Assam, India. Curr Sci 78(1421):1423

Chetia M, Singh SK, Bora K, Kalita H, Saikia LB, Goswami DC, Srivastava RB, Sarma HP (2008) Groundwater arsenic contamination in three blocks of Golaghat District in Assam. J Indian Water Works Assoc 150–155

Das B, Hazarika P, Saikia G, Kalita H, Goswami DC, Das HB, Dube SN, Dutta RK (2007) Removal of iron from groundwater by ash: a systematic study of a traditional method. J Hazard Mater 141:834–841

Dash RR, Bhanu Prakash EVP, Kumar P, Mehrotra I, Sandhu C, Grischek T (2010) River bank filtration in Haridwar, India: removal of turbidity, organics and bacteria. Hydrogeol J. https://doi.org/10.1007/s10040-010-0574-4

Davidson J (1993) Successful acid mine drainage and heavy metal site bioremediation. In: Moshiri GA (ed) Constructed wetlands for water quality improvement. Lewis Publishers, London, 632p

Enmark G, Nordborg D (2007) Arsenic in the groundwater of the Brahmaputra floodplains, Assam, India: source, distribution and release mechanisms. Minor Field Study 131, Uppsala University, Uppsala, Sweden

Hiscock KM, Grischek T (2002) Attenuation of groundwater pollution by bank filtration. J Hydrol 266(3–4):139–144

http://www.wateraid.org/documents/plugin_documents/drinkingwater.pdf. Retrieved on 7 Feb 2012

http://www.sos.arsenic.net/english/environment/flurosis.html. http://www.thehindubusinessline.com/industry-aneconomy/economy/article2766556.ece?homepage=true&ref=wl_home

Hubbard B, Sarisky J, Gelting R, Baffigo V, Seminario R, Centurion C (2011) A community demand-driven approach towards sustainable water and sanitation infrastructure development. Int J Hyg Environ Health 214:326–334

Internet (2010). http://www.scienceinafrica.co.za/2002/january/wetland.htm. Accessed on 20 Sep 2010

Jeffrey P, Jefferson B (2001) Water recycling—how is it feasible? Filtr Sep 38(4):26–29

Khatri KB, Vairavamoorthy K (2007a) Challenges for urban water supply & sanitation in the developing countries, Delft, The Netherlands

Khatri KB, Vairavamoorthy K (2007b) Challenges for urban water supply and sanitation in the developing countries. UNESCO-IHE

Kotoky P, Tamuli U, Borah GC, Baruah MK, Sarmah BK, Paul AB, Bhattacharyya KG (2010) A fluoride zonation map of the Karbi Anglong district Assam, India. Fluoride 43(2):113–115

Madrigal R, Alpizar F (2011) Determinants of performance of community based drinking water organizations. World Dev 39(9):1663–1675

Mahanta C, Pathak N, Choudhury R, Borah P, Alam W (2009) Quantifying the spread of arsenic contamination in groundwater of the Brahamputra Floodplains, Assam, India: a treat to public health of the region. In: World environmental and water resources congress, 2009: great rivers, proceedings of world environmental and water resources congress, vol 342, p 180

Messalem (2006) Membranes for unrestricted water reuse. In: Hlavinek P, Kukharchyk T, Marsalek J, Mahrikova I (eds) Integrated urban water resources management, Part of the series NATO Security through Science Series, pp 313–320

Mooley D, Parthasarathy B (1984) Fluctuations in all-India summer monsoon rain-fall during 1871–1978. Clim Change 6:287–301

NASSCOM (2011). http://www.nasscom.in/sites/default/files/researchreports/TERI-NASSCOM_Green_ICT_Executive%20_Summary_Dec2011.pdf. Retrieved on 7 Feb 2012

Nema P, Ojha CSP, Kumar A, Khanna P (2001) Techno-economic evaluation of soil-aquifer treatment using primary effluent at Ahmedabad, India. Water Res 35(9):2179–2190

O'Reilly K, Dhanju R (2012) Hybrid drinking water governance: community participation and ongoing neoliberal reforms in rural Rajasthan, India. Geoforum. https://doi.org/10.1016/j.geoforum.2011.12.001

Paul AB (2000) Slow poisoning: fluoride in groundwater of Karbianglong district, Souvenir. Assam Public Health Engineering Association

Sakthivadivel R (2007) The groundwater recharge movement in India. In: Giordano M, Villholth K (eds) The agricultural groundwater revolution. Opportunities and threats to development. CAB International Publishing, Colombo, Sri Lanka

Sandhu C, Grischek T, Kumar P, Ray C (2010) Potential for riverbank filtration in India. Clean Technol Environ Policy 13(2):295–316

Sarmah AC (2011) Groundwater crises in Assam and mitigation measures. In: Proceedings of the UGC sponsored national seminar on groundwater contamination, its challenges to human health and mitigation measures, 9–10 Dec 2011

Shah A (2001) Efficacy of the small water harvesting structures in a dryland region in India: implications for crop productivity. In: International conference on rain-water catchment systems; rainwater international 2001, Mannheim, Germany

Singh AK (2004) Arsenic contamination in groundwater of Northeastern India. In: Proceedings of the national seminar on hydrology. Roorkee

Singh S, Khanikar D, Borpujari M (2011) Concentration of fluoride in ground water with special reference to its genesis and status in and around Karbi Anglong District of Assam. In: Proceedings of the CGWB regional workshop on iron, fluoride & arsenic contamination in groundwater & its mitigation measures in northeastern states, 27 Mar 2011

SOS (2000) Retrieved from the url http://www.sos-arsenic.net/flurosis

Sushella AK (2007) A treatise on fluorosis. Fluorosis Research and Rural Development Foundation, New Delhi, p 55

Towards drinking water security in India, lessons from the field. From URL: http://www.wsp.org/wsp/sites/wsp.org/files/publications/WSP_Compendium_Water.pdf

Series Removal of Heavy Metal and Aromatic Compound from Contaminated Groundwater Using Zero-Valent Iron (ZVI)

Selvaraj Ambika and M. Nambi Indumathi

Abstract In this study, it is proposed to couple the two processes, i.e. phenol oxidation and chromium reduction. During this coupling process, it is hypothesized that the ferric iron generated from the chromium reduction process acts as the electron acceptor and catalyst for the Fenton's phenol oxidation process. The Ferrous iron formed from the Fenton reactions during phenol oxidation can be reused for the chromium reduction, and thus the iron can be made to recycle between the two reactions changing back and forth between ferrous and ferric forms. Two sizes of iron, millimetre and micron, were used in this experiment, and their optimum dosages were about 2 g/l and 20 mg/L, respectively; Cr(VI) concentration was maintained as 2 ppm and phenol concentration was about 5 ppm throughout this experiment. In case of mmZVI, 100% Cr(VI) removal was taken place at 7 h and considering mZVI, it was about 6 h, respectively. H_2O_2 was optimized as 1.5 ml for mmZVI and 1 ml for mZVI. Using mmZVI with 1.5 ml H_2O_2, for pH 4, 7 and 10, the reaction time required for the complete removal was 60, 150 and 270 min. Using mZVI with 1 ml H_2O_2, it was about 90, 240 and 390 min for pH 4, 7 and 10, respectively, series removal of phenol and Cr(VI) started with phenol reduction, and this experiment was continued for three cycles. It was also observed that the time taken for Cr(VI) reduction gets decreased in the series removal system than the individual system. The phenol oxidation process which converted some of the Fe^{3+} to Fe^{2+} sustained the chromium reduction for a longer time. The Cr(VI) reduction oxidizes Fe^0 to Fe^{2+}/Fe^{3+} and thus enabling the phenol oxidation. This cycles the iron between the two processes and sustains the barrier wall and expected to increase its lifespan.

Keywords Zero-valent iron · Chromium removal · Phenol removal
Fenton oxidation process · Iron speciation

S. Ambika (✉) · M. N. Indumathi
Department of Civil Engineering, Indian Institute of Technology, Madras, India
e-mail: ambikame@gmail.com

M. N. Indumathi
e-mail: indunambi@iitm.ac.in

© Springer International Publishing AG 2018
A. K. Sarma et al. (eds.), *Urban Ecology, Water Quality and Climate Change*,
Water Science and Technology Library 84,
https://doi.org/10.1007/978-3-319-74494-0_31

1   Introduction

Chromium is one of the most widely used metals in industry, such as in metal finishing, petroleum refining, iron and steel industries, textile manufacturing and pulp production. The effluents of these industries contain large quantities of chromium-laden wastewater. Hexavalent chromium (Cr(VI)) compounds are considered to be highly toxic, carcinogenic and mutagenic to living organisms, while trivalent chromium (Cr(III)) is generally only toxic to plants at very high concentrations and is less toxic or non-toxic to animals (Ansaf et al. 2016; Costa 2003; Anderson 1997). The discharge of Cr(VI) into surface water is regulated to below 0.05 mg/L both by the Indian standards and USEPA. Hexavalent chromium (Cr (VI)) is found together with a variety of aromatic compounds, including phenol, naphthalene and trichloroethylene (Ambika and Nambi 2016; Patterson et al. 1997; Reid et al. 1994). Oil refineries, petrochemical plants, ceramic plants, steel plants, coal conversion processes, phenolic resin industries and pharmaceutical industries all discharge high concentrations of phenol and phenolic compounds. Phenol and phenolic compounds cause skin rashes to carcinogenic effects (Ambika et al. 2016a, b).

1.1   Chromium and Phenol in Groundwater

Cr(VI) and phenol coexistence can be introduced into aquatic environments from the mixed discharges of the aforementioned industries, or from a single industry discharge (Ambika and Nambi 2014a, b, 2015; Nambi and Ambika 2012). Exposure to phenol can cause profound problems to human health and the environment. Skin irritation and off-flavour contribution in drinking and food processing waters are often attributed to phenol. Due to their toxic effects, including permeation of cellular membranes and cytoplasmic coagulation, phenolic contaminants can damage sensitive cells (Ambika et al. 2016a, b). The European Council Directive and Indian standards have set a limit of 0.5 μg/L to regulate phenol concentration in drinking waters (EC 1980). Wastewater containing Cr(VI) and phenols requires careful treatment before being discharged into a receiving water body.

During the last decades, several studies were made aiming the removal of organic compounds and/or chromium(VI) with fungi (Denizli et al. 2004; Choi et al. 2002; Taseli and Gokcay 2005), yeasts (Sheeja and Murugesan 2002; Hamed et al. 2004), bacteria (Quintelas and Tavares 2001, 2002; Edgehill 1996), consortia of bacteria (Chirwa and Wang 2000), anaerobic activated sludge (Aksu et al. 2001) and activated carbon (Streat et al. 1995; Hamdaoui et al. 2003). Utilization of organic chemicals as carbon sources for Cr(VI) reduction by microbes using pure cultures can produce toxic intermediates (Chirwa and Wang 2000), anaerobic activated sludge and activated carbon (Streat et al. 1995; Hamdaoui et al. 2003). We are in need of finding robust technology to speed up the pollutant removal process and to overcome the shortcomings of existing methods.

1.2 Limitations of In Situ Remediation Technologies

Pump and treat groundwater remediation methods have proved to be expensive and in many cases ineffective at achieving the proposed level of cleanup. An alternative to pump and treat groundwater remediation is the use of reactive barriers (Ambika et al. 2016b; Blowes et al. 1995). The use of iron as a reactive material in permeable reactive barriers (PRBs) was pioneered in the 1990s and has been the subject of considerable research and development since that date. A variety of PRBs have been developed and installed at contaminated sites to treat a range of inorganic, organic and radioactive contaminants (USEPA 2002; Mackay et al. 1993). Whereas in the residential area and deep contaminated area (>20 m), this PRB wall technology has the problem in excavation. The exhausted iron has to be periodically replaced by complete excavation and refilling which doubles the cost with every replacement. There is a need to design maintenance-free systems at the field scale in order to keep it cost effective. Permeable reactive barriers (PRBs) using nZVI (nano zero-valent iron) as a reductive medium, because of its high surface to volume ratio, can prolong the life of the PRB thus reducing the overall cost.

1.3 Reactions Involved in Chromium and Phenol Removal

Application of zero-valent iron (ZVI) to detoxify Cr(VI) is a well-proved technique (Ambika and Nambi 2016). The reactivity of iron is strongly influenced by its surface characteristics and aqueous chemistry of ground water. Cr(VI) reduction kinetics are inclined to be affected by pH, nZVI concentration, nZVI active surface, electrolyte concentration and other ionic species and contaminants. Reactions for Cr (VI) reduction and immobilization include,

$$Cr_2O_7^{2-} + 3Fe_{(s)}^0 + 14H^+ \rightarrow 2Cr^{3+} + 3Fe^{2+} + 7H_2O$$

$$Cr_2O_7^{2-} + 6Fe_{(s)}^{2+} + 14H^+ \rightarrow 2Cr^{3+} + 6Fe^{2+} + 7H_2O.$$

In which toxic or carcinogenic Cr(VI) is reduced to less toxic Cr(III) form, which readily precipitates as $Cr(OH)_3$ or as a solid in the form of $(Fe_x, Cr_{1-x})(OH)_3$. Phenol can be removed by Fenton's oxidation process using a suitable electron acceptor. Typically, oxygen is used as electron acceptor which converts phenol to carbon dioxide and water. Supplying oxygen into a contaminated aquifer for complete removal of organic compounds is an energy intensive and hence expensive process. In the so-called Fenton process, H_2O_2 is activated by Fe^{2+} to generate strong reactive species to decompose organic chemicals. According to the classical Fenton radical mechanism, a series of complex reactions occur in water during the Fenton process (Ambika and Nambi 2014a, b, 2015; Ambika et al. 2016a, b).

$$Fe^{2+} + H_2O_2 \rightarrow Fe^{3+} + HO^* + OH$$

$$Fe^{3+} + H_2O_2 \rightarrow Fe^{2+} + OOH^* + H^+.$$

2 Hypothesis

In this study, it is proposed to couple the two processes, i.e. phenol oxidation and chromium reduction. During this coupling process, it is hypothesized that the ferric iron generated from the chromium reduction process acts as the electron acceptor and catalyst for the Fenton's phenol oxidation process. The ferrous iron formed from the Fenton reactions during phenol oxidation can be reused for the chromium reduction, and thus, the iron can be made to recycle between the two reactions changing back and forth between ferrous and ferric forms.

$$Fe^0 \rightarrow Fe^{2+} + 2e$$

$$Fe^0 + Cr^{6+} \rightarrow Fe^{2+} + 2e$$

$$Fe^{2+} + Cr^{6+} \rightarrow Cr3+ + Fe^{3+}$$

$$Fe^{3+} + H_2O_2 + phenol \rightarrow Fe^{2+} + CO_2 + H_2O$$

This cycling of iron between the two redox reactions is hypothesized to be a continuous process without having to regenerate the ZVI constantly by external addition. Hence, the entire process will be a maintenance-free system and can be easily implemented in the field. Moreover, there will not be any build-up of iron in the water which may lead to other problems. The hypothesis has to be tested for success and optimal conditions identified so that the remedial technology can be adopted on field.

3   Materials and Methods

3.1   Chemicals and Reagents

Potassium chromate ($K_2Cr_2O_7$, 99.8%), hydrochloric acid (HCl), sodium hydroxide (NaOH), diphenylcarbazide, glacial acetic acid, 1,10-Phenanthroline, acetone ($\geq$ 99.5%), sulfuric acid (95–98%), ferrous ammonium sulphate ($Fe(NH_4)_2$ $(SO4)_2 \cdot 6H_2O$) and ethanol (C_2H_6O, >99.7%) were all of analytical grade. All the reagents were prepared in deionized water.

3.2   Synthesis of mZVI

Iron filings were collected from the welding workshop at Velachery, Chennai. To remove the oil and grease present in the iron filings, the filings were washed several times with hexane and acetone. After washing, the iron filings were kept in the oven for drying. The dried iron filings were again washed with acetone, and the filings were kept dry in an oxygen-free environment by purging nitrogen gas. Less than 2 mm iron filings were grounded to micron size using Fritsch P-5 High Energy Planetary Ball Mills and then sequentially washed with ethanol and acetone several times to remove the oxide layer and then stored in ethanol the oxygen-free nitrogen environment until required.

3.3   Batch Experiments

In order to determine the removal of Cr(VI) using Fe^0, batch experiments were conducted. A typical batch reactor included a 100 mL solution with the desired dose of iron nanoparticles (5, 10, 15, 20 mg/L) for each pH of 2, 4, 6, 8 and 10. The batch bottles were sealed with rubber and aluminium caps and mixed on a shaker table at 150 rpm. The Cr(VI) concentration of 2 mg/L was based on research carried out at Perungudi dumpsite, and it was found that the hexavalent chromium concentration varied from 0.24 to 1.74 mg/L. The pH was adjusted by using 0.01 N HCl and 0.01 N NaOH to maintain acidic and alkaline conditions, respectively. Phenol removal experiments were carried out by keeping the initial concentration as 5 mg/L by Fenton's advanced oxidation method, which includes iron dosage and hydrogen peroxide. Simultaneous removal of phenol and Cr(VI) was conducted in a 300 ml reagent bottle, with series input of phenol and Cr(VI).

3.4   Analytical Methods

At the desired reaction time, the samples were collected, and supernatant and precipitated iron particles were separated. The supernatant was filtered using Nupore cellulose syringe filter of size 0.45 μm for the measurements of Cr(VI) and Fe(II) concentration. Cr(VI) and Fe(II) concentrations were measured using the diphenylcarbohydrazide method and phenanthroline method at the wavelength of 540 and 510 nm, respectively (Standard Methods for the Examination of Water and Wastewater). Liquid phase phenol analysis was done by direct photometric method (5530D Direct photometric method).

4   Results and Discussion

4.1   Optimization of Iron Dosage

As the ZVI dosage increased, the reduction of Cr(VI) increased, and thus, the concentration of Cr(VI) decreased in aqueous phase. Figure 1a, b represents that in case of mmZVI, 100% Cr(VI) removal was taken place at 25, 17, 7 and 6 h for the dosage of 0.5, 1, 2 and 3 g/l, respectively. Consider in gmZVI, the time taken for the complete removal of Cr(VI) was 13, 9, 6 and 5.8 h for the dosages of 10, 15 and 20 and 40 mg/L, respectively. Cr(VI) removal pathway involves initial reduction of Cr(VI) to Cr(III), followed by adsorption of Cr(III) onto the surfaces or by precipitation of Cr(III)/Fe(III) hydroxides (Ansaf et al. 2016). All these depend on the available surface area, so that it can offer more electrons and offer more chance for the adsorption and precipitation.

4.2   Cr Removal for Various pH

Figure 2a, b showed that using mmZVI and mZVI, it is evident that the reduction rate of Cr(VI) was greatly reduced under alkaline conditions. The reductive remediation of chromate in aqueous media is strongly affected by the passivation of the metal surface with the consequent loss of reactivity (Ansaf et al. 2016).

When pH is less than 4.5, iron corrosion is mostly driven by H_2O reduction; when pH is larger than 4.5, the extent of iron corrosion depends on the properties of oxide films on Fe^0 (the solubility of Fe and the speciation of the contaminant—Cr). So at initial low pH values, iron is readily soluble and oxide films will not form immediately after the start of the experiments; keeping the low pH can avoid the formation of oxide film thus increasing the reduction rate of Cr(VI) (Ambika and Nambi 2016).

Fig. 1 **a** Optimization of mmZVI dosage. **b** Optimization of mZVI dosage

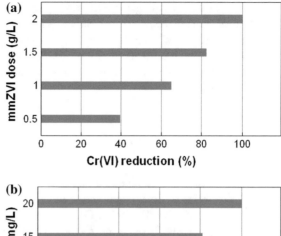

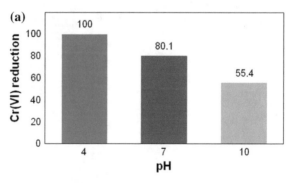

Fig. 2 **a** Cr(VI) removal using mmZVI for various pH. **b** Cr(VI) removal using mZVI for various pH

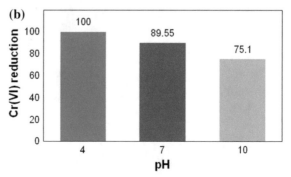

But the increase in pH value will promote the Cr(III) adsorption and co-precipitation (Liu 2009; Rivero-Huguet and Marshall 2009a, b). There are two main reasons for the observed pH dependence. First, the iron oxidation and Cr(VI) reduction reactions are favoured at low pH. Second, in the sorption of Cr(VI), an anion is preferred at low pH because the hydrous surface is more positively charged at lower pH (Li et al. 2008).

4.3 Phenol Removal for Various pH

From Fig. 3a, b, it was obtained that using mmZVI with 1.5 ml H_2O_2, for pH 4, 7 and 10, the reaction time required for the complete removal was 60, 150 and 270 min. Using mZVI with 1 ml H_2O_2, it was about 90, 240 and 390 min for pH 4, 7 and 10, respectively. Though mmZVI also can do a complete removal of phenol, mZVI showed very less reaction time for the complete degradation of phenol, because of its surface to volume ratio increment. As the specific surface area–volume ratio increases, mZVI worked very faster than the mmZVI (Ambika et al. 2016b).

Fig. 3 **a** Phenol removal using 2 g/l mmZVI, 1.5 ml H_2O_2 **b** Phenol removal using 20 mg/L mZVI, 1 ml H_2O_2

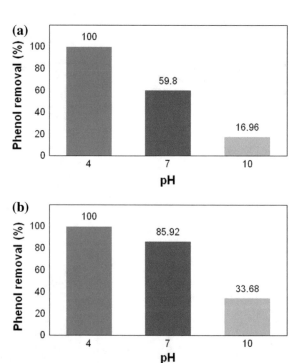

4.4 Phenol and Cr Removal for Various pH

The experiment was started with an initial concentration of mmZVI as 2 g/l and mZVI as 20 mg/L. The initial phenol concentration is kept as 5 ppm. The phenol oxidation experiment was initiated by the addition of 1.5 ml H_2O_2 solution for mmZVI and 1 ml for mZVI and represented in Fig. 4a–d.

During the initial phase of the reaction, Fe^{2+} is produced from ZVI. The colour changes from brown to orangish brown could indicate the oxidation state of iron in the solution. Figure 4a, b represents phenol and Cr(VI) removal at pH 4, 7 and 10 with 1.5 ml H_2O_2, the complete removal was achieved at 90, 240 and 390 min, respectively; whereas using 1 ml H_2O_2, the complete removal of phenol could not be obtained. In this reaction, build-up of Fe(II) was found, since iron corrosion and Fenton reaction and Fenton-like reactions (Ambika et al. 2016a, b).

After this, Cr(VI) of 2 mg/L was added to the system. It was observed that Cr (VI) was completely removed by using Fe^{2+} in the system at 60, 150 and 240 min for pH 4 and 7. It can be assured by the decrease in Fe(II) concentration in the solution.

Phenol of concentration 5 ppm was again added to check if the iron present sustains the second cycle of phenol reduction. Another 1.5 ml of H_2O_2 is added to initiate Fenton's process. It was observed that the ferric was reduced to ferrous (assured by build-up of Fe(II)) at same time phenol was oxidized and completely removed at 90, 250 and 410 min; a little lag for pH 7 and 10 was found. This was followed by a second cycle of Cr(VI) reduction with no addition of extra iron. Here also, a little lag was found for pH 7 and 10; the same time lag was obtained for the successive addition of Cr(VI) and phenol in the system without any addition of iron. The concentration of Fe(II) was measured to confirm the cycling of iron during the phenol and chromium removals. It was observed that the ferrous ions were formed and accumulated from mmZVI.

For the same experiment done using mZVI with 1 ml H_2O_2 addition, from Fig. 4b, the reaction time of 60, 150 and 270 min was found for complete removal of phenol and 45, 80 and 150 min for further addition of Cr(VI) removal for pH 4, 7 and 10, respectively. For the second cycle of phenol, time lag was found only for pH 10 and for Cr(VI), for both pH 7 and 10, time lag were found. But for the successive third trial of phenol and Cr(VI) removal, the same reaction time like second trial was obtained.

Figure 5a, b represents the Fe(II) build-up in the Cr(VI)-phenol system for various pH. As the pH decreases, the concentration of Fe(II) increased, and this was the reason for the decline in removal of Cr(VI) and phenol removal in the system as the pH increases.

Considering the Cr(VI) removal time in simultaneous series removal of phenol and Cr(VI) system, it was taking lesser time compared to the system which consisted only Cr(VI). The reason for this phenomenon may be the readily available Fe (II), which was produced during the phenol degradation.

Fig. 4 **a** Phenol removal
using mmZVI for various pH.
b Cr(VI) removal using
mmZVI for various pH.
c Phenol using mZVI for
various pH. **d** Cr(VI) removal
using mZVI for various pH

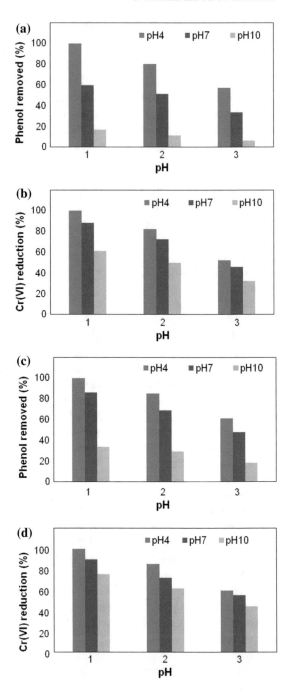

Fig. 5 **a** Fe(II) concentration in phenol-Cr(VI) removal system using mmZVI at various pH. **b** Fe(II) concentration in phenol-Cr(VI) removal system using mZVI at various pH

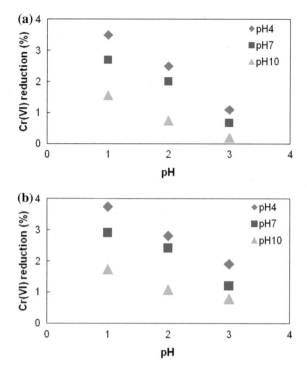

5 Conclusion

In this study, Cr(VI) was reduced to Cr(III) by oxidation of Fe^0 to Fe(III), and phenol was removed by Fenton's advanced oxidation process. In series removal process, end product of Cr(III) reduction Fe^{2+}/Fe^{3+} will act as catalyst to generate hydroxide radicals in presence of H_2O_2. Optimum pH of Cr(VI) effective removal was investigated by conducting different batch experiments with varying pH values.

From the batch experiments investigation, optimum dosage of mmZVI and mZVI was found as 2 g/l and 20 mg/L, respectively, and pH as 4. mZVI showed better Cr(VI) removal percentage compared to mmZVI, because of its enormous surface area. 92% of its total removal was obtained using 20 mg/L of mZVI and 20 g/l of mmZVI. Nearly 100 times less consumption of iron, we can obtain using mZVI instead of mmZVI, and this will considerably reduce the PRBW (permeable reactive barrier wall) thickness and amount of iron in solution and precipitation in real-world scenario. As the pH increased, the removal of Cr(VI) and phenol decreased, because of passive layer formation on the iron surface and the reduction in the rate of corrosion of iron. Increase in iron concentration resulted in the improvement in the Cr(VI) and phenol removal efficiency, due to increase in surface area for the effective electron transfer and precipitation of newly formed species.

Batch experiments were conducted to optimize the adding of H_2O_2 to remove 5 mg/L of phenol by maintaining the 2 g/l of mmZVI and 20 mg/L of mZVI dosage and found that 1.5 and 1 ml of H_2O_2 are required for its complete removal. It can be inferred that the cyclic removal of Cr(VI) and phenol experiments was possible by maintaining all the previously established optimum conditions, for single compound removal with no further addition of Fe^0/H_2O_2. It was also observed that the time taken for Cr(VI) reduction get decreased in the series removal system than the individual system. It can be inferred that series removal of Cr(VI) and phenol in the further cycles had a time delay than the first trial of the experiment. It can be inferred that in in situ PRBW application of this technology, even though the ZVI exhausted at its designed lifetime, the phenol oxidation process which converted some of the Fe^{3+} to Fe^{2+} sustained the chromium reduction for a longer time. The Cr(VI) reduction oxidizes Fe^0 to Fe^{2+}/Fe^{3+} and thus enabling the phenol oxidation. This cycles the iron between the two processes and sustains the barrier wall. The lifetime of the barrier can be prolonged much longer if recirculation of the effluent is adopted.

References

Ambika S, Nambi IM (2014a) ZVI mediated removal of Cr(VI) and phenol: a sustainable treatment technology coupling chemical Redox system and Fentons AOP. In: *Environmental and Molecular Mutagenesis,* Supplement: Environmental Mutagenesis and Genomics Society, 45th Annual Meeting, pp S52–S52

Ambika S, Nambi IM (2014b) A novel permeable reactive barrier (PRB) for simultaneous and rapid removal of heavy metal and organic matter—a systematic chemical speciation approach on sustainable technique for Pallikarani marshland remediation. In: AGU Fall Meeting, San Francisco, USA (AGUFM.B21B0048S). http://adsabs.harvard.edu/abs/2014AGUFM.B21B0048S

Ambika S, Nambi IM (2015) Sustainable permeable reactive barrier (PRB) for synchronized removal of heavy metal and organic matter for wetland remediation—a systematic chemical speciation approach. Goldschmidt2015 abstracts http://goldschmidt.info/2015/uploads/abstracts/finalPDFs/2839.pdf

Ambika S, Nambi IM (2016) Optimized synthesis of methanol-assisted nZVI for assessing reactivity by systematic chemical speciation approach at neutral and alkaline conditions. J Water Process Eng 13:107–116

Ambika S, Nambi IM, Senthilnathan J (2016a) Low temperature synthesis of highly stable and reusable CMC-Fe^{2+}(-nZVI) catalyst for the elimination of organic pollutants. Chem Eng J 289:544–553

Ambika S, Devasena M, Nambi IM (2016b) Synthesis, characterization and performance of high energy ball milled meso-scale zero valent iron in Fenton reaction. J Environ Manage 181:9847–9855

Anderson RA (1997) Chromium as an essential nutrient for humans. Regul Toxicol Pharmacol 26: S35–S41

Ansaf KV, Ambika S, Nambi IM (2016) Performance enhancement of zero valent iron based systems using depassivators: optimization and kinetic mechanisms. Water Res 102:436–444

Blowes DW, Ptacek CJ, Cherry JA, Gillham RW, Robertson WD (1995) The feoenvironment, New Orleans, LA, pp 1588–1607, 24–26 Feb 1995

Chirwa EN, Wang YT (2000) Simultaneous chromium(VI) reduction and phenol degradation in an anaerobic consortium of bacteria. Water Res 33:2376–2384

Choi SH, Moon S-H, Gu MB (2002) Biodegradation of chlorophenols using the cell-free culture broth of *Phanerochaete chrysosporium* immobilized in polyurethane foam. J Chem Technol Biotechnol 77:999–1004

Costa M (2003) Potential hazards of hexavalent chromate in our drinking water. Toxicol Appl Pharmacol 188:1–5

Denizli A, Cihangir N, Rad AY, Taner M, Alsancak G (2004) Removal of chlorophenols from synthetic solutions using Phanerochaete chrysosporium. Process Biochem 39:2025–2030

Edgehill RU (1996) Degradation of pentachlorophenol (PCP) by Arthrobacter strain ATCC 33790 in biofilm culture. Water Res 30:357–363

Hamdaoui O, Naffrechoux E, Tifouti L, Pétrier C (2003) Effects of ultrasound on adsorption–desorption of p-chlorophenol on granular activated carbon. Ultrason Sonochem 10:109–114

Hamed TA, Bayraktar E, Mehmetoglu U, Mehmetoglu T (2004) The biodegradation of benzene, toluene and phenol in a two-phase system. Biochem Eng J 19:137–146

Li X, Cao J, Zhang W (2008) Stoichiometry of Cr(VI) immobilization using nanoscalezerovalent iron (nZVI): a study with high-resolution X-ray photoelectron spectroscopy (HR-XPS). Ind Eng Chem Res 47:2131–2139

Liu J, Wang C, Shi J, Liu H, Tong Y (2009) Aqueous Cr(VI) reduction by electrodeposited zero-valent iron at neutral pH: acceleration by organic matters. J Hazard Mater 163:370–375

Mackay DM, Feenstra S, Cherry JA (1993) In: Neretnieks I (ed) Proceedings of the workshop on contaminated soils risks and remedies, Stockholm, pp 35–47, 6–28 Oct 1993

Nambi IM, Ambika S (2012) Sustainable treatment technology using nanostructured iron for combined removal of heavy metal and organic chemicals. In: Proceedings from the sixth international conference on environmental science and technology, vol I. American Science Press, Houston. (ISBN: 9780976885351) http://www.aasci.org/conference/env/2012/Table-of-Contents-EST2012-Proc-I.pdf

Patterson RR, Fendorf S, Fendorf M (1997) Reduction of hexavalent chromium by amorphous iron sulfide. Environ Sci Technol 31:2039–2044

Quintelas C, Tavares T (2001) Removal of chromium(VI) and cadmium(II) from aqueous solution by a bacterial biofilm supported on granular activated carbon. Biotechnol Lett 23:1349–1353

Quintelas C, Tavares T (2002) Lead(II) and iron(II) removal from aqueous solution: biosorption by a bacterial biofilm supported on granular activated carbon. J Resour Environ Biotechnol 3: 196–202

Reid VM, Wyatt KW, Horn JA (1994) A new angle on groundwater remediation. Civil Eng 64:56–58. EC: *Official Journal of the European Communities*, no. 80/779 (1980)

Rivero-Huguet M, Marshall WD (2009a) Reduction of hexavalent chromium mediated by micro- and nano-sized mixed metallic particles. J Hazard Mater 169:1081–1087

Rivero-Huguet M, Marshall WD (2009b) Reduction of hexavalentchromiummediated by micron- and nano-scale zero-valent metallic particles. J Environ Monit 11:1072–1079

Sheeja RY, Murugesan T (2002) Studies on biodegradation of phenol using response surface methodology. J Chem Technol Biotechnol 77:1219–1230

Streat M, Patrick JW, Camporro Perez MJ (1995) Sorption of phenol and p-chlorophenol from water using conventional and novel activated carbon. Water Res 29:467–472

Taseli BK, Gokcay CF (2005) Degradation of chlorinated compounds by *Penicillium camemberti* in batch and up-flow column reactors. Process Biochem 40:917–923

U.S. Environmental Protection Agency (2002) Economic analysis of the implementation of permeable reactive barriers for remediation of contaminated groundwater, EPA/600/R-02/034; U.S. EPA, Washington, DC

Spatial Decision Support System for Groundwater Quality Management Using Geomatics

M. V. S. S. Giridhar, B. V. Nageswra Rao and G. K. Viswanadh

Abstract The quality of groundwater is very important in evaluating its utility for agriculture, domestic and industrial purposes. In the present study, it is proposed to develop 'spatial decision support system (SDSS) for ground water quality management at village level using Geomatics' for verifying the number of borewells required, based on the population in the village and their spatial distribution for the decision makers, NGOs, government authorities, etc. Further, the proposed SDSS will also help in carrying out the study and analysis of changes in water quality parameters with respect to pre- and post-monsoon seasons and their consequent impacts on human health. Keeping these points in view, a spatial decision support system (SDSS) has been formulated and developed in GIS environment for taking appropriate decisions for groundwater quality management at village level. Nandyal Mandal, Kurnool district, Andhra Pradesh, India, with 20 villages has been considered for this study. Nitrates and sulphates are present in 140 samples out of 310 samples collected in post-monsoon season data samples. These samples need to further laboratory analysis for calculating the concentration of nitrates and sulphates in all 140 samples located in various villages. The borewell having ID No. 133401332 located at DPEP School in Ayyalur village is having excess TDS, alkalinity, hardness, and presence of nitrates and sulphates, which shows that the borewell is not suitable for drinking purpose.

Keywords Groundwater · GIS · Nandyal Mandal · Geomatics
Fluorides · Chlorides · Total hardness · TDS

M. V. S. S. Giridhar (✉) · B. V. Nageswra Rao
Centre for Water Resources, Institute of Science and Technology, JNTUH, Hyderabad, India
e-mail: mvssgiridhar@gmail.com

G. K. Viswanadh
Civil Engineering, JNTUH College of Engineering, JNTUH, Hyderabad, India

© Springer International Publishing AG 2018   429
A. K. Sarma et al. (eds.), *Urban Ecology, Water Quality and Climate Change*,
Water Science and Technology Library 84,
https://doi.org/10.1007/978-3-319-74494-0_32

1 Introduction

The chief merit of groundwater is its availability in some quantity almost in every human settlement. Even in areas where there is abundant surface water supply, groundwater is playing an increasingly crucial role in supplementing surface water. In areas where surface water availability is very less, groundwater is the only source for survival. The GIS could be the front end to a groundwater modelling simulation devised to fully capture the contaminant. Areas that have been overpumped of groundwater can subside, and when near the sea, this may invite flooding. Groundwater quality mapping will also assist the planners to take suitable action to improve the groundwater quality in contaminated regions. 'The spatial variation in chemical and physiochemical parameters of groundwater were used to identify suitable zone for pumping for both domestic and irrigation purposes' (Srinivasarao 2007). GIS was effectively used for the evaluation of groundwater quality by several researchers (Shahid and Nath 2000; Phukon et al. 2004). The overlay/index approaches gained popularity particularly with the ease of use of GIS technology (Bonhamcartar 1996; Corwin et al. 1997; Fuest et al. 1998). The spatial ground-water quality index and the procedure of weighing are widely used to identify the spatial characteristics of the aquifer by many researchers (Horton 1965; Prati et al. 1971; Inhaber 1975; Provencher and Lamontagne 1979; Maha and Al-Dabbagh 1989; Sinha and Shrivastava 1994; Secunda et al. 1998; Pradhan et al. 2001; Connell and Van den Daele 2003; Babiker et al. 2005; Chachadi and Lobo-Ferreira 2005; Stigter et al. 2006; Ckakraborthy et al. 2007; Mamadou et al. 2010). Many approaches have been developed to extract aquifer vulnerability such as process-based method, statistical methods and overlay/index methods (Zhang et al. 1996; Tesoriero et al. 1998). Ramakrishnaiah et al. (2008) used the spatial groundwater quality index was used for ranking, which reflected the influence of different parameters. The results obtained by such techniques were used to point out the groundwater from a particular region is good or not (Rajankar et al. 2009). Spatial decision support system (SDSS) will serve the purpose of verifying the number of borewells required, based on the population in the village and their spatial distribution for the decision makers, NGOs, government authorities, etc. Further, SDSS will also help in carrying out the study and analysis of changes in water quality parameters with respect to pre- and post-monsoon seasons and their consequent impacts on human health. Keeping these points in view, a spatial decision support system (SDSS) has been formulated and developed in GIS environment for taking appropriate decisions for groundwater quality management at village level. Nandyal Mandal, Kurnool district, Andhra Pradesh, India, with 20 villages has been considered for this study with the following specific objectivities:

1. To prepare the spatial distribution maps of groundwater quality parameters such as fluorides, chlorides, total hardness and total dissolved solids.

2. To analyse water quality parameters, viz. fluorides, chlorides, total hardness and TDS according to the IS:10500-1991 guidelines and to decide the suitability of water for potable purpose.
3. To design and develop spatial decision support system (SDSS) for taking appropriate decisions based on relevant technology for drilling of new borewells in the village.

2 Study Area

The study area comprises Nandyal Mandal of Kurnool district, which lies between 78° 21′ 43″ and 78° 33′ 53″ East longitudes and 15° 22′ 11″ and 15° 34′ 02″ North latitudes. The Mandal comprises 20 revenue villages including 12 hamlets. Nandyal town is a municipality having a population of 151,771 and the rest of 31 habitations account for a total rural population of about 45,296 as per 2001 census. It is at an elevation of 217.00 m above the MSL.

3 Methodology

Based on the literature review and the objects identified for the study, suitable methodology is identified to carry out the present work, and they were discussed in the following sections.

3.1 Creation of Spatial Database

In order to create the database required to carry out the present study, field visit has been conducted to each village in the Mandal for the purpose of collection of latitude and longitudes of all the borewells in the village using GPS technology for all the 20 villages. The collected spatial information such as borewell code as per Rural Water Supply and Sanitation Department, Government of Andhra Pradesh, India, coding, location of each borewell in the village, village code, latitude, longitude and elevation details has been listed.

3.2 Spatial Analysis of Water Quality Parameters

In addition to the creation of spatial database creation, information pertaining to water quality parameters is essential for the present study. For this purpose, water

samples have been collected from 310 borewells located in the study area for pre-
and post-monsoon seasons for the year 2010 during the field visits. Further, the
collected water samples were analysed in the Nandyal division Laboratory, Rural
Water Supply and Sanitation Department (RWS&S), Nandyal Mandal, Kurnool
district, AP.

3.3 Design and Development of Spatial Decision Support System (SDSS)

Initially, a home page has been designed as part of SDSS to show various select
options such as state, district, mandal and village to enable the user to go to location
of his choice. The 'state' option enables the user to select the state of his choice
from the drop menu. Once a state has been selected, a pop-up window shall be
activated guiding the user to select a district. Similarly, the user is guided to select
the mandal as well as the village from the database. Further, spatial distribution
maps pertaining to various water quality parameters such as fluoride, chloride, total
hardness and total dissolved solids will be activated in the bottom of the web page,
and the user is required to click on the map icons, so that they are displayed in the

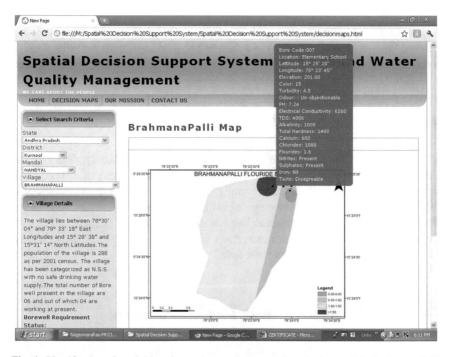

Fig. 1 Identification of spatial location and groundwater quality of a borewell in Brahmana Palli
village

centre of the SDSS web page. Further, navigation pane is facilitated to benefit the user to navigate to any part of the map to obtain the required information corresponding to the location in that village. A query option has been created at the bottom of the village description about the status of the borewells with respect to the number of borewells required. The technology used for the design and development of SDSS is IIS web server, C#.net language, ASP.Net web technology and JavaScript as client side scripting (Fig. 1).

4   Results and Discussion

Water samples were collected from all the 310 borewells for pre- and post-monsoon seasons for the year 2010–11. Spatial location of each borewell has been collected using Global Positioning System technology. The collected samples were processed and analysed with respect to physical and chemical parameters such as colour, odour, pH, electrical conductivity, alkalinity, fluorides, chlorides, total hardness, calcium, nitrates, sulphates, iron and total dissolved solids for each borewell. Subsequently, these water quality parameters were compared and analysed with the guidelines as per the drinking water standards given in IS code for Nandyal Mandal at village level. Further, spatial maps were prepared in GIS environment for fluorides, chlorides, total hardness and total dissolved solids and were developed at village level. Village level water quality information and its suitability for drinking purpose have been explained for two villages (Table 1).

4.1   Ayyalur Village

The population of the village is 3475 as per 2001 census. The village has been categorized as PC-2 with an average supply of safe drinking water at the rate of 19.62 lpcd. The total number of borewells present in the village is 28 and out of which 20 are working. Water samples were collected for all the 20 borewells for pre- and post-monsoon season for the year 2010 from the Nandyal division Laboratory, Rural Water Supply and Sanitation Department (RWS&S), Nandyal. Highest and lowest elevation observed as 228.00 m and 209.00 m from MSL.

4.1.1   Physical Parameters

Physical water quality parameters such as colour and turbidity with minimum and maximum values are 5 and 20 in Hazen units, 2.90 and 5.20 in NTU scale, respectively, and are found to be within the safe limits as per Indian Standard drinking water specifications IS:10500-1991. Odour is found to be not objectionable for all the borewells in the village. The minimum and maximum pH value in all

Table 1 The analytical results showing quality of groundwater during post-monsoon season (2010–11) in the study area

S. No.	Borewell code	Location of the borewell	Colour	Turbidity	pH	Electrical conductivity	Total dissolved solids	Alkalinity	Total hardness	Calcium	Chlorides	Fluorides	Nitrites	Sulphates	Iron	Taste
Name of the village: Ayyaluru																
1	033	Urdu School	20	4.2	7.61	1460	934	230	320	135	251	0.4	Nil	Trace	Nil	Agreeable
2	002	Sivaalyam	10	3.8	7.56	1120	717	170	250	110	193	0.8	Nil	Nil	Nil	Agreeable
3	031	Chavidi	15	4.6	7.43	2100	1344	330	470	220	362	1	Present	Trace	Nil	Agreeable
4	004	Masidpuram Rasta	15	4.9	7.42	2160	1382	345	480	240	372	0.6	Present	Trace	Nil	Agreeable
5	021	Durgamma Temple—Main village	5	2.9	7.41	1740	1114	270	380	180	300	0.8	Nil	Nil	Nil	Agreeable
6		Pakkiraiah (Velugudu)	15	4.8	7.65	3170	2029	500	710	305	547	1	Present	Present	Nil	Agreeable
7	009	Chennakesava Temple	10	4.2	7.45	3150	2016	500	700	300	543	0.4	Present	Present	Nil	Agreeable
8	008	Harijanwada	5	2.9	7.46	1250	800	200	280	125	215	0.6	Nil	Trace	Nil	Agreeable
9	027	New Church	10	4.1	7.37	2090	1338	334	460	220	360	0.8	Trace	Trace	Nil	Agreeable
10	011	Harijanwada Gangsubbarayadu	15	4.4	7.42	2490	1594	390	550	235	430	1	Present	Trace	Nil	Agreeable
11	032	DPEP School	20	5.2	7.53	4250	2720	680	950	415	733	1.2	Present	Present	Nil	Agreeable
12	017	Near Ramalayam	20	4.8	7.33	1550	992	230	340	140	267	0.8	Nil	Trace	Nil	Agreeable
13	001	Near well	10	3.9	7.6	1340	858	210	300	145	231	0.8	Nil	Nil	Nil	Agreeable
14	030	Near Nagulakatta	10	4.1	7.36	1730	1107	270	380	175	298	0.6	Trace	Trace	Nil	Agreeable
15	019	Suddalapeta	15	4.9	7.45	1320	845	210	290	140	227	0.2	Nil	Trace	Nil	Agreeable
16	028	Durgamma Temple	10	4.1	7.61	1450	928	230	320	160	250	0.8	Trace	Nil	Nil	Agreeable
17	024	Sankar Reddy/ Pakkiraiah House	5	3.1	7.59	990	634	150	220	110	171	1	Nil	Nil	Nil	Agreeable

(continued)

Table 1 (continued)

S. No.	Borewell code	Location of the borewell	Colour	Turbidity	pH	Electrical conductivity	Total dissolved solids	Alkalinity	Total hardness	Calcium	Chlorides	Fluorides	Nitrites	Sulphates	Iron	Taste
Name of the village: Bheemavaram																
18	001	Kothakottala Jamal Shab	15	4.9	7.13	3780	2419	480	950	420	773	0.4	Trace	Trace	Nil	Agreeable
19	004	Kothakottala Dastagiri	20	6	7.04	4320	2764	550	1100	500	883	0.6	Trace	Trace	Nil	Agreeable
20	008	Maddilati House	15	4.5	7.24	2510	1606	320	630	310	513	0.4	Trace	Trace	Nil	Agreeable
21	003	Yerraguntla Rasta	15	5.1	7.21	2410	1542	300	610	300	492	0.6	Trace	Trace	Nil	Agreeable
22	007	Elementary School	15	4.4	7.17	1730	1107	220	440	210	353	1	Nil	Nil	Nil	Agreeable
23	011	Elementary School back side														
24	009	Near Ratnamaiah House	10	2.9	7.25	2390	1529	300	600	295	488	1.4	Trace	Trace	Nil	Agreeable
25	006	Housing colony Balaiah	15	4.4	7.23	3040	1945	385	780	390	621	1.6	Trace	Trace	Nil	Agreeable
26	010	MPE School	10	3.8	7.08	2960	1894	380	740	370	605	1.6	Trace	Trace	Nil	Agreeable
27	005	Near Devaraju House	15	4.9	7.2	3270	2092	420	820	410	668	1.4	Trace	Trace	Nil	Agreeable

the borewells are noticed as 7.33 and 7.65, and are found within in the limits as per specifications. Hence, water can be used for potable purpose as far as pH parameter is concerned. Total dissolved solids (TDS) have been measured from the values of electrical conductivity and found as 634 and 2720 as minimum and maximum for this village. Borewells with ID Nos. 133401332, 133401323 and 133401309 have TDS values as 2720, 2029 and 2016 mg/lit; these values are more than the permissible standards. Hence, these three borewells are not suggestible for drinking purpose, and the same has been reflected in the figures in GIS environment.

4.1.2   Chemical Parameters

The total alkalinity (as $CaCo_3$) values with minimum and maximum are found to be 150 and 680, respectively, for all the borewells within the village. The borewell located at DPEP School is having the excess alkalinity value (680), which is more than the allowable standards and is found to be not safe for the drinking purpose. Total hardness values were found more than permissible limits in borewells with ID Nos. 133401332, 133401323 and 133401309 with the hardness values 950, 710 and 700, respectively, and the same has been represented in the figure. Chlorides present in all the borewells have been observed and found to be within the limits, and the same was depicted in the figure. Flourides present in all borewells are observed as within the recommendations of IS:10500-1991, and the same was shown in the figure. Nitrates and sulphates are traced in all the borewells except borewell located at Durgamma Temple, Sankar Reddy/Pakkiraiah House, Sivaalyam, Near well and Durgamma Temple—Main village. The borewell having ID No. 133401332 located at DPEP School is having excess TDS, alkalinity, hardness, and presence of nitrates and sulphates, which shows that the borewell is not suitable for drinking purpose.

4.2   Bheemavaram Village

The population of the village is 2171 as per 2001 census. The village has been categorized as PC-4 with an average supply of safe drinking water at the rate of 35.06 lpcd. The total number of borewell present in the village is 13 and out of which 10 are working. Water samples were collected for all the 10 borewells for pre- and post-monsoon season for the year 2009–10 from the Nandyal division Laboratory, Rural Water Supply and Sanitation Department (RWS&S), Nandyal. Highest and lowest elevation observed as 210.00 and 207.00 m from MSL.

4.2.1 Physical Parameters

Minimum and maximum for the water quality parameter colour is 10 and 20 in Hazen units and for turbidity as 2.90 and 6.90 in NTU scale, respectively. Colour and turbidity for all the borewells in this village are found to be safe for potable purpose. Odour is found to be not objectionable for all the borewells. pH value was observed in all the borewells and found to be 7.04 and 7.25 as minimum and maximum; hence, water can be used for potable purpose as pH parameter is concerned. Total dissolved solids (TDS) have been measured from the values of electrical conductivity and found as 1107 and 2764 as minimum and maximum for this village. Borewells with ID Nos. 133400501 and 133400504 have TDS values as 2419 and 2764 mg/l; these values are more than the permissible standards of physical and chemical quality as per Indian Standard drinking water specifications IS:10500-1991. Hence, these two borewells are not suggestible for drinking purpose, and the same has been reflected in the figure prepared spatially in GIS environment.

4.2.2 Chemical Parameters

Alkalinity values were observed for all the borewells and found these values are within the permissible limits. Total hardness values were found more than permissible limits in borewells with ID Nos. 133400501, 133400504, 133400505 and 133400506. The water quality parameter, i.e. chlorides, has been observed and found that all the values are within the specifications. Borewells located at Housing colony Balaiah House and MPE School are found as 1.60 which is more than IS:10500-1991 with respect to fluorides, and the same was shown in the figure. Nitrates and sulphates are traced in all the borewells except borewell located at Elementary School.

4.3 Design and Development of Spatial Decision Support System (SDSS) for Groundwater Management at Village Level

Spatial decision support system (SDSS) will serve the purpose of verifying the number of borewells required, based on the population in the village and their spatial distribution for the decision makers, NGOs, government authorities, etc. Further, SDSS will also help in carrying out the study and analysis of changes in water quality parameters with respect to pre- and post-monsoon seasons and their consequent impacts on human health. Keeping these points in view, a spatial decision support system (SDSS) has been designed and developed in GIS environment for taking appropriate decisions for groundwater quality management at village level. Nandyal Mandal, Kurnool district, Andhra Pradesh, India, with 20 villages has been considered for this study. The technology used for the design and development of SDSS is IIS web server, C#.net language, ASP.Net web technology and JavaScript as client side scripting. Home page has been designed as part of SDSS to show various select options such as state, district, mandal and village to enable the user to go to the location of his choice. The 'state' option enables the user to select the state of his choice from the drop menu. Once a state has been selected, a pop-up window shall be activated guiding the user to select a district as given in spatial map. Similarly, the user is guided to select the mandal as well as the village from the database as displayed in spatial maps. Further, spatial distribution maps pertaining to various water quality parameters such as fluoride, chloride, total hardness and total dissolved solids will be activated in the bottom of the web page, and the user is required to click on the map icons, so that they are displayed in the centre of the SDSS web page. Further, navigation pane is facilitated to benefit the user to navigate to any part of the map to obtain the required information corresponding to the location in that village. A query option has been created at the bottom of the village description about the status of the borewells with respect to the number of borewells required and can be seen in spatial map in GIS environment.

5 Conclusions

The maximum values for various water quality parameters such as colour, turbidity, pH, electrical conductivity, total dissolved solids, alkalinity, total hardness, calcium, chlorides and fluorides were observed as 35, 16.8, 7.84, 6830,1257, 10005, 4300, 2050, 3976, and 2 for post-monsoon season in the study area. The minimum values for various water quality parameters such as colour, turbidity, pH, electrical conductivity, total dissolved solids, alkalinity, total hardness, calcium, chlorides and fluorides were observed as 2, 0.3, 6.07, 200, 408, 35, 140, 16, 110, and 0.1 for post-monsoon season in the study area. Nitrates and sulphates are present in 140 samples out of 310 samples collected in Nandyal Mandal in post-monsoon season data samples. These samples need further laboratory analysis for calculating the concentration of nitrates and sulphates in all 140 samples located in various villages. The borewell having ID No. 133401332 located at DPEP School in Ayyalur village is having excess TDS, alkalinity, hardness and presence of nitrates and sulphates; therefore, the borewell is not suitable for drinking purpose. Except borewell having ID No. 133400710 located near Darga in Brahmanapalli village, all other borewells are having excess total dissolved solids (TDS) for this village, and the same has been reflected in the maps prepared spatially in GIS environment. The borewell having ID No. 133401201015 located at Village of Pedda Kottala is having excess colour, TDS, alkalinity, hardness, chlorides and presence of nitrates and sulphates, which indicated that the borewell is not suitable for drinking purpose. The borewell having ID No. 133401002001 located at Elementary School

Pandurangapuram Village is having excess colour, TDS, alkalinity, hardness, chlorides and presence of nitrates and sulphates, which indicated that the borewell is not suitable for drinking purpose.

References

Babiker IS, Mohammed MAA, Hiyama T, Kato K (2005) A GIS-based DRASTIC model for assessing aquifer vulnerability in Kakamigahara heights" Gifu Prefecture, central Japan. Sci Total Environ 345:127–140

Bonhamcarter GF (1996) Geographic information systems for geoscientists: modeling with GIS computer methods in the geosciences. Elsevier, Pergamon, vol 13, pp 1–50

Chachadi AG, Lobo-Ferreira JP (2005) Assessing aquifer vulnerability to sea-water intrusion using GALDIT method: part 2—GALDIT indicator descriptions. IAHS and LNEC. In: Proceedings of the fourth inter celtic colloquium on hydrology and management of water resources. Universidade do Minho, Guimarães, Portugal

Ckakraborthy S, Paul PK, Sikdar PK (2007) Assessing aquifer vulnerability to arsenic pollution using DRASTIC and GIS of North Bengal Plain: a case study of English Bazar Block, Malda District, West Bengal, India. J Spat Hydrol 7(1):101–121

Connell LD, Van den Daele G (2003) A quantitative approach to aquifer vulnerability mapping. J Hydrol 276:71–88

Corwin DL, Vaughan PJ, Loague K (1997) Modeling nonpoint source pollutants in the vadose zone with GIS. Environ Sci Technol 31(8):2157–2175

Fuest S, Berlekamp J, Klein M, Matthies M (1998) Risk hazard mapping of groundwater

Horton RK (1965) An index number system for rating water quality. J Water Pollut Control Fed 37:300–305

Inhaber H (1975) An approach to a water quality index for Canada. Water Resour 9:821–833

Maha ASMS, Al-Dabbagh (1989) Water quality index of groundwater wells at Horan area (western Iraq). Regional characterization of water quality. In: Proceedings of the Baltimore symposium IAHS publication no. 182, 1989, pp 99–108

Phukon P, Phukan S, Das P, Sarma B (2004) Multicriteria evaluation in GIS environment for groundwater resource mapping in Guwahati City Areas, Assam

Pradhan SK, Patnaik D, Rout SP (2001) Groundwater quality index for groundwater around a phosphatic fertilizers plant. Indian J Environ Prot 21(4):355–358

Prati L, Pavanello R, Pesarin F (1971) Assessment of surface water quality by a single index of pollution. Water Resour 5:741–751

Provencher M, Lamontagne J (1979) A method for establishing a water quality index for different uses. Quebec: Gouvernment du Quea' bec, Ministea' re des richesses naturelles, le Service de la qualitea' des eaux, Bibliotequea' nationale du Quea' bec

Rajankar PN, Gulhane SR, Tambekar DH, Ramteke DS, Wate SR (2009) Water quality assessment of groundwater resources in Nagpur Region (India) based on WQI. http://www.e-journals.net

Ramakrishnaiah CR, Sadashivaiah C, Ranganna G (2008) Assessment of water quality index for the groundwater in Tumkur Taluk, Karnataka State, India. http://www.e-journals.net

Samake M, Tang Z, Hlaing W, M'Bue I, Kasereka K (2010) Assessment of groundwater pollution potential of the Datong basin, northern China. J Sustain Develop 3(2):140–152

Secunda S, Collin ML, Melloul AJ (1998) Groundwater vulnerability assessment using a composite model combining DRASTIC with extensive agricultural land use in Israel's Sharon Region. J Environ Manage 54:39–57

Shahid S, Nath SK (2000) GIS integration of remote sensing and electrical sounding data for hydrogeological exploration. J Spat Hydrol 2(1):1–12

Sinha DK, Shrivastava AK (1994) Water quality index for river Sai at Rae Bareli for the pre monsoon period and after the onset of monsoon. Indian J Environ Prot 14(5):340–345

Srinivasarao Y (2007) Groundwater quality suitable zones identification: application of GIS, Chittoor area, Andhra Pradesh. Environ Geol 53(1):201–210

Stigter TY, Ribeiro L, Carvalho Dill AMM (2006) Evaluation of an instrinsic and a specific vulnerability assessment method in comparison with groundwater salinisation and nitrate contamination levels in two agricultural regions in the south of Portugal. J Hydrogeol 14(1–2):79–99. https://doi.org/10.1007/s10040-004-0396-3

Tesoriero AJ, Inkpet EL, Voss FD (1998) Assessing groundwater vulnerability using logistic regression. In: Proceedings for the source water assessment and protection 98 conference, Dallas, TX, pp 157–165

Zhang R, Hamerlinck JD, Gloss SP, Munn L (1996) Determination of non point-source pollution using GIS and numerical models. J Environ Qual 25:411–418

Printed in the United States
By Bookmasters